Advances in
VIRUS RESEARCH

VOLUME **76**

Natural and Engineered Resistance to
Plant Viruses, Part II

Advances in VIRUS RESEARCH

VOLUME 76

Natural and Engineered Resistance to Plant Viruses, Part II

Edited by

JOHN P. CARR
Department of Plant Sciences
University of Cambridge, U.K.

GAD LOEBENSTEIN
Agricultural Research Organization
Bet Dagan, Israel

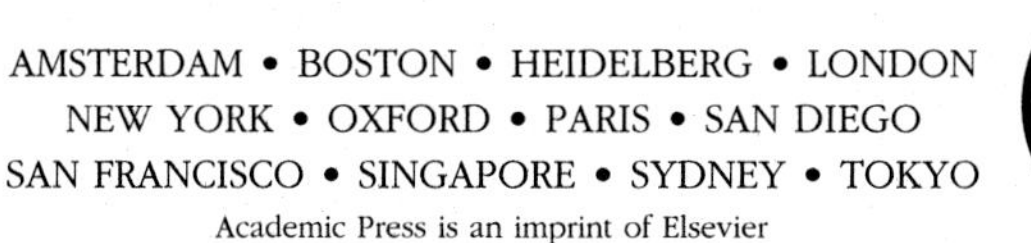

Academic Press is an imprint of Elsevier

32 Jamestown Road, London, NW1 7BY, UK
Radarweg 29, PO Box 211, 1000 AE Amsterdam, The Netherlands
30 Corporate Drive, Suite 400, Burlington, MA 01803, USA
525 B Street, Suite 1900, San Diego, CA 92101-4495, USA

First edition 2010

Library of Congress Cataloging-in-Publication Data

A catalog record for this book is available from the Library of Congress

British Library Cataloguing-in-Publication Data

A catalogue record for this book is available from the British Library

ISBN: 978-0-12-374525-5
ISSN: 0065-3527

For information on all Academic Press publications
visit our website at elsevierdirect.com

Printed and bound in USA
10 11 12 10 9 8 7 6 5 4 3 2 1

CONTENTS

PREFACE

We present the second of two volumes of *Advances in Virus Research* dealing with natural and engineered resistance to plant viruses. Viruses constitute the single largest group of causative agents of emerging and novel plant diseases and thus represent a serious threat to crop production throughout the world. Nevertheless, and as described in our preface to Volume 75, plant viruses also provide an apparently never-ending source of fundamental insights into pathogen and host biology. In this volume, our authors address topics that must be better understood in order to foster future developments in basic and applied plant virology. These range from virus epidemiology and virus/host coevolution and the control of vector-mediated transmission through to systems biology investigations of virus–cell interactions. Other chapters cover the current status of signaling in natural resistance and the potential for a revival in the use of cross-protection (a method for exploitation of strain–strain interactions to provide protection of crops against viral diseases), as well as future opportunities for the deployment of the underutilized but highly effective crop protection strategy of pathogen-derived resistance. As with the previous volume, our aim in editing these thematic volumes of *Advances in Virus Research* is to contribute to the development of much-needed new approaches to disease control by stimulating further research on natural and engineered resistance, and by encouraging younger scientists to join in this important quest.

Once again, we want to thank Professor Karl Maramorosch for giving us the opportunity to edit these thematic volumes. We are also profoundly grateful to our contributors for their hard work in preparing these insightful and up-to-date review chapters. We also thank the technical staff of the *Advances*, particularly Mr. Ezhilvijayan Balakrishnan and Ms. Narmada Thangavelu from Chennai, India.

December 2009

John Peter Carr and Gad Loebenstein
Editors

CHAPTER 1

The Coevolution of Plants and Viruses: Resistance and Pathogenicity

Aurora Fraile and **Fernando García-Arenal**

Contents

Abstract

Virus infection may damage the plant, and plant defenses are effective against viruses; thus, it is currently assumed that plants and viruses coevolve. However, and despite huge advances in understanding the mechanisms of pathogenicity and virulence in viruses and the mechanisms of virus resistance in plants, evidence in support of this hypothesis is surprisingly scant, and refers almost only to the virus partner. Most evidence for coevolution derives from the study

Centro de Biotecnología y Genómica de Plantas (UPM-INIA) and E.T.S.I. Agrónomos, Universidad Politécnica de Madrid, Campus de Montegancedo, Pozuelo de Alarcón (Madrid), Spain,
E-mail: fernando.garciaarenal@upm.es

Advances in Virus Research, Volume 76
ISSN 0065-3527, DOI: 10.1016/S0065-3527(10)76001-2

of highly virulent viruses in agricultural systems, in which humans manipulate host genetic structure, what determines genetic changes in the virus population. Studies have focused on virus responses to qualitative resistance, either dominant or recessive but, even within this restricted scenario, population genetic analyses of pathogenicity and resistance factors are still scarce. Analyses of quantitative resistance or tolerance, which could be relevant for plant–virus coevolution, lag far behind. A major limitation is the lack of information on systems in which the host might evolve in response to virus infection, that is, wild hosts in natural ecosystems. It is presently unknown if, or under which circumstances, viruses do exert a selection pressure on wild plants, if qualitative resistance is a major defense strategy to viruses in nature, or even if characterized genes determining qualitative resistance to viruses did indeed evolve in response to virus infection. Here, we review evidence supporting plant–virus coevolution and point to areas in need of attention to understand the role of viruses in plant ecosystem dynamics, and the factors that determine virus emergence in crops.

ACRONYMS AND NAMES OF VIRUSES

BaMMV	*Barley mild mosaic virus*
BaYMV	*Barley yellow mosaic virus*
BCMV	*Bean common mosaic virus*
BDMV	*Bean dwarf mosaic virus*
BYDV	*Barley yellow dwarf virus*
BYMV	*Bean yellow mosaic virus*
CMV	*Cucumber mosaic virus*
CYDV	*Cereal yellow dwarf virus*
GRSV	*Groundnut ringspot virus*
LMV	*Lettuce mosaic virus*
MNSV	*Melon necrotic spot virus*
PMMoV	*Pepper mild mottle virus*
PSbMV	*Pea seed-borne mosaic virus*
PVMV	*Pepper veinal mottle virus*
PVX	*Potato virus X*
PVY	*Potato virus Y*
RRSV	*Raspberry ringspot virus*
RYMV	*Rice yellow mottle virus*
SMV	*Soybean mosaic virus*
TCSV	*Tomato chlorotic spot virus*
TCV	*Turnip crinkle virus*
TEV	*Tobacco etch virus*
TMV	*Tobacco mosaic virus*

ToMV	*Tomato mosaic virus*
TSWV	*Tomato spotted wilt virus*
TuMV	*Turnip mosaic virus*
ZYMV	*Zucchini yellow mosaic virus*

I. INTRODUCTION

Pathogens are able to infect a host and, as a result of infection, they cause damage to this host. The appropriate terminology for these two central properties of pathogens has lead to much discussion in the phytopathological literature since Vanderplank (1968) defined aggressiveness as the quantitative negative effect of a pathogen on its host and virulence as the capacity of a pathogen to infect a particular host genotype. However, in other areas of biology, including animal pathology and evolutionary biology, virulence is defined as the detrimental effect of parasite infection on host fitness (e.g., Read, 1994; Woolhouse *et al.*, 2002), that is, virulence is related to the damage that parasite infection causes to the host, and the capacity to infect a host is named infectivity (Gandon *et al.*, 2002; Tellier and Brown, 2007). In spite of other conventions in the phytopathological literature, the American Phytopathological Society defines pathogenicity as the ability of a pathogen to cause disease on a particular host (i.e., a qualitative property), and virulence as the degree of damage caused to the host (i.e., a quantitative property), assumed to be negatively correlated to host fitness (D'Arcy *et al.*, 2001). These definitions are more in line with those used by other scientists interested in the biology of hosts and pathogens, and will be used in this review, except that for gene-for-gene (GFG) and matching-allele (MA) interactions we will retain the usual terminology of avirulence/virulence genes or factors.

Because pathogen infection reduces their fitness, hosts have developed different defense strategies to avoid or limit infection and to compensate for its costs (Agnew *et al.*, 2000). In plants, the two major defense mechanisms are resistance (defined as the ability of the host to limit parasite multiplication) and tolerance (defined as the ability of the host to reduce the damage caused by parasite infection) (Clarke, 1986). Host defenses may have a negative effect on parasite fitness. Hence, hosts and parasites may modulate the dynamics and genetic structure of each other's populations, and hosts and pathogens may coevolve, defining coevolution as the process of reciprocal, adaptive genetic change in two or more species (Woolhouse *et al.*, 2002).

Woolhouse *et al.* (2002) point to three conditions that are required for host–pathogen coevolution: (1) reciprocal effects of the relevant traits of the interaction (e.g., defense and pathogenicity) on the fitness of the two species (i.e., pathogens and hosts), (2) dependence of the outcome of the host–pathogen interaction on the combinations of host and pathogen genotypes involved, and (3) genetic variation in the relevant host and pathogen traits. If these three conditions are met, demonstrating coevolution requires in addition to show changes in genotype frequencies in both the host and the pathogen populations in the field (Woolhouse *et al.*, 2002). Although it is currently assumed that plants coevolve with their pathogens, this evidence is available for few plant–pathogen systems, and derives from analysis of the interaction of plants with bacteria, fungi, and oomycetes in their natural habitats (Burdon and Thrall, 2009; Salvaudon *et al.*, 2008). To our knowledge, analyses of genotype changes of plants and their infecting viruses in natural populations have not been reported, and there is no specific demonstration of plant–virus coevolution. However, evidence consistent with coevolution of plants and pathogens has accumulated for more than 50 years (e.g., Flor, 1971; Salvaudon *et al.*, 2008; Thompson and Burdon, 1992), deriving in a large part from agricultural systems, and this includes a considerable body of data from plant–virus interactions. In this review, we discuss the available evidence in support of coevolution in plant–virus systems; a major goal will be to pinpoint research areas in need of attention.

II. VIRUS INFECTION AND HOST DEFENSES RECIPROCALLY AFFECT THE FITNESS OF HOST AND VIRUS

Selection for resistance in plants, and for pathogenicity in viruses, would occur only if pathogenicity and resistance would negatively affect the fitness of plants and viruses, respectively. It is widely assumed that pathogen infection decreases the fitness of the infected host, that is, that pathogens are virulent, and that resistance decreases the fitness of the pathogen. However, direct evidence of these two assumptions for plants and viruses is surprisingly scarce, probably due to a limited interest till recent times of plant virologists in virulence evolution, on the one hand, and to difficulties in estimating experimentally the fitness of any organisms and linking these estimates to its evolution in nature (Kawecki and Ebert, 2004).

For animal pathogens, the effect of infection on host fitness, that is, virulence, is usually estimated as increased host mortality (Frank, 1996). This assumes that a reduction in lifespan conveys a decrease in fecundity and, hence, in fitness. But this is not obvious in many plant species, particularly domesticates, which are semelparous, that is, reproduce only once during their lives. Also, most plant pathogens do not cause an immediate increase in host mortality and their effect on host fitness depends on the pathogen life history (Barrett *et al.*, 2008). Hence, virulence on plants is most often estimated as the effect of pathogen infection on the plant's fecundity (i.e., viable seed production) or on one of its correlates, as plant size or biomass, or even symptom severity, the most commonly used correlate of virulence (Jarozs and Davelos, 1995). However, the relationship between fecundity and biomass or symptom severity may be nonlinear and depend on both genetic and environmental factors (e.g., Pagán *et al.*, 2007; Schürch and Roy, 2004), and this relationship has been analyzed only seldom for plant viruses (Agudelo-Romero *et al.*, 2008; Pagán *et al.*, 2008). Thus, the assumption that plant viruses decrease the fitness of their hosts rests mostly on the severity of the symptoms induced by virus infection on crops, and on the effects of infection on crop productivity, what may not be relevant for plant–virus coevolution. Moreover, although several reports of experiments showing that virus infection can decrease the fitness of wild plants under controlled conditions (e.g., Friess and Maillet, 1996; Kelly, 1994; Pagán *et al.*, 2007), there is little evidence that plant viruses have any effect on plant fitness in natural ecosystems, and it has been proposed that most often viruses would be mutualistic symbionts of plants (Roossinck, 2005; Wren *et al.*, 2006). This hypothesis rests on the interesting observation that in wild hosts growing in nonagricultural ecosystems, virus infection most often does not cause any obvious symptom, at odds with what is known to occur in crops. But estimates of the effect of virus infection on wild plants fitness are presently scarce. The negative effect of virus infection on plant fitness in nature has been best documented for BYDV and CYDV on wild grasses in California (Malmstrom *et al.*, 2006; Power and Mitchell, 2004). Interestingly, virus infection, in addition to direct fitness costs, has important indirect costs as it may reduce the competitive ability of the infected plants, a phenomenon (apparent competition) that may also occur among genotypes of the same species (Pagán *et al.*, 2009). Virus infection has also been shown to increase mortality and to reduce fecundity in wild cabbage in southern England (Maskell *et al.*, 1999), and to reduce lifespan of wild pepper in its natural habitats in Mexico (our unpublished results). Other reports suggest that the effect of virus infection on the population dynamics of wild plants will vary largely according to site or population (Pallett *et al.*, 2002). On the other hand, virus infection may be beneficial for plants, as shown by an increase of tolerance to abiotic stress in virus-infected plants as compared with uninfected controls (Xu *et al.*, 2008), or by a decreased herbivory on tymovirus-infected *Kennedia rubicunda* in Australia (Gibbs, 1980). Thus, it is obvious that the effects of virus infection on plant fitness in natural ecosystems may vary largely according to the specific virus–host interaction and, probably, according to the environment, a subject that requires further attention by virologists with an interest in ecology and evolution.

For parasites, fitness is also best estimated as fecundity, that is, production of new infections per unit time (Anderson and May, 1982).

However, for plant viruses, fitness is usually estimated as within-host multiplication rates (e.g., Sacristán *et al.*, 2005) or, when different genotypes are compared, as competitive ability (e.g., Carrasco *et al.*, 2006; Elena *et al.*, 2006; Fraile *et al.*, 2010). Because resistance results in a decrease of within-host virus multiplication, it is assumed that resistance decreases virus fitness. This assumption implicitly considers that rates of between-host transmission positively correlate with rates of within-host multiplication. Indeed, for viruses transmitted by aphids, both nonpersistently and persistently, it has been shown that transmission efficiency is positively correlated with virus accumulation in source tissues (Barker and Harrison, 1986; Escriu *et al.*, 2000; Foxe and Rochow, 1975; Jiménez-Martínez and Bosque-Pérez, 2004; Pirone and Megahed, 1966). Whether or not this correlation holds for other mechanisms of horizontal transmission, or for seed transmission, remains to be analyzed.

In summary, although direct evidence is far less common than might be expected, it supports that in the case of virulent virus–plant interactions traits related to pathogenicity have a negative effect on the plant's fitness, and traits related to defense have a negative effect on the virus fitness.

III. THE OUTCOME OF PLANT–VIRUS INTERACTIONS DEPENDS ON THE PLANT AND VIRUS GENOTYPES INVOLVED

For the last 50 years, different theoretical analyses aimed at understanding and modeling host–pathogen coevolution have been published. All these analyses assume that the outcome of the host–pathogen interaction is determined by the combination of host and pathogen genotypes involved. Two major models of host–parasite interaction determining the success of infection have been proposed: the GFG and the MA models, which have been applied mostly to plant and animal systems, respectively. Genetic and molecular genetic evidence support both these models to explain plant–virus interactions (Kang *et al.*, 2005a; Maule *et al.*, 2007; Sacristán and García-Arenal, 2007).

In plant–pathogen systems, pathogenicity has been most often related and analyzed as conforming to GFG interactions, first described in the flax–flax rust system (Flor, 1955). According to this model, the interaction of specific products of the plant and pathogen genotypes determines an incompatible interaction (Fig. 1), that is, host defenses are triggered and infection is limited. Plant proteins encoded by resistance genes (R proteins) recognize corresponding proteins of the pathogen, encoded by avirulence genes (*AVR*). Recognition can be either through a direct R–AVR interaction or, more often, via multiprotein interactions, including AVR–host protein

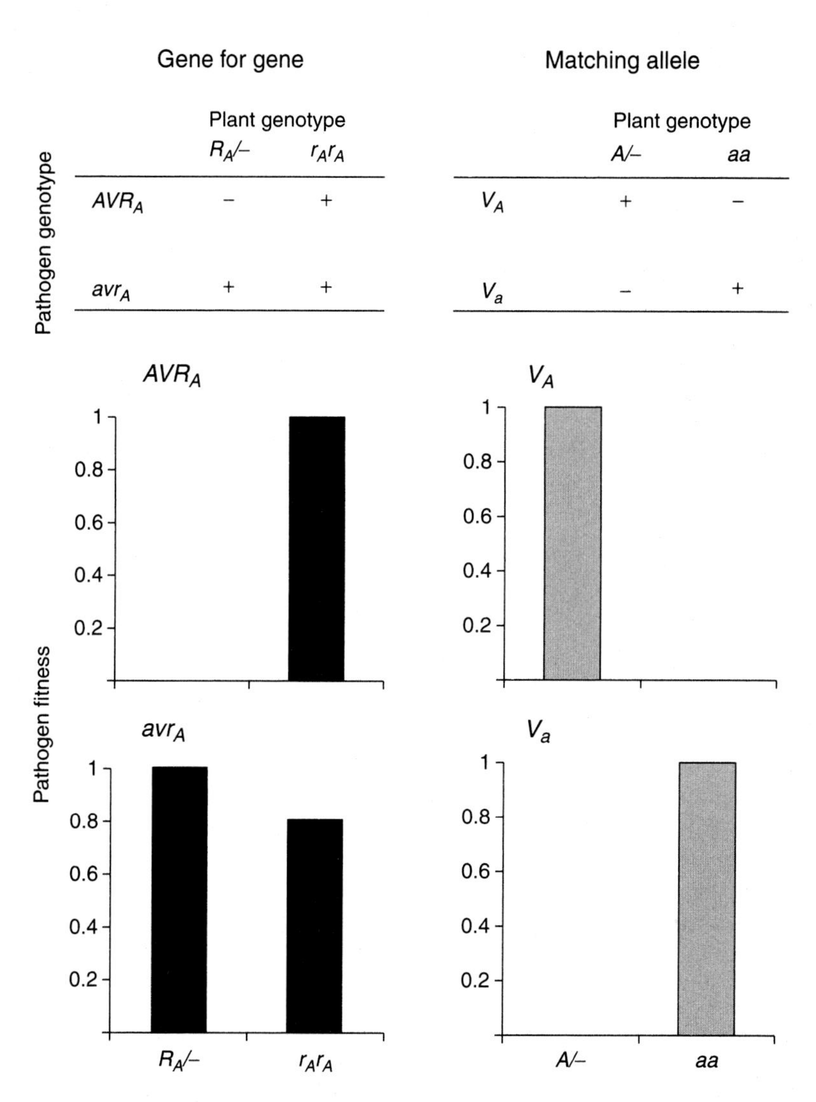

FIGURE 1 Models of host–pathogen coevolution for a diploid host and a haploid pathogen species. Left panel: gene-for-gene model. The product of the dominant resistance allele at locus A (R_A) in the host allows recognition of the product of the avirulence gene A (AVR_A) in the pathogen, triggering defenses and limiting infection (–). If the plant genotype is homozygous for the recessive susceptibility allele r_A, or the pathogen genotype has the virulence allele avr_A, the pathogen is not recognized, defenses are not triggered and infection occurs (+). In the resistant host genotype (R_A/–), the relative fitness of the avirulent pathogen genotype (AVR_A) is near zero, while that of the virulent one (avr_A) is considered as 1. In the susceptible host genotype (r_Ar_A), the virulent pathogen genotype has a lower relative fitness than the avirulent genotype (cost of pathogenicity). Right panel: matching-allele model.

complexes, modified/unmodified host targets of AVR, and/or adapter proteins that mediate binding, stabilize, or localize R (Friedman and Baker, 2007; Jones and Dangl, 2006; McDowell and Simon, 2006; Moffett, 2009). The recognition of AVR by the host triggers defense responses leading to limitation of multiplication and spread of the pathogen which remains localized at the infection site, and the resistance response is often associated to a hypersensitive response (HR), often involving localized host cell death. In the absence of the *AVR* allele in the pathogen or of the *R* allele in the host, the parasite is not recognized by the host, resistance is not triggered, and the host is infected, resulting in a compatible interaction. Accordingly, a key feature of the GFG model is that universal pathogenicity occurs, that is, there are pathogen genotypes able to infect all host genotypes (Agrawal and Lively, 2002).

Pathogen recognition by plant genotypes resulting in a HR was first described by Holmes (1937) for the interaction of TMV with *Nicotiana* spp. Ever since, polymorphisms for resistance to different viruses have been described in many plant species. About 51% of characterized resistance factors have a monogenic dominant inheritance, and are most often manifested as a HR (Kang *et al.*, 2005a; Khetarpal *et al.*, 1998). Twelve dominant genes conferring resistance to viruses expressed either as HR or as extreme resistance (ER; i.e., virus multiplication is limited to the initially infected cells without an apparent necrotic local lesion) have been cloned and sequenced (Table I) (Maule *et al.*, 2007; Palukaitis and Carr, 2008). All encode members of the NB–LRR class of R proteins (Dangl and Jones, 2001) that localize to the cytoplasm consistent with the lifestyle of viruses (Maule *et al.*, 2007). Viral genotypes that break down a defense response (i.e., that no longer elicit a HR or ER) have been described for many resistance factors (e.g., García-Arenal and McDonald, 2003; Janzac *et al.*, 2009), that is, there are polymorphisms in the virus population for pathogenicity. The viral *AVR* genes responsible for eliciting the defense reaction, or for resistance breaking, have been identified in many instances (García-Arenal and McDonald, 2003; Janzac *et al.*, 2009;

The product of allele *A* at a certain locus in the host genotype interacts with the product of the virulence allele V_A in the pathogen, allowing infection (+), while this interaction does not occur with the product of allele V_a, resulting in a lack of susceptibility (−) or resistance. Similarly, the product of allele *a* in the host interacts with the product of allele V_a in the pathogen, allowing infection, but not with the product of allele V_A. In a pure matching-allele model, in the host genotype *A*/− the relative fitness of the pathogen genotype with allele V_A is 1, while that of the pathogen genotype with allele V_a is 0, the opposite being true in the host genotype *aa*, and there are no fitness penalties for pathogenicity. Here alleles *A* and *a* are represented as dominant and recessive, respectively, but this is not a requirement of the model.

TABLE I Characterized genes conferring quantitative, genotype-specific resistance to viruses

	Protein	Plant species	Virus targets	AVR factor	References
Dominant genes					
Cloned					
N	TIR–NB–LRR	*Nicotiana tabacum*	Tobamoviruses	Replicase/ helicase	Padgett *et al.* (1997) and Whitham *et al.* (1994)
Rx1	CC–NB–LRR	*Solanum andigena*	PVX	Coat protein	Bendahmane *et al.* (1997, 1999)
Rx2	CC–NB–LRR	*Solanum acaule*	PVX	Coat protein	Bendahmane *et al.* (1997, 2000)
Sw-5	CC–NB–LRR	*Solanum lycopersicum*	TSWV, TCSV, GRSV	Movement protein	Bromonschenkel *et al.* (2000)
HRT	CC–NB–LRR	*Arabidopsis thaliana*	TCV	Coat protein	Cooley *et al.* (2000)
RCY1	CC–NB–LRR	*Arabidopsis thaliana*	CMV	Coat protein	Takahashi *et al.* (2002)
Y-1	TIR–NB–LRR	*Solanum tuberosum*	PVY		Vidal *et al.* (2002)
Tm-2/Tm-2^2	CC–NB–LRR	*Solanum peruvianum*	ToMV, TMV	Movement protein	Lanfermeijer *et al.* (2003, 2005), Weber and Pfitzner (1998), and Weber *et al.* (2004)

(*continued*)

TABLE I *(continued)*

	Protein	Plant species	Virus targets	AVR factor	References
Rsv1	CC–NB–LRR	*Glycine max*	SMV	P3 protein	Hajimorad *et al.* (2005) and Hayes *et al.* (2004)
RT4-4	TIR–NB–LRR	*Phaseolus vulgaris*	CMV	Replicase/ helicase	Seo *et al.* (2006)
PvVTT1	TIR–NB–LRR	*Phaseolus vulgaris*	BDMV	Nuclear shuttle protein	Garrido-Ramirez *et al.* (2000) and Seo *et al.* (2007)
RTM1	Lectin-like	*Arabidopsis thaliana*	TEV	Coat protein	Chisholm *et al.* (2000) and Decroocq *et al.* (2009)
RTM2	Small heat-shock protein	*Arabidopsis thaliana*	TEV	Coat protein	Decroocq *et al.* (2009) and Whitham *et al.* (2000)
Tm-1	TIM barrel structure	*Solanum habrochaites*	TMV, ToMV	Replicase	Ishibashi *et al.* (2007) and Meshi *et al.* (1988)
Mapped to complex loci					
Tsw		*Capsicum*	TSWV	NSs protein	Margaria *et al.* (2007)
L^1, L^2, L^3, L^4	CC–NB–LRR	*Capsicum*	Tobamovirus	Coat protein	Tomita *et al.* (2008)
I	TIR–NB–LRR	*Phaseolus vulgaris*	BCMV		Vallejos *et al.* (2006)

Recessive genes					
Cloned					
pvr2^i + *pvr6*	*pvr2*: eIF4E	*Capsicum annuum*	PVMV, TEV	–	Caranta *et al.* (1996) and Ruffel *et al.* (2006)
pvr1/*pvr2*^{i}	eIF4E	*Capsicum chinense*	PVMV, PVY, TEV	VPg	Charron *et al.* (2008), Kang *et al.* (2005b), and Ruffel *et al.* (2002, 2006)
nsv	eIF4E	*Cucumis melo*	MNSV	3′-UTR	Díaz *et al.* (2004) and Nieto *et al.* (2006)
rym4/5	eIF4E	*Hordeum vulgare*	BaMMV, BaYMV	VPg	Kanyuka *et al.* (2004) and Stein *et al.* (2005)
mol^1/*mol*^{2}	eIF4E	*Lactuca sativa*	LMV	VPg and CI	Nicaise *et al.* (2003) and Roudet-Tavert *et al.* (2007)
rymv-1	eIF(iso)4G	*Oryza sativa* and *Oryza glaberrima*	RYMV	VPg	Albar *et al.* (2003, 2006)
sbmv1^i	eIF4E	*Pisum sativum*	PSbMV, BYMV	VPg	Bruun-Rasmussen *et al.* (2007), Gao *et al.* (2004), and Johansen *et al.* (2001)
pot-1	eIF4E	*Solanum habrochaites*	PVY, TEV	VPg	Ruffel *et al.* (2005) and Schaad *et al.* (2000)

Kang *et al.*, 2005a; Maule *et al.*, 2007). Virtually all classes of virus-encoded protein have been shown to have the potential to be AVR factors in different plant–virus systems (e.g., Table I). Thus, monogenic dominant resistance of plants to viruses, expressed as HR or as ER, conforms to a GFG model of host–pathogen interaction.

The other major model of host–pathogen interaction is the MA model. Its key feature is that infection requires a specific match between host and parasite genes (Fig. 1). Hence, it is at odds with the GFG model since "recognition" of the pathogen by the host leads to susceptibility rather than to resistance. In a pure MA system, pathogenicity on all host genotypes (i.e., "universal pathogenicity") cannot exist, an important difference with the GFG model (Agrawal and Lively, 2002). Although data from plant–pathogen systems have been mostly analyzed under the GFG model, the MA model could better fit some types of interactions. One obvious instance is recessive resistance to plant viruses. In contrast to resistance to cellular plant pathogens, a high percentage (35%) of monogenic resistance of plants to viruses is recessive (Kang *et al.*, 2005a; Khetarpal *et al.*, 1998). Polymorphisms for recessive resistance are in fact polymorphisms for impaired susceptibility, and this type of resistance is most often expressed as immunity at the cell level (Díaz-Pendón *et al.*, 2004; Kang *et al.*, 2005a; Maule *et al.*, 2007). Several recessive resistance genes have been cloned and sequenced (Table I) (Truniger and Aranda, 2009), and in all instances they encode translation initiation factors, either eIF4E, eIF4G, or their isoforms. Polymorphisms for pathogenicity on recessive resistant host genotypes have been described for many systems, and the viral gene products responsible for the expression or the breakdown of the resistance have been identified in most cases as the viral genome-linked protein (VPg), which is thought to interact with the initiation factors for cap-independent translation to occur (Maule *et al.*, 2007; Palukaitis and Carr, 2008; Truniger and Aranda, 2009). One notable exception is the system MNSV–melon, in which the pathogenicity determinant has been mapped to the 3′-UTR of the viral genomic RNA (Díaz *et al.*, 2004). It has been proposed that interaction of the 3′-UTR of MNSV and the eIF4E is required for messenger circularization and translation (Truniger *et al.*, 2008). Thus, available information is compatible with the adequacy of the MA model to explain plant–virus interactions determined by recessive resistance systems.

Resistance to viruses in plants may also be polygenically inherited. Polygenic resistance is usually expressed as quantitative or partial resistance, in which within-host multiplication of the virus is reduced. Few QTLs for virus resistance have been mapped (Maule *et al.*, 2007), but polygenic resistance has been used in crop breeding for virus disease control, and virus genotypes overcoming these resistances have been reported (García-Arenal and McDonald, 2003; Khetarpal *et al.*, 1998).

Also, quantitative resistance of Arabidopsis to CMV depended on host–virus genotype × genotype interaction (Pagán *et al.*, 2007). Thus, there is evidence showing that the outcome of the host–virus interaction again depends on the specific plant and virus genotypes.

The other major defense strategy of plants against pathogens is tolerance. Tolerance has received considerably less attention from scientists than resistance, and its use for viral disease control has been limited by obvious difficulties of breeding for increased tolerance, as phenotype evaluation can only be done at later stages of the plant's life cycle. Also, the mechanisms of tolerance are poorly understood, but they may be related to the ability of the plant to modify its life history program upon infection (Pagán *et al.*, 2008). The genetic control of tolerance may be monogenic or, most often, polygenic (Clarke, 1986). Tolerance depends on the interacting virus and plant genotypes (Pagán *et al.*, 2008). Accordingly, virus genotypes able to overcome tolerance in crops have been described and, at least in one case, ZYMV in melon, tolerance-breaking genotypes have been shown to become prevalent in the virus population after the extensive use of tolerant varieties (Desbiez *et al.*, 2002, 2003). This was a quite unexpected finding, because as tolerance does not affect the within-host multiplication of the virus it was traditionally considered not to exert a selection pressure upon it. However, recent theoretical analyses have shown that tolerance, through reducing virulence, will select for virus genotypes with an increased within-host multiplication, as far as virulence and multiplication are linked (van den Bosch *et al.*, 2006), which may explain these observations.

In summary, no matter the type of plant defense against virus infection, there is evidence that the outcome of the plant–virus interaction depends on the interacting plant and virus genotypes, thus fulfilling this condition for plant–virus coevolution to occur.

IV. GENETIC VARIATION OF RESISTANCE AND PATHOGENICITY

For evolution to occur there must be genetic variation for the relevant trait. In the previous section we have shown that the outcome of plant–virus interactions depends on the specific plant and virus genotypes, and this is evidence for genetic variability in resistance or tolerance and pathogenicity. Analysis of the patterns of variability of plant genes determining resistance, and of the viral genes determining pathogenicity, further supports the hypothesis that plants and viruses coevolve. The available evidence derives mostly from analyses of interactions resulting in qualitative resistance, either dominant or recessive, and thus conforming to the GFG or MA models.

A. Variability of resistance and pathogenicity under the gene-for-gene model

GFG interactions between plant and pathogens have been much analyzed and, in recent years, knowledge on the structure of R and AVR proteins, on their molecular variation and on the mechanisms underlying recognition, has made enormous progress. This is also the case for GFG plant–virus interactions. However, evidence supporting plant–virus coevolution is sparse, comes from different pathosystems, and detailed analyses of the variation of R and AVR in the same system are lacking.

1. *R*-gene variability

Most molecularly characterized genes that determine dominant qualitative resistance to viruses are involved in GFG-like plant–virus interactions and resistance is expressed as either an ER or a HR. Exceptions to this include the Arabidopsis genes *RTM1* and *RTM2*, which confer resistance to systemic colonization by TEV, and the tomato gene *Tm1* that encodes an inhibitor of ToMV replication (Chisholm *et al.*, 2000; Ishibashi *et al.*, 2007; Whitham *et al.*, 2000). All resistance genes that determine an ER or HR reaction upon virus inoculation encode proteins (R proteins) that contain nucleotide-binding site (NB) and leucine-rich repeat (LRR) domains, with either TIR or CC domains at their N-terminal regions (Table I). No function other than resistance is known for this protein class, and many copies of NB–LRR protein-encoding genes occur in plant genomes (Dangl and Jones, 2001; Friedman and Baker, 2007).

Most NB–LRR *R* genes to plant pathogens occur in complex loci, formed by tightly linked homologous genes (Hulbert *et al.*, 2001). This is also the case for *R* genes targeting different viruses in different plant species, for instance, the *N* gene of resistance to TMV in *Nicotiana tabacum* (Whitham *et al.*, 1994), the *Rx1* and *Rx2* genes of resistance to PVX in *Solanum tuberosum* and *S. acaule* (Bendahmane *et al.*, 1999, 2000), the *HRT* gene of resistance to TCV in *Arabidopsis thaliana* (Dempsey *et al.*, 1997; McDowell *et al.*, 1998), the *Y-1* gene of resistance to PVY in *S. tuberosum* (Vidal *et al.*, 2002), the *Rsv1* gene of resistance to SMV in soybean (Hayes *et al.*, 2004), or the *L* gene of resistance to tobamoviruses in *Capsicum* spp. (Tomita *et al.*, 2008). Often, resistance genes to viruses in complex loci are allelic or tightly linked to resistance genes against other pathogens or herbivores, as is the case for *Rx2* and *Gpa3* (*Globodera pallida*) and *R1* (*Phytophthora infestans*) in potato or *HRT* and *RPP8* (*Hyaloperonospora parasitica*) in Arabidopsis (Bendahmane *et al.*, 2000; Cooley *et al.*, 2000). Duplications and recombination through unequal crossover seem to be a major mechanism in the evolution of *R* genes and a way to generate new specificities (Friedman and Baker, 2007). Indeed, reported mutation rates at *R* genes are high; thus, the frequency of reversion to susceptibility to

TMV in an *Nn* population of tobacco is ~1/2000 (Whitham *et al.*, 1994), which is better explained by unequal crossing over causing deletions between repetitive sequences than by point mutations, considering the spontaneous rate of nucleotide substitutions in eukaryotes (Drake *et al.*, 1998). Recombination may occur between linked genes within complex loci (e.g., Hayes *et al.*, 2004; Whitham *et al.*, 1994) or between unlinked disease resistance genes, as shown for *Rx1*, *Rx2*, and *Gpa2*, which locate at different chromosomes in the potato genome (Bendahmane *et al.*, 2000). It has been shown that virus infection promotes recombination in the host plant genome with transgeneration effects (Kovalchuk *et al.*, 2003; Molinier *et al.*, 2006). Hence, the appearance of new recognition specificities through recombination in *R* genes could be favored by infection and play an important part in plant–virus coevolution (Friedman and Baker, 2007), a hypothesis to be analyzed.

In addition to recombination, point mutation is another major source of genetic variation in *R* genes. Evidence for diversifying selection, compatible with plant–pathogen coevolution, has been reported for *R* genes conferring resistance to cellular plant pathogens (e.g., Allen *et al.*, 2004; Dodds *et al.*, 2006; Mauricio *et al.*, 2003; Parniske *et al.*, 1997; Rose *et al.*, 2004; Wang *et al.*, 1998). Diversifying selection affects mostly the LRR domain, which is associated to specificity of recognition in R. Diversifying selection has not been demonstrated for *R* genes to viruses, but characterization of the single-gene locus alleles *lptm2*/*Tm2*^{2} of *S. peruvianum*, and *Tm2* of *S. habrochaites*, in which alleles *Tm2* and *Tm2*^{2} confer resistance to different genotypes of ToMV, showed that the resistance alleles differ from one another and from the susceptibility allele *lptm2* by a reduced number of mostly nonsynonymous nucleotide changes (Lanfermeijer *et al.*, 2003, 2005).

Under the GFG model of host–pathogen coevolution, fitness costs of resistance and pathogenicity are required for stable polymorphism at these traits to occur in host and pathogen populations (Fig. 1). Costs of resistance have been much reviewed (Bergelson and Purrington, 1996; Bergelson *et al.*, 2001; Brown, 2003; Mauricio, 1998): evidence is controversial and there is none for any resistance factor to viruses. However, there is indirect evidence that resistance may be costly: artificial evolution of *Rx* by introducing random point mutations in the LRR domain resulted in the appearance of variants with new specificities, which showed enlarged recognition of PVX genotypes or even of distantly related viruses (Farnham and Baulcombe, 2006). Since alleles with these enlarged recognition abilities do not occur in nature, these results suggest that fitness penalties constrain the evolution of *R* genes.

Conclusions on adaptive evolution of R proteins should consider several traits that show that interactions under the GFG model are certainly more complex than originally considered. On the one hand, the process of plant–pathogen recognition itself may involve other proteins than R and AVR, so that recognition of AVR is indirect involving multiprotein complexes. Indirect recognition may be more common than direct R–AVR interaction, and it has been shown to occur in all characterized plant–virus systems (Caplan *et al.*, 2008a,b; Jeong *et al.*, 2008; Ren *et al.*, 2000; Tameling and Baulcombe, 2007). It has been proposed that the mode of R–AVR recognition, direct or indirect, will affect the evolution of *R* and *AVR*, indirect recognition resulting in balancing selection in *AVR* and *R* (Van der Hoorn *et al.*, 2002), but evidence in support of this hypothesis is scant, and there is none from plant–virus systems. Also, *R* genes often determine the unspecific recognition of different virus species within the same genus, rather than specific recognition of viral genotypes, for example, the *Sw5* of *S. lycopersicum* determines resistance against different tospoviruses, the *I* gene of *Phaseolus vulgaris* determines resistance to different potyviruses, while the *N* gene of *Nicotiana* determines resistance to different tobamovirus (Bromonschenkel *et al.*, 2000; Fisher and Kyle, 1994; Padgett and Beachy, 1993). Thus, it is not known which species within these genera exerted a selection on the host plant leading to the appearance of resistance, or if resistance to other virus species is due to shared structures in AVR or just coincidental. Last, resistance/susceptibility alleles may be more complex than originally considered from inheritance analyses, and rather than resulting from variations in single-gene loci, be due to rearrangements resulting in gain/loss of several cistrons encoding NB–LRR proteins, as shown for *Rsv1* (Hayes *et al.*, 2004) or as proposed to explain the evolution of the *L* locus within the genus *Capsicum* (Tomita *et al.*, 2008). How all these traits of GFG systems would affect selection pressures of viruses on plants and, hence, the evolution of *R* genes, remains to be explored, but certainly should be considered both when analyzing evidence apparently in support of plant–virus coevolution and when developing theoretical models of host–pathogen coevolution.

2. *AVR*-gene variability

The first AVR factor identified in a plant pathogen was the capsid protein (CP) of TMV, which elicits the HR defense response triggered by the *N′* resistance gene in *Nicotiana* spp. (Knorr and Dawson, 1988; Saito *et al.*, 1987). Since then, it has been shown that many other virus-encoded proteins may act as an AVR factor on different R proteins (see Table I). For example, within the genus *Tobamovirus*, the CP of TMV is the AVR factor for *N′*; the p50 helicase domain of the RNA-dependent polymerase (RdRp) of TMV is the AVR for the *N* gene in *Nicotiana*, and the movement protein of ToMV is the AVR for *Tm2* and $Tm2^2$ in tomato (Meshi *et al.*, 1989; Padgett *et al.*, 1997; Weber and Pfitzner, 1998). Within the *Potyvirus* genus, the NIa protease of PVY is the AVR factor for *Ry* in potato

(Mestre *et al.*, 2003), the P3 protein of SMV elicits *Rsv1* in soybean (Hajimorad *et al.*, 2005), or the cylindrical inclusion helicase of TuMV elicits *TuRB01* of *Brassica* (Jenner *et al.*, 2000). Within the *Cucumovirus* genus the CP of CMV is the AVR for *RCY1* in *Arabidopsis* and the 2a protein for *RT4-4* in *Phaseolus* (Seo *et al.*, 2006; Takahashi *et al.*, 2002). Further examples can be found in Kang *et al.* (2005a) or Maule *et al.* (2007). Because it is necessary that virus-encoded proteins interact with host factors for completion of the virus life-cycle within the infected host, they can be considered as pathogenicity effectors, as is the case for AVR factors of cellular plant pathogens (Jones and Dangl, 2006).

While AVR of cellular plant pathogens may avoid recognition by R through a large array of mechanisms including point mutations, recombination and even the deletion of AVR itself (Friedman and Baker, 2007; Sacristán and García-Arenal, 2007), obviously this cannot be the case for plant viruses, which have small genomes encoding multifunctional proteins. Changes in recognition of viral AVR by R proteins depend on one to few amino acid substitutions (Harrison, 2002; Maule *et al.*, 2007). For many R factors only one or few *avr* genotypes have been reported, with no evidence of diversifying selection on *AVR*/*avr*. Thus, virulence on *N* is extremely rare, occurring only in tobamovirus species with a restricted geographical distribution (García-Arenal and McDonald, 2003). Similarly, although pathogenicity on *Rx* is due to mutations at two positions in PVX CP (Goulden *et al.*, 1993), in nature only one strain, PVX-HB, with limited distribution, has been reported with these mutations and phenotype (see García-Arenal and McDonald, 2003). On the other hand, different mutations in the CP of TMV led to breakage of *N′*-mediated resistance in tobacco. In an elegant series of papers, Culver and colleagues have analyzed a large set of TMV CP variants, obtained by site-directed mutagenesis, that totally or partly overcame *N′*-mediated resistance, and determined that the maintenance of the CP three-dimensional structure is essential for *N′* elicitation (see Culver, 2002). Similarly, field isolates of PMMoV that overcome partially or totally L^3 resistance in pepper, inducing systemic necroses or mosaics, respectively, differ from the AVR genotype in few nucleotide substitutions resulting in different amino acid changes in the CP that may destabilize its three-dimensional structure. Interestingly, although different resistance-breaking genotypes have been characterized, CP mutations resulting in resistance breakage occur only in certain combinations, and the different resistance-breaking genotypes have different geographical distributions, having been reported either in the Mediterranean or in Japan (Berzal-Herranz *et al.*, 1995; Hamada *et al.*, 2002, 2007; Tsuda *et al.*, 1998), which suggests that only certain evolutionary pathways may lead to pathogenicity on L^3. Also, in TSWV different genotypes virulent on pepper carrying the resistance gene *Tsw* have been reported in different areas of the Mediterranean basin, and most of the few nucleotide substitutions in the *AVR* gene encoding Nss protein resulted in amino acid substitutions (Margaria *et al.*, 2007).

Limited polymorphism in resistance-breaking genotypes suggests that there are fitness penalties associated with increased pathogenicity. Although experimental estimates of putative costs of pathogenicity are few, evidence for these costs derives from several systems. Thus, there is evidence for selection against PVX CP mutants pathogenic on *Rx* (Goulden *et al.*, 1993), and no field isolate of PVY has been reported to overcome *Ry* in potato, although resistance-breaking mutants in the NIa protein were obtained experimentally (Mestre *et al.*, 2003). Also, fitness penalties could relate to functions other than virus multiplication: RRSV strains overcoming *Irr* resistance in raspberry had a decreased transmission both by nematodes and through the seed in alternate hosts (Hanada and Harrison, 1977; Murant *et al.*, 1968). TuMV genotypes overcoming *TuRB01* resistance in rape were outcompeted by avirulent ones in susceptible hosts. Assays were done with engineered *avr* mutants with no second-site mutations, thus providing evidence for a cost due to a pleiotropic effect of the *avr* mutation (Jenner *et al.*, 2002a). Fitness costs have also been reported for TuMV mutants overcoming a second resistance gene, *TuRB04* (Jenner *et al.*, 2002b). At odds with other reports, the data in Jenner *et al.* (2002a,b) allow us to estimate the fitness of virulent mutants relative to avirulent ones, which shows values of about 0.50 (Sacristán and García-Arenal, 2007). Similarly, competition experiments among field PMMoV isolates virulent or avirulent on L^3 resistance in pepper also showed high differences in relative fitness; the fitness of *avr* isolates being on average 0.47 relative to that of *AVR* isolates (Fraile *et al.*, 2010). Interestingly, evidence for fitness penalties was also provided by the dynamics of *avr/AVR* genotypes of pepper-infecting tobamoviruses in the field, as compared with the relative acreage of the different *L* alleles deployed over a period of more than 20 years (Fraile *et al.*, 2010); to our knowledge, the only long term analysis of avr/AVR dynamics for a plant–virus system.

Consistent evidence for pathogenicity-associated fitness penalties in plant viruses contrasts strikingly with the conflicting results for fungi and oomycetes (Sacristán and García-Arenal, 2007) and may be highly relevant for the analysis of the durability of the resistance of crops to viruses.

B. Variability of resistance and pathogenicity under the matching-allele model

Recessive resistance (i.e., lack of susceptibility) to plant viruses may be best interpreted under the MA model. Mutations leading to resistance can be countered by mutations in the virus, thus restoring compatibility on the mutated host gene. All evidence on the structure of VPg proteins and on the mutations resulting in overcoming recessive resistance strongly suggest a direct interaction between the eIF4E/eIF(iso)4G and the VPg, required for infection (Charron *et al.*, 2008; Hebrard *et al.*, 2008; Truniger and Aranda, 2009). Although interest in recessive resistance is more recent than on dominant resistance, and fewer pathosystems have been analyzed in detail for the variation of either the resistance or the pathogenicity determinants, they have provided the best evidence so far for plant–virus coevolution.

The system for which more information is available is the pepper–PVY interaction determined by the *pvr2* locus of *Capsicum*, encoding eIF4E, and the virus VPg. In the only large-scale survey of the variation of a gene encoding resistance to a virus, Charron *et al.* (2008) have reported 10 *pvr2* alleles in a worldwide survey of accessions of *Capsicum annuum*. The most common allele, with a 0.4 frequency, $pvr2^{+}$, determines susceptibility to PVY and to another pepper-infecting potyvirus, TEV, while alleles $pvr2^{1}$–$pvr2^{9}$, determine resistance, with different specificities toward different PVY and TEV genotypes. *pvr2* alleles conferring resistance differ from $pvr2^{+}$ by 1–4 amino acid substitutions at nine positions in two domains of eIF4E, polymorphic sites being located in the surface of the protein, and there is evidence for diversifying selection at these domains. Amino acid substitutions resulting in resistance impair the physical interaction of eIF4E and the VPg of incompatible virus genotypes. On the virus side, up to 11 amino acid changes in the central region of PVY VPg have been described to determine pathogenicity on resistance alleles of *pvr2*, and, again, there is evidence that positive selection on these sites leads to diversification of the VPg. Overcoming one *pvr2* allele does, or does not, confer pathogenicity on other alleles, pending on the specific mutations (Ayme *et al.*, 2007; Moury *et al.*, 2004), as corresponds to a MA interaction. Thus, data on the pepper/PVY system provides evidence that the direct interaction *pvr2*/VPg drives the coevolution between resistance and pathogenicity leading to diversifying selection at both genes.

The MA model predicts that polymorphisms for pathogenicity and resistance will be maintained by negative frequency-dependent selection, with no need of resistance or pathogenicity costs (Fig. 1) (Agrawal and Lively, 2002). It has been argued that pure GFG and MA models are extremes of a continuum, in the inside of which the MA model should be modified to admit partial infection, that is, the parasite infects, but reproduces less effectively, and the host suffers less intensely from parasitism than in a "full infection." Within this continuum, costs of resistance and pathogenicity would exist as a function of the degree of success of partial infection (Agrawal and Lively, 2002; Parker, 1994). Functional assay of the 10 eIF4E variants in yeast failed to detect differences in the efficiency of cap-dependent mRNA translation (Charron *et al.*, 2008), strongly suggesting that there is no fitness penalty for PVY resistance, although this may not be universal for eIF4E-mediated resistance to potyviruses (e.g., Kang *et al.*, 2005a). The relative fitness of different VPg mutants overcoming allele *pvr2*^{3} was analyzed in pepper genotypes homozygous either for *pvr2*^{3} or for the susceptibility allele *pvr2*$^+$, and in the susceptible host *Nicotiana clevelandii*. The various pathogenic PVY genotypes differed in fitness in all three hosts, but some of them were as fit in susceptible pepper and *N. clevelandii* plants as the nonpathogenic wild type (Ayme *et al.*, 2006). Again, in this respect the *pvr2*/PVY system corresponds to a modified MA model.

Another well-characterized system is the rice–RYMV interaction. Recessive resistance in rice to RYMV is conferred by *rymv1*, encoding eIF(iso)4G (Albar *et al.*, 2006). Different mutations in RYMV VPg determine pathogenicity on the different resistance alleles at *rymv1*. Mutations at five amino acid positions in the central region of the VPg are involved in overcoming *rymv1–2*, and there is evidence of diversifying selection at these positions (Pinel-Galzi *et al.*, 2007). A high percentage (~17%) of field isolates of RYMV from Africa were pathogenic on either allele *rymv1–2* or *rymv1–3*, and fewer (~5%) on both (Traoré *et al.*, 2006). No fitness penalty for pathogenicity on *rymv1–2* was found in passage competition experiments with nonpathogenic genotypes (Sorho *et al.*, 2005).

V. COSTS OF PATHOGENICITY AND RESISTANCE DURABILITY

The use of resistance bred into cultivars is a preferred strategy for the control of infectious diseases of plants. However, the advantages of resistance for the control of plant pathogens are countered by the common short life of the resistant variety, as the protection conferred by the resistance factor may be lost due to the increase in frequency of resistance-breaking genotypes in the pathogen's population (Kang *et al.*, 2005a; Khetarpal *et al.*, 1998; Maule *et al.*, 2007). Hence, a major interest in the study of plant–pathogen coevolution is to understand the factors that lead to resistance breakage, and to predict the durability of the protection conferred by resistance factors. As resistance durability, by definition (Johnson, 1979), can only be known *a posteriori*, this is rather a difficult task.

It has long been observed that, on average, the life of resistance factors deployed against viruses is considerably longer than that deployed against cellular plant pathogens. Thus, the life of resistance factors deployed against fungi and oomycetes was of 7.3 years (average for 27 host–pathogen systems from data in McDonald and Linde, 2002), while for viruses it was of 12.8 years (average for 25 host–pathogen systems from data in García-Arenal and McDonald, 2003) (Fig. 2). While earlier analyses identified the inheritance and mechanisms of the resistance as a

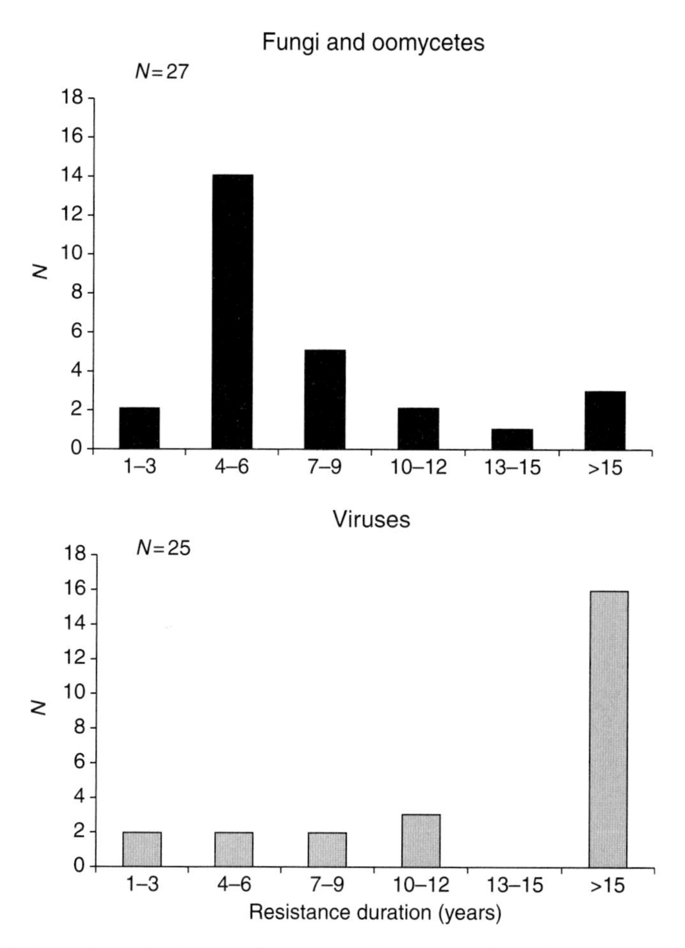

FIGURE 2 Duration of resistance factors bred into crops. The distribution of the effective duration of resistance factors to fungi and oomycetes over 27 pathosystems, or to viruses over 25 pathosystems is shown. Mean duration of resistance was 7.3 $\pm$ 4.0 and 12.8 $\pm$ 4.9 years (mean $\pm$ standard deviation) to fungi and oomycetes and to viruses, respectively. Median duration values were 4–6 and over 15 years for resistance to fungi and oomycetes and to viruses, respectively. Data are from García-Arenal and McDonald (2003) and McDonald and Linde (2002).

major factor in its durability (Fraser, 1990), the capacity of the virus to evolve as a factor determining resistance durability received more attention later on. Thus, the appearance of resistance-breaking virus genotypes on 10 monogenic dominant resistance factors was partially related to the number of amino acid substitutions required to convert the avirulence factor into a virulent one (Harrison, 2002). The analysis of the effective life of 50 resistance factors (including monogenic dominant, monogenic recessive, and polygenic) in relation to a compound risk index based on life history traits affecting the evolutionary potential of the virus, indicated a relationship between evolutionary potential and resistance durability (García-Arenal and McDonald, 2003). While this analysis suggested a broad relationship between the evolutionary potential of the virus and resistance durability, it failed to explain why different resistance factors deployed to control the same virus species in the same or in different host species often had different effective lives. Janzac *et al.* (2009) have re-examined the relationship between virus evolvability and resistance durability, not finding a relationship between the risk index proposed by García-Arenal and McDonald (2003) and resistance duration for a set of 14 dominant and 5 recessive monogenic resistance factors in 20 pathosystems. They neither found a relationship between the nucleotide diversity of the genes encoding for the avirulence factors in the different virus species considered. However, they found a significant association between the evolutionary constraints on avirulence, measured as a relationship between nucleotide substitutions at nonsynonymous and synonymous sites (d_N/d_S ratios) and resistance durability. While in both studies association and correlation between virus factors and resistance durability was always weak, regardless of statistical signification, it was clearly determined that the inheritance of the resistance was not a factor in its durability (García-Arenal and McDonald, 2003; Janzac *et al.*, 2009).

Thus, current evidence points to the evolvability of the avirulence factor itself as a predictor to the durability of a resistance factor. For many analyzed pathosystems, resistance-breaking virus isolates have been reported to occur in the field without becoming prevalent in the virus population (García-Arenal and McDonald, 2003; Janzac *et al.*, 2009). Hence, all evidence suggests that the cost of pathogenicity may be a major determinant of the durability of a resistance. It is important to consider that functional constraints on protein evolution, as uncovered by the d_N/d_S ratio, would explain only in part the costs of pathogenicity. This cost, that is, the selection against unnecessary avirulence in the absence of resistance, will depend also on other evolutionary factors, some of which will be determined by intrinsic traits of the small genomes of viruses. An example would be constraints to recombination due to multifunctionality of genes, or due to epistatic interaction both within and among genes (Escriu *et al.*, 2007; Lefeuvre *et al.*, 2007; Martin *et al.*, 2005; Sanjuán *et al.*, 2004). Other factors will be related to the virus life history rather than to the virus genome structure. This will be the case for effective population size or gene and genome flow between viral populations. Thus, the relationship between costs of pathogenicity and resistance durability is complex, and much in need of further study.

VI. CONCLUDING REMARKS

Although it is currently assumed that plants and viruses coevolve, evidence in support of this hypothesis is quite weak. A major conclusion of this review is that research in several areas related to plant–virus coevolution is badly needed and should be encouraged.

A major limitation of current knowledge is that most data taken as evidence for plant–virus coevolution derive from the analysis of highly virulent viruses infecting crops, and is mostly limited to the virus side, that is, to the evolution of the virus population in response to the use of resistance factors in the crop directed at controlling virus-induced diseases. Few data are available on the evolution of the host in response to virus infection. The occurrence in crops and their wild relatives of resistance factors effective against viruses is usually taken as evidence of virus–host coevolution, but those factors could have evolved under the pressure of other pathogens or herbivores. It has been proposed that diversification of identified virus taxa, which are those infecting crops, occurred after the expansion of agriculture, and was driven by agriculture-associated ecological changes (Fargette *et al.*, 2008; Gibbs *et al.*, 2008). Little is known of virus–plant interactions in wild ecosystems, or on whether these interactions are pathogenic or mutualistic. There is an urgent need of studies on the occurrence of viruses in wild plants and on the effect of virus infection on wild plant fitness. Such studies are a prerequisite to analyze plant–virus coevolution and put it in a similar ground to current knowledge on the coevolution of plants and cellular pathogens, which has been carried on for decades.

In recent years, huge progress has been achieved in understanding the molecular aspects of plant–virus interactions, as determined by both dominant and recessive resistance (Moffett, 2009; Truniger and Aranda, 2009). Other defense mechanisms, such as quantitative resistance or tolerance, have received little attention. Population genetic analyses of resistance and pathogenicity factors are still scarce, and more effort is needed also in this area. Notably, there are few instances of the joint analysis of resistance and pathogenicity in the same plant–virus system, which is a drawback to derive general conclusions. Analyses such as those published for the pepper–PVY and rice–RYMV systems are in urgent need for GFG-like interactions.

Current knowledge on the molecular mechanisms underlying virus recognition by plants, defense reactions, and pathogenicity factors should be considered by scientists involved in theoretical analyses on host–pathogen coevolution. Ecological and epidemiological factors are currently being incorporated into theoretical models (e.g., Tellier and Brown, 2007, 2009), but this is not yet the case for mechanistic aspects such as broad recognition of groups of taxa (i.e., reduced specificity of recognition) or recognition involving multiprotein complexes. Theoreticians should also consider the peculiarities of viruses as pathogens and as evolving entities, for instance how would high pathogenicity costs resulting from limited evolvability affect current models of host–pathogen coevolution.

If the analysis of plant–pathogen coevolution is a promising area of research, with deep academic and applied consequences, this is even more so for the specific case of plant–virus coevolution, a field still in its infancy. We hope that this review will contribute to drive the attention of scientists to this most interesting field.

ACKNOWLEDGMENTS

We want to thank Dr. Soledad Sacristán for critical reading of a previous version of this review. This work was in part supported by grant AGL2008-02458, Plan Nacional de I+D+i, Spain.

REFERENCES

Agnew, P., Koella, J. C., and Michalakis, Y. (2000). Host life history responses to parasitism. *Microbes Infect.* **2:**891–896.

Agrawal, A. F., and Lively, C. M. (2002). Infection genetics: Gene-for-gene versus matching-alleles models and all points in between. *Evol. Ecol. Res.* **4:**79–90.

Agudelo-Romero, P., de la Iglesia, F., and Elena, S. E. (2008). The pleiotropic cost of host-specialization in Tobacco etch potyvirus. *Infect. Genet. Evol.* **8:**806–814.

Albar, L., Ndjiondjop, M. N., Esshak, Z., Berger, A., Pinel, A., Jones, M., Fargette, D., and Ghesquiere, A. (2003). Fine genetic mapping of a gene required for Rice yellow mottle virus cell-to-cell movement. *Theor. Appl. Genet.* **107:**371–378.

Albar, L., Bangratz-Reyser, M., Hebrard, E., Ndjiondjop, M., Jones, M., and Ghesquière, A. (2006). Mutations in the eIF(iso)4G translation initiation factor confer high resistance to *Rice yellow mottle virus. Plant J.* **47:**417–426.

Allen, R. L., Bittner-Eddy, P. D., Greeville-Briggs, L. J., Meitz, J. C., Rehmany, A. P., Rose, L. E., and Beynon, J. L. (2004). Host–parasite coevolutionary conflict between Arabidopsis and downy mildew. *Science* **306:**1957–1960.

Anderson, R. M., and May, R. M. (1982). Coevolution of hosts and parasites. *Parasitology* **85:**411–426.

Ayme, V., Souche, S., Caranta, C., Jacquemond, M., Chadoeuf, J., Palloix, A., and Moury, B. (2006). Different mutations in the genome-linked protein VPg of *Potato virus Y* confer virulence on the pvr2^3 resistance in pepper. *Mol. Plant Microbe Interact.* **19:**557–563.

Ayme, V., Petit-Pierre, J., Souche, S., Palloix, A., and Moury, B. (2007). Molecular dissection of the potato virus Y VPg virulence factor reveals complex adaptations to the pvr2 resistance allelic series in pepper. *J. Gen. Virol.* **88:**1594–1601.

Barker, H., and Harrison, B. D. (1986). Restricted distribution of potato leafroll virus antigen in resistant potato genotypes and its effect on transmission of the virus by aphids. *Ann. Appl. Biol.* **109:**595–604.

Barrett, L. G., Thrall, P. H., Burdon, J. J., and Linde, C. C. (2008). Life history determines genetic structure and evolutionary potential of host–parasite interactions. *Trends Ecol. Evol.* **23:**678–685.

Bendahmane, A., Kanyuka, K. V., and Baulcombe, D. C. (1997). High resolution and physical mapping of the *Rx* gene for extreme resistance to potato virus X in tetraploid potato. *Theor. Appl. Genet.* **95:**153–162.

Bendahmane, A., Kanyuka, K., and Baulcombe, D. C. (1999). The *Rx* gene from potato controls separate virus resistance and cell death responses. *Plant Cell* **11:**781–792.

Bendahmane, A., Quercy, M., Kanyuka, K., and Baulcombe, D. C. (2000). *Agrobacterium* transient expression system as a tool for the isolation of disease resistant genes: Application to the *Rx2* locus in potato. *Plant J.* **21:**73–81.

Bergelson, J., and Purrington, C. B. (1996). Surveying patterns in the cost of resistance in plants. *Am. Nat.* **148:**536–558.

Bergelson, J., Dwyer, G., and Emerson, J. J. (2001). Models and data on plant-enemy coevolution. *Annu. Rev. Genet.* **35:**469–499.

Berzal-Herranz, A., de la Cruz, A., Tenllado, F., Díaz-Ruíz, J. R., López, L., Sanz, A. I., Vaquero, C., Serra, M. T., and García-Luque, I. (1995). The *Capsicum* L^3 gene-mediated resistance against the tobamoviruses is elicited by the coat protein. *Virology* **209:**498–505.

Bromonschenkel, S. H., Frary, A., and Tanksley, S. D. (2000). The broad-spectrum tospovirus resistance gene *Sw-5* of tomato is a homolog of the root-knot nematode resistance gene. *Mi. Mol. Plant Microbe Interact.* **13:**1130–1138.

Brown, J. K. M. (2003). A cost of disease resistance: Paradigm or peculiarity? *Trends Genet.* **19:**667–671.

Bruun-Rasmussen, M., Moller, I. S., Tulinius, G., Hansen, J. K. R., Lund, O. S., and Johansen, I. E. (2007). The same allele of translation initiation factor 4E mediates resistance against two *Potyvirus* spp. in *Pisum sativum*. *Mol. Plant Microbe Interact.* **20:**1075–1082.

Burdon, J. J., and Thrall, P. H. (2009). Coevolution of plants and their pathogens in natural habitats. *Science* **324:**755–756.

Caplan, J., Padmanabhan, M., and Dinesh-Kumar, S. P. (2008). Plant NB–LRR immune receptors: From recognition to transcriptional reprogramming. *Cell Host Microbe* **3:**126–135.

Caplan, J. L., Mimillapalli, P., Burch-Smith, T. M., Czymmek, H., and Dinesh-Kumar, S. P. (2008). Chloroplastic protein NRIP1 mediates innate immune receptor recognition of a viral effector. *Cell* **132:**449–462.

Caranta, C., Palloix, A., GebreSelassie, K., Lefebvre, V., Moury, B., and Daubeze, A. M. (1996). A complementation of two genes originating from susceptible *Capsicum annuum* lines confers a new and complete resistance to pepper veinal mottle virus. *Phytopathology* **86:**739–743.

Carrasco, P., Daros, J. A., Agudelo-Romero, P., and Elena, S. F. (2006). A real-time RT-PCR assay for quantifying the fitness of tobacco etch virus in competition experiments. *J. Virol. Methods* **139:**181–188.

Charron, C., Nicolaï, M., Gallois, J.-L., Robaglia, C., Moury, B., Palloix, A., and Caranta, C. (2008). Natural variation and functional analyses provide evidence for co-evolution between plant eIF4E and potyviral VPg. *Plant J.* **54:**56–68.

Chisholm, S. T., Mahajan, S. K., Whitham, S. A., Yamamoto, M. L., and Carrington, J. C. (2000). Cloning of the Arabidopsis *RTM1* gene, which controls restriction of long-distance movement of tobacco etch virus. *Proc. Natl. Acad. Sci. USA* **97:**489–494.

Clarke, D. D. (1986). Tolerance of parasites and disease in plants and its significance in host–parasite interactions. *Adv. Plant Pathol.* **5:**161–198.

Cooley, M. B., Pathirana, S., Wu, H. J., Kachroo, P., and Klessig, D. F. (2000). Members of the Arabidopsis *HRT/RPP8* family of resistance genes confer resistance to both viral and oomycete pathogens. *Plant Cell* **12:**663–676.

Culver, J. N. (2002). Tobacco mosaic virus assembly and disassembly: Determinants in pathogenicity and resistance. *Annu. Rev. Phytopathol.* **40:**287–308.

Dangl, J. L., and Jones, J. D. G. (2001). Plant pathogens and integrated defence responses to infection. *Nature* **411:**826–833.

D'Arcy, C. J., Eastburn, D. M., and Schumann, G. L. (2001). Illustrated glossary of plant pathology. *Plant Health Instr.* doi: 10.1094/PHI-I-2001-0219-01.

Decroocq, V., Salvador, B., Sicard, O., Glasa, M., Cosson, P., Svanella-Dumas, L., Revers, F., Garcia, J. A., and Candresse, T. (2009). The determinant of Potyvirus ability to overcome the RTM resistance of *Arabidopsis thaliana* maps to the N-terminal region of the coat protein. *Mol. Plant Microbe Interact.* **22:**1302–1311.

Dempsey, D. A., Pathirana, M. S., Wobbe, K. K., and Klessig, D. F. (1997). Identification of an *Arabidopsis* locus required for resistance to turnip crinkle virus. *Plant J.* **11:**301–311.

Desbiez, C., Wipf-Scheibel, C., and Lecoq, H. (2002). Biological and serological variability, evolution and molecular epidemiology of *Zucchini yellow mosaic virus* (ZYMV, Potyvirus) with special reference to Caribbean islands. *Virus Res.* **85:**5–16.

Desbiez, C., Gal-On, A., Girard, M., Wipf-Scheibel, C., and Lecoq, H. (2003). Increase in *Zucchini yellow mosaic virus* symptom severity in tolerant zucchini cultivars is related to a point mutation in P3 protein and is associated with a loss of relative fitness on susceptible plants. *Phytopathology* **93:**1478–1484.

Díaz, J. A., Nieto, C., Moriones, E., Truniger, V., and Aranda, M. (2004). Molecular characterization of a *Melon necrotic spot virus* strain that overcomes the resistance in melon and non-host plants. *Mol. Plant Microbe Interact.* **17:**668–675.

Díaz-Pendón, J. A., Truniger, V., Nieto, C., García-Mas, J., Bendahmane, A., and Aranda, M. (2004). Advances in understanding recessive resistance to plant viruses. *Mol. Plant Pathol.* **5:**223–233.

Dodds, P. N., Lawrence, G. J., Catanzariti, A.-M., Teh, T., Wang, C.-I., Ayliffe, M. A., Kobe, B., and Ellis, J. G. (2006). Direct protein interaction underlies gene-for-gene specificity and coevolution of the flax resistance genes and flax rust avirulence genes. *Proc. Natl. Acad. Sci. USA* **103:**8888–8893.

Drake, J. W., Charlesworth, B., Charlesworth, D., and Crow, J. F. (1998). Rates of spontaneous mutation. *Genetics* **148:**1667–1686.

Elena, S. F., Carrasco, P., Darós, J. A., and Sanjuán, R. (2006). Mechanisms of genetic robustness in RNA viruses. *EMBO Rep.* **7:**168–173.

Escriu, F., Perry, K. L., and García-Arenal, F. (2000). Transmissibility of *Cucumber mosaic virus* by *Aphis gossypii* correlates with viral accumulation and is affected by the presence of its satellite RNA. *Phytopathology* **90:**1068–1072.

Escriu, F., Fraile, A., and García-Arenal, F. (2007). Constraints to genetic exchange support gene coadaptation in a tripartite RNA virus. *PLoS Pathog.* **3:**e8.

Fargette, D., Pinel-Galzi, A., Sereme, D., Lacombe, S., Hebrard, E., Traore, O., and Konate, G. (2008). Diversification of *Rice yellow mottle virus* and related viruses spans the history of agriculture from the Neolithic to the present. *PLoS Pathog.* **4:**e1000125.

Farnham, G., and Baulcombe, D. C. (2006). Artificial evolution extends the spectrum of viruses that are targeted by a disease-resistance gene from potato. *Proc. Natl. Acad. Sci. USA* **103:**18828–18833.

Fisher, M. L., and Kyle, M. M. (1994). Inheritance of resistance to potyviruses in *Phaseolus vulgaris* L. III. Cosegregation of phenotypically similar dominant response to nine potyviruses. *Theor. Appl. Genet.* **89:**818–823.

Flor, H. H. (1955). Host–parasite interactions in flax—Its genetics and other implications. *Phytopathology* **45:**680–685.

Flor, H. H. (1971). Current status of the gene-for-gene concept. *Annu. Rev. Phytopathol.* **9:**275–296.

Foxe, M. J., and Rochow, W. F. (1975). Importance of virus source leaves in vector specificity of barley yellow dwarf virus. *Phytopathology* **65:**1124–1129.

Fraile, A., Pagan, I., Anastasio, G., Saez, E., and Garcia-Arenal, F. (2010). High fitness penalties associated with increased pathogenicity in a plant virus. (in press).

Frank, S. A. (1996). Models of parasite virulence. *Q. Rev. Biol.* **71:**37–78.

Fraser, R. S. S. (1990). The genetics of resistance to plant viruses. *Annu. Rev. Phytopathol.* **28:**179–200.

Friedman, A. R., and Baker, B. J. (2007). The evolution of resistance genes in multi-protein plant resistance systems. *Curr. Opin. Genet. Dev.* **17:**493–499.

Friess, N., and Maillet, J. (1996). Influence of cucumber mosaic virus infection on the intraspecific competitive ability and fitness of purslane (*Portulaca oleracea*). *New Phytol.* **132:**103–111.

Gandon, S., van Baalen, M., and Jansen, V. A. A. (2002). The evolution of parasite virulence, superinfection, and host resistance. *Am. Nat.* **159:**658–669.

Gao, Z., Johansen, E., Eyers, S., Thomas, C. L., Noel Ellis, T. H., and Maule, A. J. (2004). The potyvirus recessive resistance gene, *sbm1*, identifies a novel role for translation initiation factor eIF4E in cell-to-cell trafficking. *Plant J.* **40:**376–385.

García-Arenal, F., and McDonald, B. A. (2003). An analysis of the durability of resistance to plant viruses. *Phytopathology* **93:**941–952.

Garrido-Ramirez, E. R., Sudarshana, M. R., Lucas, W. J., and Gilbertson, R. L. (2000). Bean dwarf mosaic virus BV1 protein is a determinant of the hypersensitive response and avirulence in *Phaseolus vulgaris*. *Mol. Plant Microbe Interact.* **13:**1184–1194.

Gibbs, A. J. (1980). A plant virus that partially protects its wild legume host against herbivores. *Intervirology* **13:**42–47.

Gibbs, A. J., Ohshima, K., Phillips, M. J., and Gibbs, M. J. (2008). The prehistory of potyviruses: Their initial radiation was during the dawn of agriculture. *PLoS ONE* **3:**e2523.

Goulden, M. G., Köhm, B. A., Santa Cruz, S., Kavanagh, T. A., and Baulcombe, D. A. (1993). A feature of the coat protein of potato virus X affects both induced virus resistance in potato and viral fitness. *Virology* **197:**293–302.

Hajimorad, M. R., Eggenberger, A. L., and Hill, J. H. (2005). Loss and gain of elicitor function of *Soybean mosaic virus* G7 provoking *Rsv1*-mediated lethal systemic hypersensitive response maps to P3. *J. Virol.* **79:**1215–1222.

Hamada, H., Takeuchi, S., Kiba, A., Tsuda, S., Hikichi, Y., and Okuno, T. (2002). Amino acid changes in *Pepper mild mottle virus* coat protein that affect L^3 gene-mediated resistance in pepper. *J. Gen. Plant Pathol.* **68:**155–162.

Hamada, H., Tomita, R., Iwadate, Y., Kobayashi, K., Minemura, I., Takeuchi, S., Hikichi, Y., and Suzuki, K. (2007). Cooperative effect of two amino acid mutations in the coat protein of *Pepper mild mottle virus* overcomes L^3-mediated resistance in *Capsicum* plants. *Virus Genes* **34:**205–214.

Hanada, K., and Harrison, B. D. (1977). Effects of virus genotype and temperature on seed transmission of nepoviruses. *Ann. Appl. Biol.* **85:**79–92.

Harrison, B. D. (2002). Virus variation in relation to resistance breaking plants. *Euphytica* **124:**181–192.

Hayes, A. J., Jeong, S. C., Gore, M. A., Yu, Y. G., Buss, G. R., Tolin, S. A., and Maroof, M. A. S. (2004). Recombination within a nucleotide-binding site/leucine-rich-repeat gene cluster produces new variants conditioning resistance to soybean mosaic virus in soybeans. *Genetics* **166:**493–503.

Hebrard, E., Pinel-Galzi, A., and Fergette, D. (2008). Virulence domain of the RYMV genome-linked viral protein VPg towards rice *rymv1-2*-mediated resistance. *Arch. Virol.* **153:**1161–1164.

Holmes, F. O. (1937). Genes affecting response of *Nicotiana tabacum* hybrids to tobacco mosaic virus. *Science* **85:**104–105.

Hulbert, S. C., Webb, C. A., Smith, S. M., and Sun, Q. (2001). Resistance gene complexes: Evolution and utilization. *Annu. Rev. Phytopathol.* **39:**285–312.

Ishibashi, K., Masuda, K., Naito, S., Meshi, T., and Ishikawa, M. (2007). An inhibitor of viral RNA replication is encoded by a plant resistance gene. *Proc. Natl. Acad. Sci. USA* **104:**13833–13838.

Janzac, B., Fabre, F., Palloix, A., and Moury, B. (2009). Constraints on evolution of virus avirulence factors predict the durability of corresponding plant resistances. *Mol. Plant Pathol.* **10:**599–610.

Jarozs, A. M., and Davelos, A. I. (1995). Effects of disease in wild plant populations and the evolution of pathogen aggressiveness. *New Phytol.* **129:**371–387.

Jenner, C. E., Sánchez, F., Nettleship, S. B., Foster, G. D., Ponz, F., and Walsh, J. A. (2000). The cylindrical inclusion gene of *Turnip mosaic virus* encodes a pathogenic determinant to the *Brassica* resistance gene *TuRB01*. *Mol. Plant Microbe Interact.* **13:**1102–1108.

Jenner, C. E., Wang, X., Ponz, F., and Walsh, J. A. (2002). A fitness cost for *Turnip mosaic virus* to overcome host resistance. *Virus Res.* **86:**1–6.

Jenner, C. E., Tomimura, K., Oshima, K., Hughes, S. L., and Walsh, J. A. (2002). Mutations in *Turnip mosaic virus* P3 and cylindrical inclusion proteins are separately required to overcome two *Brassica napus* resistance genes. *Virology* **300:**50–59.

Jeong, R. D., Chandra-Shekara, A. C., Kachroo, A., Klessig, D. F., and Kachroo, P. (2008). HRT-mediated hypersensitive response and resistance to *Turnip crinkle virus* in *Arabidopsis* does not require the function of TIP, the presumed guardee protein. *Mol. Plant Microbe Interact.* **21:**1316–1324.

Jiménez-Martínez, E. S., and Bosque-Pérez, N. A. (2004). Variation in barley yellow dwarf virus transmission efficiency by *Rhopalosiphum padi* (Homoptera: Aphididae) after acquisition from transgenic and nontransformed wheat genotypes. *J. Econ. Entomol.* **97:**109–127.

Johansen, I. E., Lund, O. S., Hjulsager, C. K., and Laurse, J. (2001). Recessive resistance in *Pisum sativum* and *Potyvirus* pathotype resolved in a gene-for-cistron correspondence between host and virus. *J. Virol.* **75:**6609–6614.

Johnson, R. (1979). The concept of durable resistance. *Phytopathology* **69:**198–199.

Jones, J. D. G., and Dangl, J. L. (2006). The plant immune system. *Nature* **444:**323–329.

Kang, B. C., Yeam, I., and Jahn, M. M. (2005). Genetics of virus resistance. *Annu. Rev. Phytopathol.* **43:**581–621.

Kang, B. C., Yeam, I., Frantz, J. D., Murphy, J. F., and Jahn, M. M. (2005). The *pvr1* locus in pepper encodes a translation initiation factor eIF4E that interacts with *Tobacco etch virus* VPg. *Plant J.* **41:**392–405.

Kanyuka, K., McGrann, G., Alhudaib, K., Hariri, D., and Adams, M. J. (2004). Biological and sequence analysis of a novel European isolate of *Barley mild mosaic virus* that overcomes the barley *rym5* resistance gene. *Arch. Virol.* **149:**1469–1480.

Kawecki, T. J., and Ebert, D. (2004). Conceptual issues of local adaptation. *Ecol. Lett.* **7:**1225–1241.

Kelly, S. E. (1994). Viral pathogens and the advantage of sex in the perennial grass *Anthoxantum odoratum*: A review. *Philos. Trans. R. Soc. Lond. B* **346:**295–302.

Khetarpal, R. K., Maisonneuve, B., Maury, Y., Chalhoub, B., Dianant, S., Lecoq, H., and Varma, A. (1998). Breeding for resistance to plant viruses. *In* "Plant Virus Disease Control" (A. Hadidi, R. K. Khetarpal, and H. Koganzawa, eds.), pp. 14–33. American Phytopathological Society Press, St. Paul, MN.

Knorr, D. A., and Dawson, W. O. (1988). A point mutation in the tobacco mosaic virus capsid protein gene induces hypersensitivity in *Nicotiana sylvestris*. *Proc. Natl. Acad. Sci. USA* **85:**170–174.

Kovalchuk, I., Kovalchuk, O., Kalck, V., Boyko, V., Filkowski, J., Heinlein, M., and Hohn, B. (2003). Pathogen-induced systemic plant signal triggers DNA rearrangements. *Nature* **423:**760–762.

Lanfermeijer, F. C., Dijkhuis, J., Sturre, M. J. G., and Hille, J. (2003). Cloning and characterization of the durable tomato mosaic virus resistance gene *Tm-2*^{2} from *Lycopersicon esculentum*. *Plant Mol. Biol.* **52:**1037–1049.

Lanfermeijer, F. C., Warmink, J., and Hille, J. (2005). The products of the broken *Tm-2* and the durable *Tm-2*^{2} resistance genes from tomato differ in four amino acids. *J. Exp. Bot.* **56:**2925–2933.

Lefeuvre, P., Lett, J. M., Raynaud, B., and Martin, D. P. (2007). Avoidance of protein fold disruption in natural virus recombinants. *PLoS Pathog.* **3:**e181.

Malmstrom, C. M., Stoner, C. J., Brandenburg, S., and Newton, L. A. (2006). Virus infection and grazing exert counteracting influences on survivorship of native bunchgrass seedlings competing with invasive exotics. *J. Ecol.* **94:**264–275.

Margaria, P., Ciuffo, M., Pacifico, D., and Turina, M. (2007). Evidence that the nonstructural protein of *Tomato spotted wilt virus* is the avirulence determinant in the interaction with resistant pepper carrying the *Tsw* gene. *Mol. Plant Microbe Interact.* **20:**547–558.

Martin, D. P., van der Walt, E., Posada, D., and Rybicki, E. P. (2005). The evolutionary value of recombination is constrained by genome modularity. *PLoS Genet.* **1:**475–479.

Maskell, L. C., Raybould, A. F., Cooper, J. I., Edwards, M. L., and Gray, A. J. (1999). Effects of turnip mosaic virus and turnip yellow mosaic virus on the survival, growth and reproduction of wild cabbage (*Brassica oleracea*). *Ann. Appl. Biol.* **135:**401–407.

Maule, A., Caranta, C., and Boulton, M. (2007). Sources of natural resistance to plant viruses: Status and prospects. *Mol. Plant Pathol.* **8:**223–231.

Mauricio, R. (1998). Costs of resistance to natural enemies in field populations of the annual plant *Arabidopsis thaliana*. *Am. Nat.* **151:**20–28.

Mauricio, R., Stahl, E. A., Korves, T., Tian, D. C., Kreitman, M., and Bergelson, J. (2003). Natural selection for polymorphism in the disease resistance gene Rps2 of *Arabidopsis thaliana*. *Genetics* **163:**735–746.

McDonald, B. A., and Linde, C. (2002). Pathogen population genetics, evolutionary potential, and durable resistance. *Annu. Rev. Phytopathol.* **40:**340–379.

McDowell, J. M., and Simon, S. A. (2006). Recent insights into R gene evolution. *Mol. Plant Pathol.* **7:**437–448.

McDowell, J. M., Dhandaydham, M., Long, T. A., Aarts, M. G. M., Goff, S., Holub, E. B., and Dangl, J. L. (1998). Intragenic recombination and diversifying selection contribute to the evolution of downy mildew resistance at the *RPP8* locus of *Arabidopsis*. *Plant Cell* **10:**1861–1874.

Meshi, T., Motoyoshi, F., Adachi, A., Watanabe, Y., and Okada, Y. (1988). Two concomitant base substitutions in the putative replicase genes of tobacco mosaic virus confer the ability to overcome the effects of tomato resistance gene, *Tm-1*. *EMBO J.* **7:**1575–1581.

Meshi, T., Motoyoshi, F., Maeda, T., Yoshikawa, S., Watanabe, Y., and Okada, Y. (1989). Mutations in the tobacco mosaic virus 30-kD protein gene overcome *Tm-2* resistance in tomato. *Plant Cell* **1:**515–522.

Mestre, P., Brigneti, G., Durrant, M. C., and Baulcombe, D. C. (2003). Potato virus Y NIa protease activity is not sufficient for elicitation of *Ry*-mediated disease resistance in potato. *Plant J.* **36:**755–761.

Moffett, P. (2009). Mechanisms of recognition in *R* gene mediated resistance. *Adv. Virus Res.* **75:**1–33.

Molinier, J., Ries, G., Zipfel, C., and Hohn, B. (2006). Transgeneration memory of stress in plants. *Nature* **442:**1046–1049.

Moury, B., Morel, C., Johansen, E., Guilbaud, L., Souche, S., Ayme, V., Caranta, C., Palloix, A., and Jacquemond, M. (2004). Mutations in *Potato virus Y* genome-linked protein determine virulence toward recessive resistances in *Capsicum annuum* and *Lycopersicon hirsutum*. *Mol. Plant Microbe Interact.* **17:**322–329.

Murant, A. F., Taylor, C. E., and Chambers, J. (1968). Properties, relationships and transmission of a strain of raspberry ringspot virus infecting raspberry cultivars immune to the common Scottish strain. *Ann. Appl. Biol.* **61:**175–186.

Nicaise, V., German-Retana, S., Sanjuan, R., Dubrana, M. P., Mazier, M., Maisonneuve, B., Candresse, T., Caranta, C., and LeGall, O. (2003). The eukaryotic translation initiation factor 4E controls lettuce susceptibility to the potyvirus *Lettuce mosaic virus. Plant Physiol.* **132;**1272–1282.

Nieto, C., Morales, M., Orjeda, G., Clepet, C., Monfort, A., Sturbois, B., Puigdomenech, P., Pitrat, M., Caboche, M., Dogimont, C., García-Mas, J., Aranda, M. A., *et al.* (2006). An *eIF4E* allele confers resistance to an uncapped and non-polyadenylated RNA virus in melon. *Plant J.* **48:**452–462.

Padgett, H. S., and Beachy, R. N. (1993). Analysis of a tobacco mosaic virus strain capable of overcoming N gene-mediated resistance. *Plant Cell* **5:**577–586.

Padgett, H. S., Watanabe, Y., and Beachy, R. N. (1997). Identification of the TMV replicase sequence that activates the *N* gene mediated hypersensitive response. *Mol. Plant Microbe Interact.* **10:**709–715.

Pagán, I., Alonso-Blanco, C., and García-Arenal, F. (2007). The relationship of within-host multiplication and virulence in a plant–virus system. *PLoS ONE* **2:**e786.

Pagán, I., Alonso-Blanco, C., and García-Arenal, F. (2008). Host responses in life-history traits and tolerance to virus infection in *Arabidopsis thaliana. PLoS Pathog.* **4:**e1000124.

Pagán, I., Alonso-Blanco, C., and García-Arenal, F. (2009). Differential tolerance to direct and indirect density-dependent costs of viral infection in *Arabidopsis thaliana. PLoS Pathog.* **5:**e1000531.

Pallett, D. W., Thurston, M. I., Cortina-Borja, M., Edwards, M. L., Alexander, M., Mitchell, E., Raybould, A. F., and Cooper, J. I. (2002). The incidence of viruses in wild *Brassica rapa* ssp. *sylvestris* in southern England. *Ann. Appl. Biol.* **141:**163–170.

Palukaitis, P., and Carr, J. P. (2008). Plant resistance responses to viruses. *J. Plant Pathol.* **90:**153–171.

Parker, M. A. (1994). Pathogens and sex in plants. *Evol. Ecol.* **8:**560–584.

Parniske, M., HammondKosack, K. E., Golstein, C., Thomas, C. M., Jones, D. A., Harrison, K., Wulff, B. B. H., and Jones, J. D. G. (1997). Novel disease resistance specificities result from sequence exchange between tandemly repeated genes at the Cf-4/9 locus of tomato. *Cell* **91:**821–832.

Pinel-Galzi, A. S., Rakotomalala, M., Sangu, E., Sorho, F., Kanyeka, Z., Traore, O., Sereme, D., Poulicard, N., Rabenantoandro, Y., Sere, Y., Konate, G., Ghesquiere, A., *et al.* (2007). Theme and variations in the evolutionary pathways to virulence of an RNA plant virus species. *PLoS Pathog.* **3:**e180.

Pirone, T. P., and Megahed, E. (1966). Aphid transmissibility of some purified viruses and viral RNAs. *Virology* **30:**630–637.

Power, A. G., and Mitchell, C. E. (2004). Pathogen spillover in disease epidemics. *Am. Nat.* **164:**S79–S89.

Read, A. F. (1994). The evolution of virulence. *Trends Microbiol.* **2:**73–76.

Ren, T., Qu, F., and Morris, T. J. (2000). *HRT* gene function requires interaction between a NAC protein and viral capsid protein to confer resistance to *Turnip crinkle virus. Plant Cell* **12:**1917–1925.

Roossinck, M. J. (2005). Symbiosis versus competition in plant virus evolution. *Nat. Rev. Microbiol.* **3:**917–924.

Rose, L. E., Bittner-Eddy, P. D., Langley, C. H., Holub, E. B., Michelmore, R. W., and Beynon, J. L. (2004). The maintenance of extreme amino acid diversity at the disease resistance gene, Rpp 13, in *Arabidopsis thaliana. Genetics* **166:**1517–1527.

Roudet-Tavert, G., Michon, T., Walter, J., Delaunay, T., Redondo, E., and Le Gall, O. (2007). Central domain of a potyvirus VPg is involved in the interaction with the host translation initiation factor eIF4E and the viral protein HcPro. *J. Gen. Virol.* **88:**1029–1033.

Ruffel, S., Dussault, M. H., Palloix, A., Moury, B., Bendahmane, A., Robaglia, C., and Caranta, C. (2002). A natural recessive resistance gene against potato virus Y in pepper corresponds to the eukaryotic initiation factor 4E (eIF4E). *Plant J.* **32:**1067–1075.

Ruffel, S., Gallois, J. L., Lesage, M., and Caranta, C. (2005). The recessive potyvirus resistance gene *pot-1* is the tomato orthologue of the pepper *pvr2*-eIF4E gene. *Mol. Genet. Genomics* **274:**346–353.

Ruffel, S., Gallois, J. L., Moury, B., Robaglia, C., Palloix, A., and Caranta, C. (2006). Simultaneous mutations in translation initiation factors eIF4E and eIF(iso)4E are required to prevent pepper veinal mottle virus infection of pepper. *J. Gen. Virol.* **87:**2089–2098.

Sacristán, S., and García-Arenal, F. (2007). The evolution of virulence and pathogenicity in plant pathogen populations. *Mol. Plant Pathol.* **9:**369–384.

Sacristán, S., Fraile, A., Malpica, J. M., and García-Arenal, F. (2005). An analysis of host adaptation and its relationship with virulence in *Cucumber mosaic virus*. *Phytopathology* **95:**827–833.

Saito, T., Meshi, T., Takamatsu, N., and Okada, Y. (1987). Coat gene sequence of tobacco mosaic virus encodes host response determinant. *Proc. Natl. Acad. Sci. USA* **84:**6074–6077.

Salvaudon, L., Giraud, T., and Shykoff, J. A. (2008). Genetic diversity in natural populations: A fundamental component of plant–microbe interactions. *Curr. Opin. Plant Biol.* **11:**135–143.

Sanjuán, R., Moya, A., and Elena, S. F. (2004). The contribution of epistasis to the architecture of fitness in an RNA virus. *Proc. Natl. Acad. Sci. USA* **101:**15376–15379.

Schaad, M. C., Anderberg, R. J., and Carrington, J. C. (2000). Strain-specific interaction of the tobacco etch virus NIa protein with the translation initiation factor eIF4E in the yeast two-hybrid system. *Virology* **273:**300–306.

Schürch, S., and Roy, B. A. (2004). Comparing single- vs. mixed-genotype infections of *Mycosphaerella graminicola* on wheat: Effects of pathogen virulence and host tolerance. *Evol. Ecol.* **18:**1–14.

Seo, Y.-S., Rojas, M. R., Lee, J.-Y., Lee, S.-W., Jeon, J.-S., Ronald, P., Lucas, W. J., and Gilbertson, R. L. (2006). A viral resistance gene from common bean functions across plant families and is up-regulated in a non-virus-specific manner. *Proc. Natl. Acad. Sci. USA* **103:**11856–11861.

Seo, Y.-S., Jeon, J. S., Rojas, M. R., and Gilbertson, R. L. (2007). Characterization of a novel Toll/interleukin-1 receptor (TIR)-TIR gene differentially expressed in common bean (*Phaseolus vulgaris* cv. Othello) undergoing a defence response to the geminivirus *Bean dwarf mosaic virus*. *Mol. Plant Pathol.* **8:**151–162.

Sorho, F., Pinel, A., Traoré, O., Bersoult, A., Ghesquière, A., Hébrard, E., Konaté, G., Séré, Y., and Fargette, D. (2005). Durability of natural and transgenic resistances to *Rice yellow mottle* virus. *Eur. J. Plant Pathol.* **112:**349–359.

Stein, N., Perovic, D., Kumlehn, J., Pellio, B., Stracke, S., Streng, S., Ordon, F., and Graner, A. (2005). The eukaryotic translation initiation factor 4E confers multiallelic recessive *Bymovirus* resistance in *Hordeum vulgare*. *Plant J.* **42:**912–922.

Takahashi, H., Miller, J., Nonaki, Y., Takeda, M., Shah, J., Hase, S., Ikegami, M., Ehara, Y., and Dinesh-Kumar, S. P. (2002). *RCY1*, an *Arabidopsis thaliana RPP8/HRT* family resistance gene, conferring resistance to cucumber mosaic virus requires salicylic acid, ethylene and a novel signal transduction mechanism. *Plant J.* **32:**655–667.

Tameling, W. I. L., and Baulcombe, D. C. (2007). Physical association of the NB-LRR resistance protein Rx with a Ran GTPase-activating protein is required for extreme resistance to *Potato virus X*. *Plant Cell* **19:**1682–1694.

Tellier, A., and Brown, J. K. M. (2007). Stability of genetic polymorphism in host–parasite interactions. *Proc. R. Soc. Lond. B* **274:**809–817.

Tellier, A., and Brown, J. K. M. (2009). The influence of perenniality and seed banks on polymorphism in plant–parasite interactions. *Am. Nat.* **174:**769–779.

Thompson, J. N., and Burdon, J. J. (1992). Gene-for-gene coevolution between plants and parasites. *Nature* **360:**121–126.

Tomita, R., Murai, J., Miura, Y., Ishikara, H., Liu, S., Kubotera, Y., Honda, A., Hatta, R., Kuroda, T., Hamada, H., Sakamoto, M., Munemura, I., *et al.* (2008). Fine mapping and DNA fiber FISH analysis locates the tobamovirus resistance gene L^3 of *Capsicum chinense* in a 400-kb region of R-like genes cluster embedded in highly repetitive sequences. *Theor. Appl. Genet.* **117:**1107–1118.

Traoré, O., Pinel, A., Hébrard, E., Gumedzoé, M. Y. D., Fargette, D., Traoré, A. S., and Konaté, G. (2006). Occurrence of resistance-breaking isolates of *Rice yellow mottle virus* in west and central Africa. *Plant Dis.* **90:**259–263.

Truniger, V., and Aranda, M. A. (2009). Recessive resistance to plant viruses. *Adv. Virus Res.* **75:**119–159.

Truniger, V., Nieto, C., Gonzalez-Ibeas, D., and Aranda, M. A. (2008). Mechanism of plant eIF4E-mediated resistance against a Carmovirus (Tombusviridae): Cap-independent translation of a viral RNA controlled in cis by an (a)virulence determinant. *Plant J.* **56:**716–727.

Tsuda, S., Kirita, M., and Watanabe, Y. (1998). Characterization of a pepper mild mottle tobamovirus strain capable of overcoming the L^3 gene-mediated resistance, distinct from the resistance-breaking Italian isolate. *Mol. Plant Microbe Interact.* **11:**327–331.

Vallejos, C. E., Astua-Monge, G., Jones, V., Plyler, T. R., Sakiyama, N. S., and Mackenzie, S. A. (2006). Genetic and molecular characterization of the *I* locus of *Phaseolus vulgaris*. *Genetics* **172:**1229–1242.

Van den Bosch, F., Akudibilah, G., Seal, S., and Jeger, M. (2006). Host resistance and the evolutionary response of plant viruses. *J. Appl. Ecol.* **43:**506–516.

Van der Hoorn, R. A. L., De Wit, P. J. G. M., and Joosten, M. H. A. J. (2002). Balancing selection favors guarding resistance proteins. *Trends Plant Sci.* **7:**67–71.

Vanderplank, J. E. (1968). Disease Resistance in Plants. Academic Press, New York, NY.

Vidal, S., Cabrera, H., Andersson, R. A., Fredriksson, A., and Valkonen, J. P. T. (2002). Potato gene *Y-1* is an *N* gene homolog that confers cell death upon infection with *potato virus Y*. *Mol. Plant Microbe Interact.* **7:**717–727.

Wang, G. L., Ruan, D. L., Song, W. Y., Sideris, S., Chen, L., Pi, L. Y., Zhang, S., Zhang, Z., Fauquet, C., Gaut, B. S., Whalen, M. C., and Ronald, P. C. (1998). *Xa21D* encodes a receptor-like molecule with a leucine-rich repeat domain that determines race-specific recognition and is subject to adaptive evolution. *Plant Cell* **10:**765–779.

Weber, H., and Pfitzner, A. J. P. (1998). $Tm2^2$ resistance in tomato requires recognition of the carboxy terminus of the movement protein of tobacco mosaic virus. *Mol. Plant Microbe Interact.* **11:**498–503.

Weber, H., Ohnesorge, S., Silber, M. V., and Pfitzner, A. J. P. (2004). The *Tomato mosaic virus* 30 kDa movement protein interacts differentially with the resistance genes *Tm-2* and $Tm\text{-}2^2$. *Arch. Virol.* **149:**1499–1514.

Whitham, S., Dinesh-Kumar, S. P., Choi, D., Hehl, R., Corr, C., and Baker, B. (1994). The product of the tobacco mosaic virus resistance gene *N*: Similarity to Toll and the interleukin-1 receptor. *Cell* **78:**1101–1115.

Whitham, S. A., Anderberg, R. J., Chisholm, S. T., and Carrington, J. C. (2000). Arabidopsis *RTM2* gene is necessary for specific restriction of tobacco etch virus and encodes an unusual small heat shock-like protein. *Plant Cell* **12:**569–582.

Woolhouse, M. E. J., Webster, J. P., Domingo, E., Charlesworth, B., and Levin, B. R. (2002). Biological and biomedical implications of the co-evolution of pathogens and their hosts. *Nat. Genet.* **32:**569–577.

Wren, J. D., Roossinck, M. J., Nelson, R. S., Scheets, K., Palmer, M. W., and Melcher, U. (2006). Plant virus biodiversity and ecology. *PLoS Biol.* **4:**314–315.

Xu, P., Chen, F., Mannas, J. P., Feldman, T., Sumner, L. W., and Roossinck, M. J. (2008). Virus infection improves drought tolerance. *New Phytol.* **180:**911–921.

CHAPTER 2

Assessment of the Benefits and Risks for Engineered Virus Resistance

Jeremy R. Thompson* and **Mark Tepfer*,†**

Contents

* Plant Virology Group, ICGEB Biosafety Outstation, Via Piovega 23, 31056 Ca' Tron di Roncade, Italy
† Institut Jean-Pierre Bourgin UMR1318, INRA-Versailles, Versailles Cedex, France,
E-mail: Mark.Tepfer@versailles.inra.fr

Advances in Virus Research, Volume 76
ISSN 0065-3527, DOI: 10.1016/S0065-3527(10)76002-4

Abstract

Viral diseases of cultivated crops are responsible for the worldwide loss of billions of US dollars in agricultural productivity every year. Historically, this loss has been reduced or minimized principally by the implementation of specific agricultural/phytosanitary measures, and by the introduction of naturally occurring virus-resistance genes into appropriate cultivars by plant breeding. Since the first report of virus-resistant transgenic plants (VRTPs) in 1986, a remarkable diversity of virus-resistance transgenes has been developed. Despite this, to a large part due to controversy surrounding the use of genetically modified organisms, the number of commercially available VRTPs remains small. However, since the potential risks associated with VRTPs were first formulated in the early 1990s, fundamental research on plant–virus interactions and also research specifically aimed at resolving biosafety issues have greatly circumscribed the potential impact of the risks envisaged. Yet, in spite of the advances, both in strategies for creating VRTPs and in the assessment of potential risks, it remains remarkably difficult to weigh the risks/costs and benefits of different means to manage plant viral diseases, and even to make scientifically well-founded choices of the most appropriate strategy for creating VRTPs. Many of the outstanding issues concern the lack of sufficient knowledge of the breadth and durability of the resistance of VRTPs under field conditions. VRTPs will only take their appropriate place in modern agriculture when their potential users will be able to base their choices on realistic assessments of their efficacy, durability, and safety.

I. INTRODUCTION

Since genetic transformation technology was developed in the mid-1980s, the use of transgenic crops has been fraught with controversy, fuelled principally by a polarizing of public and scientific opinion (Tester, 2001). This has occurred due to a growing mutual mistrust as to, on the one hand, the true benefits of the technology, and on the other, the potential health and environmental risks involved in their use, both of which are perceived as inflated by their detractors. Here, we try to provide an overview of the development of virus-resistant transgenic plants (VRTPs), looking at both the potential advantages and disadvantages of their use. In doing this, we first take one step backwards, in order to briefly consider estimated losses caused to world agriculture and the future challenges posed by virus infection, issues which provide the justification for developing control strategies.

II. THE COST OF PLANT–VIRUS INFECTION

There is surprisingly little information in the scientific literature on the global economic losses caused by plant viruses. It has been calculated that the total annual global crop loss in 2002 was 36.5%, of which 40% was due to disease, which represented a loss of 14.1% of total crop production—translating to an estimated total economic loss of 220 billion USD (Agrios, 2005). The proportion of this total loss that is due to virus infection is not clear, but it is generally accepted that, after hyphal pathogens (fungi and oomycetes), viruses are the most damaging (Matthews, 1991). In Europe, it has been suggested that total losses due to viruses in field crops are 10–15%, and even higher in vegetables and fruits (Spence and Garcia-Arenal, 2005). This situation would be even worse were it not for the control measures already in place, including pesticide use, an absence of which would double world losses to 70% of total production (Oerke *et al.*, 1994). Further hidden losses due to virus infection also exist, such as product quality deterioration and the effects of symptomless viruses that are undetectable by present methods.

But crop losses caused by a few single viruses for a specific host in a particular geographical region have been well documented (for reviews, see Bos, 1982; Pennazio *et al.*, 1996; Waterworth and Hadidi, 1998). For instance, the annual global impact of *Tomato spotted wilt virus* (TSWV) and the viruses causing African cassava mosaic disease, which are of exceptional agronomic importance, has been estimated to be approximately 1 billion USD each (Goldbach and Peters, 1996; Taylor *et al.*, 2004). Equivalent estimates for other crop/virus/region combinations are needed to understand the balance between the costs and benefits of any virus control strategy.

The accumulated weight of qualitative and quantitative evidence provided by the scientific literature highlights the need for the continued development of strategies to minimize the impact of viruses on cultivated crops. This need is made even more acute if future pressures on the global food market are considered. These principally comprise an increase in the human population, predicted to reach 9.1 billion by 2050 (Alexandratos, 2005), increased water consumption, of which 70% is presently used for agriculture (Brown, 2005), and global warming that will alter both agricultural practices and the threat of emerging plant pests (Rodoni, 2009). The combined effects of these changes have been predicted to reduce agricultural production in developing countries by as much as 25% by 2080 (Cline, 2007). Predicted changes in the climate are likely in some latitudes to expose more cultivated varieties to a greater variety of vectors of viral diseases, while increasing other vectors' survival and reproduction rates (Yamamura and Kiritani, 1998) and reducing the effectiveness of the disease resistances of some hosts (Garrett *et al.*, 2006).

III. ANTIVIRAL STRATEGIES NOT BASED ON GENETIC ENGINEERING

The range of approaches to reduce the effect of virus infection in cultivated crops has been the subject of a number of reviews (see Agrios, 2005; Hull, 2002; Irwin *et al.*, 2000 and references therein), and will not be entered into in any depth here, except to illustrate very generally the possible options available as alternatives to engineered resistance.

A. Naturally occurring virus-resistance genes

The most effective way to prevent virus infection is by means of genetic resistance, thereby in theory (though unfortunately not always in practice) precluding the need for any other control measures. Success stories of the identification and introduction of virus-resistance genes into cultivated varieties abound (for review, see Kang *et al.*, 2005; Ritzenthaler, 2005), and yet in many cases there are no natural resistance genes available for a particular crop species against a specific virus. Eighty percent of the known naturally existing viral resistances described are monogenic, with just over half being based on dominant alleles (Kang *et al.*, 2005); the remaining potentially more durable polygenic resistances are difficult to introgress into elite crop lines, due to more complex inheritance. Difficulties can also be encountered postdeployment with the development of resistance-breaking virus strains (Garcia-Arenal and McDonald, 2003).

B. Agricultural practices

Other important control strategies are based on reducing the risk of virus infection in a cultivated crop by minimizing exposure of the crop to the virus. The International Plant Protection Convention (Anon, 1995) aims to facilitate implementation of the correct phytosanitary measures, including creation of pest-free areas within given geographical regions, so that exporting countries are able to demonstrate that any exported material is free from pests that might threaten the importing country. Despite all this, quarantine measures can be insufficient, as has been shown in Australia with *Wheat streak mosaic virus* (WSMV), *Potato virus Y* (PVY), and *Tomato yellow leaf curl virus* (TYLCV) (Rodoni, 2009). Further safeguards are often in place in the importing country in the form of quarantine and certification schemes. Even when these safeguards fail, if enough is known about a nonendemic virus, effective action can still minimize virus spread—as has been shown for *Plum pox virus* (PPV) in North America (Levy, 2006; Thompson, 2006). Failing this, if the virus is or becomes endemic, then preventive measures can include keeping source material (e.g., seeds) virus-free, with the use of micropropagation in the case of vegetatively propagated crops, combined with effective methods of detection, thereby minimizing the threat of vertical transmission (Lopez *et al.*, 2009). In addition, planting healthy source material limits horizontal transmission of the virus within the crop. Of course, these strategies do not protect the plants against pathogen invasion from other sources, an event that becomes more likely if the virus is transmitted by a highly mobile vector such as a winged insect (Irwin *et al.*, 2000), which is the case for around half of the plant virus genera (Hull, 2002). Horizontal transmission of the virus can be minimized if the vector is controlled, which can be done in some cases by the regular use of pesticides, and this can be done in the context of an integrated pest management (IPM) program (Jones, 2004), involving, among others, a forecasting system to better time applications (Qi *et al.*, 2004; Thackray *et al.*, 2004). The use of pesticides in agriculture has a long history (Agrios, 2005), yet in the last two decades there have been concerted efforts to reduce their application because of environmental concerns (van der Vlugt, 2006). Viruses, unlike other pests and pathogens, are not directly affected by either biological or chemical control. In addition, preventing the spread of those that are transmitted by vectors in a nonpersistent fashion has met with little success, due to the relatively slow action of insecticides (Irwin *et al.*, 2000), though the arrival of a new generation of insecticides may allow for more effective control (Castle *et al.*, 2009). The grower may also practice a range of IPM strategies in an effort to reduce losses, such as the removal of volunteer plants, roguing infected plants, intercropping, sowing early-maturing cultivars, use of mulches, and the introduction of natural insect predators (Irwin *et al.*, 2000; Jones, 2004).

C. Cross-protection

Another approach to the problem has been to inoculate the plant with an attenuated virus, so that subsequent infections by virulent isolates of the virus are diminished. This kind of cross-protection has been successfully employed using attenuated viruses in both perennials (Costa and Müller, 1980; Komar *et al.*, 2008) and annuals (Kajihara *et al.*, 2008; Kosaka *et al.*, 2006). However, cross-protection is relatively labor-intensive, and it may be inadvisable to voluntarily disseminate a pathogen, even if it is attenuated, at least on the crop host.

IV. TRANSGENE-MEDIATED RESISTANCE

In the absence of simply inherited natural resistance genes, there are clear agronomic and practical benefits for the adoption of virus-resistance transgenes to combat the threat of virus infection. Transgene-mediated virus resistance has been reviewed recently (Gottula and Fuchs, 2009; Morroni *et al.*, 2008; Prins *et al.*, 2008; Reddy *et al.*, 2009), and will thus be presented here only briefly. It is also important to emphasize that in our view, the existence of naturally occurring resistance genes should not preclude the development of transgenic resistance, since it would be imprudent to base important characters such as virus resistance on a single gene, and the stacking of resistance genes of all types should confer greater durability of the resistance in the field.

The first demonstration of the insertion of a foreign gene into a plant genome using *Agrobacterium* and a Ti plasmid (Barton *et al.*, 1983) was quickly put to use, as shown by the first report of a VRTP three years later (Abel *et al.*, 1986). The concept behind the approach, termed "pathogen-derived resistance" was simple—that the development of a pathogen could be interfered with by causing the host to synthesize a pathogen gene product, either RNA or protein (Sanford and Johnston, 1985). Most of the initial work centered on using transgenes encoding a viral coat protein (CP) (Fitchen and Beachy, 1993; Prins *et al.*, 2008; Wilson, 1993) which encountered varying degrees of success. For example, in 20 reports (between 1987 and 2002) using a CP transgene against infection by *Cucumber mosaic virus* (CMV) in six different hosts, the degree of resistance conferred varied from 0% to 100% (Morroni *et al.*, 2008). Nevertheless, in 1994 the summer squash line ZW-20, transformed with the CP genes of both *Zucchini yellow mosaic virus* (ZYMV) and *Watermelon mosaic virus* (WMV), became the first VRTP to be deregulated (Medley, 1994). This was followed in 1996 by transgenic squash CZW-3, which expresses the CP of CMV, ZYMV, and WMV (Acord, 1996), and the highly successful commercialization of papaya resistant to *Papaya ringspot virus* (PRSV) in Hawaii in 1998 (Gonsalves, 1998). These examples demonstrate the actual potential of the technology. In particular, the PRSV-resistant papaya story is an excellent example of a concerted technological response to a local agronomic problem, which in the long term has allowed for a return to small-scale cultivation of nontransgenic papaya varieties because of the reduced incidence of the virus, which has in turn contributed to an increase in cultivar diversity (Fuchs and Gonsalves, 2007).

Since then, there have been a remarkable number of field trials of VRTPs, including cereals, flowers, fruits, forage crops, grasses, legumes, and vegetables (Fuchs and Gonsalves, 2007). There have been further VRTPs deregulated in the United States and Canada, with two potato lines, one resistant to PVY, and another resistant to *Potato leafroll virus* (PLRV). However, there are now only two VRTPs that are still in commercial use, the PRSV-resistant papaya, and the CZW-3 squash. The first squash line, ZW-20, was in effect replaced by CZW-3, and the two potato lines were abandoned by their provider, primarily because certain food processors refused to purchase them because of perceived acceptability issues.

V. REFINEMENTS TO ENGINEERING RESISTANCE

Since the creation of the first VRTPs using viral CPs, researchers have employed a variety of different transgenes that can be roughly grouped into two broad strategies: (1) pathogen-derived resistance and (2) nonpathogen-derived resistance.

A. Further developments of pathogen-derived resistance

Once it was shown that the technique could work (Abel *et al.*, 1986), researchers began developing a variety of pathogen-derived resistance approaches without in fact having a clear understanding of the underlying mechanism(s). That was until posttranscriptional gene silencing (PTGS) was discovered, and evidence rapidly accumulated in support of its involvement in resistance (Goodwin *et al.*, 1996; Hamilton and Baulcombe, 1999; Lindbo *et al.*, 1993; Mueller *et al.*, 1995), although in at least two cases, transgenes encoding the CP of either *Tobacco mosaic virus* (TMV) (Asurmendi *et al.*, 2007; Bendahmane *et al.*, 1997, 2007) or CMV (Jacquemond *et al.*, 2001), it has been clearly demonstrated that the mechanism is based on protein and not RNA, and in others the picture is still far from clear (Baulcombe, 1996; Morroni *et al.*, 2008; Prins *et al.*, 2008).

Knowledge of PTGS and how it works has provided the main impetus for the refinements made to pathogen-derived resistance. Researchers have been able to design transgenes that generate inverted-repeat (IR) RNA molecules, providing directly the substrate for the Dicer-like (DCL) proteins that produce siRNAs from dsRNA precursors (Voinnet, 2008). There are now several methods specifically designed to facilitate IR construction—for example, pHANNIBAL (Helliwell and Waterhouse, 2003), pHELLSGATE (Wielopolska *et al.*, 2005), IR-PCR (Pawloski *et al.*, 2005), gateway vectors (Karimi *et al.*, 2007), and rolling circle amplification-mediated hairpin RNA (Wang *et al.*, 2008). Resistance has been conferred using a range of sequence lengths, from using whole virus cistrons (Kamachi *et al.*, 2007; Krubphachaya *et al.*, 2007; Tougou *et al.*, 2006) to sequences as short as 120 nucleotides (Bucher *et al.*, 2006). In this last example, IRs of three distinct tospoviruses were combined in the same construct under a single promoter. A similar approach has also been successful for the distantly related viruses *Sweet potato chlorotic stunt virus* (SPCSV) (family *Closteroviridae*) and *Sweet potato feathery mottle virus* (SPFMV) (family *Potyviridae*) (Kreuze *et al.*, 2008). At present, the lower effective size limit for such constructs designed against virus infection appears to be about 100 bp, though shorter IR sequences have been employed successfully in virus-induced gene silencing (VIGS) systems (Lacomme *et al.*, 2003; Pflieger *et al.*, 2008). However, since VIGS targets cellular mRNAs, which are generally less abundant than viral RNAs, VIGS may not require as efficient silencing to be effective. The technique of using IRs for DNA viruses has also proved successful (Bonfim *et al.*, 2007; Noris *et al.*, 2004; Ribeiro *et al.*, 2007; Vanderschuren *et al.*, 2009; Zrachya *et al.*, 2007), although questions have been raised over the potential durability of such a method in the single-stranded DNA geminiviruses (Lucioli *et al.*, 2008). The potential lack of durability of the IR approach, because of the need for high nucleotide identity (>90%) between the transgene and viral target, is its main disadvantage (de Haan *et al.*, 1992). Even shorter 21-nt siRNA constructs have been shown to work against mammalian RNA viruses (Capodici *et al.*, 2002; Kapadia *et al.*, 2003; Shin *et al.*, 2009), although there are clear problems, in that the inherently high mutation rate of the viral genome is able to evade inhibition (Das *et al.*, 2004; Wilson and Richardson, 2005).

The above methods are geared in principle to producing the shortest noncoding transgenes that confer the desired levels of resistance. In doing so, the theoretical risks associated with the use of their larger full-cistron predecessors; that of complementation, synergy and transencapsidation, and recombination between viral RNAs and transgene mRNAs mRNA recombination, have practically been eliminated. Nevertheless, it is important to remember therefore that the apparent technical convenience of IRs should not overshadow the potential of other more longstanding approaches which might provide a more durable resistance using viral cistrons coding for the CP, movement protein, and replicase (for review, see Morroni *et al.*, 2008; Prins *et al.*, 2008; Sudarshana *et al.*, 2007).

The use of significantly shorter viral target sequences has been achieved by using artificial microRNAs (amiRNAs) that target plant viruses. This technique, which has been shown to be effective against *Turnip yellow mosaic virus* (TYMV) (Lin *et al.*, 2009; Niu *et al.*, 2006), CMV (Duan *et al.*, 2008; Qu *et al.*, 2007), and *Turnip mosaic virus* (TuMV) (Lin *et al.*, 2009), exploits existing miRNA machinery to target virus-specific sequences. A better understanding of the evolution of the 21-nt target site in plants expressing amiRNAs and identification of those nucleotides critical for resistance should help in further refining the design of future constructs.

B. Virus-resistance transgenes without viral sequences

Nonpathogen-derived resistance may possess the potential to impart a broader and more durable resistance than PTGS-derived strategies, but most have lacked the latter's ease of design and technical simplicity. A nonpathogen-derived technique that has shown particular promise is the expression of intracellular antibodies (intrabodies), and more specifically single-chain *F* variable fragment (scFv) antibodies. This was first demonstrated by the expression in plants of an scFv containing the variable domains of a preselected monoclonal antibody directed against the CP of *Artichoke mottled crinkle virus* (AMCV), named F8 (Tavladoraki *et al.*, 1993). This success led to the development of an F8-derived library (Donini *et al.*, 2003) from which scFvs specific to the CP of CMV were selected, two of which conferred resistance when expressed in tomato (Villani *et al.*, 2005). Concerns that variability in the CP would limit the breadth of resistance were addressed using phage display to select scFvs that recognized a specific and highly conserved 10-amino acid motif of the RNA-dependent RNA polymerase of *Tomato bushy stunt virus* (TBSV) (Boonrod *et al.*, 2004). The Rep protein of the single-stranded DNA geminivirus, TYLCV, has also been recently targeted in *Nicotiana benthamiana* expressing scFvs (Safarnejad *et al.*, 2009). Further development of virus-specific scFvs has been somewhat hindered by the relatively high technical knowledge and practical investment required at the initial stages of scFv selection, resulting in, when compared to IR technology, a slower adoption by the scientific community. However, to its benefit, the intrabody approach provides a potentially more robust alternative to PTGS-based techniques, with regards to breadth and durability of interaction, as mentioned above, and possible off-target effects (Jackson *et al.*, 2003). However, from a biosafety point of view, since they express a protein that is not normally consumed, intrabody-expressing plants will require additional consideration from a food safety perspective.

Another field of research still in relatively initial stages is that of modulating existing host factors to interfere with the virus life cycle. This approach has been shown to be effective against *Potato spindle tuber viroid* when tobacco plants and protoplasts silenced for homologues of the tomato *VirP1* gene were resistant to infection (Kalantidis *et al.*, 2007), and in *N. benthamiana* silenced for the *HSP70* gene, where replication of TBSV was shown to be significantly reduced (Wang *et al.*, 2009). Any stably transformed plants derived from this method would still probably be considered transgenic, even if the only likely foreign element present would be the promoter.

VI. CONCEPTS OF RISK ASSESSMENT

Humans are often—but not always—uncomfortable with uncertainty. When anyone claims that something is safe, they are merely making an evaluation that an object or situation is free of unacceptable danger or risk. Scientific risk assessment (RA) seeks to evaluate more objectively the level of risk in a given situation, and consists of the identification and characterization of the two principal components of risk, namely, hazard and exposure (or likelihood) to determine the gravity of the hazard and probability that it will occur (Johnson *et al.*, 2007; Poppy, 2004; Raybould and Cooper, 2005). In the strictest sense, RA forms along with risk management and risk communication part of the wider activity of risk analysis, which tries to take into account all factors (e.g., ethical, political, economical, etc.) relevant to decision making. The confusion among these various components and the misinformation generated has contributed to the public's scepticism about the safety of GM crops.

Some of the limitations that are important to recognize in doing RA for GM crops are that unlike RA for chemical and pharmaceutical products, which have a much longer history of assessment and monitoring, our knowledge of the full scope of potential hazards may still be limited, and it may be difficult to quantify exposure (likelihood) accurately, leading to more qualitative and therefore less precise conclusions. In some cases, this has led to misinterpretation of results, and on occasion to an overestimation of the risks (Ewen and Pusztai, 1999; Losey *et al.*, 1999).

Calls advocating a more rigorous and quantitative approach have been gaining momentum both in environmental RA (Poppy, 2000, 2004; Raybould and Cooper, 2005; Wilkinson and Tepfer, 2009; Wilkinson *et al.*, 2003) and nutrition and health RA (Atherton, 2002; Cockburn, 2002; Goodman *et al.*, 2008; König *et al.*, 2004; Kuiper *et al.*, 2002; Magaña-Gomez and Calderón de la Barca, 2009), which are the two main branches pertinent to GM plants. For environmental RA, this approach involves a clear step-by-step process which begins with the identification of the appropriate risk hypotheses, and thus the components in the environment that need to be protected—known as "assessment endpoints." A tiered approach for quantitative or semiquantitative evaluation of both the exposure and hazard provides "test endpoint" values which can then be used to estimate risk, with for example a "hazard quotient" or "risk estimate matrix." It is then at this point that a decision has to be made based on established thresholds and acceptable risks. A model for the tiered approach to potential risks mediated by gene flow from crop to wild relatives has been outlined for TuMV in oilseed rape, and provides a useful framework for controlled laboratory and field trials under appropriate assumptions (Raybould and Cooper, 2005).

The guiding principle for nutritional and health RA is the concept of "substantial equivalence" (Anon, 2003). This is based on the assumption that if the characteristics and composition of a new GM food are comparable to those of a familiar food with a history of safe consumption, then the new GM food will be no less safe, assuming it is used in the same way as the conventional food. In most cases, because the obvious difference between the unmodified familiar parental variety and the GM variety are the genes that have been introduced, it has been deemed sufficient to focus on the safety of the novel protein(s) expressed (Atherton, 2002). Newly introduced proteins and their allergenic potential can be tested using a variety of parameters (Goodman *et al.*, 2008).

VII. POTENTIAL RISKS ASSOCIATED WITH VIRUS-RESISTANT TRANSGENIC PLANTS

As mentioned above, in all countries, the legal frameworks regulating environmental release of GMOs require evaluation of undesirable impacts on human or animal health and for the environment. It is remarkable that quite shortly after the first report of VRTPs in 1986, there were several articles that considered the broad issues of their potential impacts (Hull, 1990; Palukaitis, 1991; Tepfer, 1993), and it is quite reassuring that no scientifically credible new issues have been raised since then. An additional highly positive point is that research carried out since then on virus–host interactions, but also specifically on VRTPs, has clarified many of the issues raised in the early 1990s.

A. Potential food safety issues

Since nucleic acids are universal in living organisms, the only novel molecules that are expected to exist in VRTPs that could have a food safety impact are the proteins synthesized from the transgenes. From a practical perspective, this primarily concerns viral CPs, since the vast majority of VRTPs developed so far, and all but one of the VRTPs that have been deregulated or authorized for large-scale release, are ones that express CP transgenes. But it has been argued that there is in fact a long history of consumption of viral CPs in virus-infected nontransgenic plant foods that we consume regularly. Thus, although they may be transgene-encoded, CPs are in fact very much familiar parts of human and animal diets. Transgenes that express no protein, as is the case for those used in PTGS and amiRNA approaches, should in theory not pose this sort of question, although it might be argued that such RNA-based technologies could alter the plant's protein expression profile to a degree that would render it nonequivalent. Protein profiling, which is becoming routine before commercial release of a GM crop, reduces these uncertainties considerably.

B. Potential impact on the environment

Ten years ago, it was pointed out that the potential environmental impacts of VRTPs could be segregated into two groups, ones that would be mediated by changes in the phenotype of virus–host interactions, and ones in which the genotype of either the virus or its host would be modified (Teycheney and Tepfer, 1999). This distinction continues to be useful, since it corresponds in fact to potential impacts that—if they proved to be undesirable—would be expected to be reversible (phenotype-mediated) and ones that could be irreversible (genotype-mediated). In the paragraphs below, we will consider the potential risks briefly, focusing on recent results, but for a more complete overview, see Tepfer (2002) and references cited therein.

1. Potential impacts mediated by phenotypic changes in virus–host interactions

Viral CPs are remarkably selective in the RNAs they will encapsidate, but it has long been known that viral genomes can be encapsidated in particles composed partially or entirely of CP of a closely related virus. This phenomenon, known as heterologous encapsidation, occurs frequently when plants are infected with more than one virus, but also in VRTPs expressing a CP to high levels when infected with a related virus. The concern with heterologous encapsidation in VRTPs is that the genome of the infecting virus could acquire the interactions with the vectors of the transgene donor virus. In fact, this could only constitute a risk when heterologous encapsidation could confer transmission by a vector different from the usual one. So for instance, one would not expect that heterologous encapsidation of an aphid-transmitted potyviral genome in particles of another aphid-transmitted potyvirus could have any impact. In addition, in cases where heterologous encapsidation could have an impact (by conferring transmission by a novel vector), it has been shown that the problem can be solved, since it has been shown in several cases that the CP transgene can be modified so that the protein either no longer interacts with the vector, or is unable to assemble viral particles, while still conferring resistance to the target virus.

When, as frequently happens, plants are infected with more than one virus, this can lead to synergy, that is, increased accumulation of one of the viruses concerned and aggravation of the symptoms observed. Since at the time little was known about the mechanisms of synergy, in the 1990s there was serious concern that VRTPs expressing viral sequences could lead to worse symptoms when the plants are infected with a nontarget virus. Since then, it has become apparent that synergy is mediated by viral silencing suppressor proteins, such as the potyviral HC-Pro, or the cucumoviral 2b, so the only transgenes that would be of concern for synergy would be ones encoding silencing suppressors, and here it is worth noting that in a few cases the viral CP can also act as a suppressor (Genoves *et al.*, 2006; Qu *et al.*, 2003; Thomas *et al.*, 2003). Before undergoing large-scale field releases, it would be quite simple—if this had not been previously established—to determine using standard techniques if the viral protein expressed from the transgene is a silencing suppressor, in which case it would be wise to consider alternative VRTP strategies.

2. Potential impacts mediated by changes in host or virus genotype

Although in any given area of the globe, sexually compatible wild relatives of crop plants may not be present (e.g., there are no close relatives of maize in Europe, Asia, South America, etc.), it is now generally recognized that compatible relatives of nearly all crop species do occur somewhere (Ellstrand, 2003). As a result, seen from a worldwide perspective, the question of the potential impact of the introduction of transgenes into crop relatives by outcrossing is always pertinent, and in the case of VRTPs this would apply to all types of novel virus-resistance traits. It should be noted that the same issues could be raised for natural resistance genes, but in nearly all jurisdictions they escape the regulatory burden that applies to the equivalent GMOs. The environmental impact that is of concern is that a virus-resistance transgene could confer enhanced fitness on a wild crop relative, and in doing so, possibly increase its weediness or invasiveness. Determining if a virus has an effect on certain fitness parameters of a wild crop relative is comparatively easy (Bartsch *et al.*, 1996; Friess and Maillet, 1996, 1997; Fuchs *et al.*, 2004; Maskell *et al.*, 1999), but as has been discussed in detail recently, it can be a daunting task to go beyond, and answer the critical question, which is whether changed fitness parameters will lead to increased population size, the accepted proxy for increased weediness or invasiveness (Quemada *et al.*, 2008; Wilkinson and Tepfer, 2009).

In addition to the above plant-to-plant gene flow, in plants expressing viral sequences, a form of plant-to-virus gene flow can occur by recombination between the transgene mRNA and the RNA of an infecting virus. Recombination between viruses is a well-known source of genetic variability that contributes to virus evolution. And indeed, there have been several reports of recombination between the mRNA encoded by a viral transgene and the RNA of an infecting virus (Adair and Kearney, 2000; Borja *et al.*, 1999; Gal *et al.*, 1992; Greene and Allison, 1994; Turturo *et al.*, 2008; Varrelmann *et al.*, 2000). This of course raised the issue of whether recombination in transgenic plants in the field could lead to the creation of novel viral genomes, and thus to the emergence of new diseases.

A first approach to answering this question was taken by Vigne *et al.* (2004). In two small-scale field trials of grapevine rootstocks expressing the CP of *Grapevine fanleaf virus* (GFLV), the plants were exposed to natural infection by the virus. When they screened the viruses associated with the plants for recombinants, no GFLV recombinants were observed. Thus in this case the potential for messenger/viral RNA recombination did not lead to any detectable changes. The weak point of this approach is one of scale; in the absence of the detection of recombinants between the viral mRNA and the infecting viral genome, it is not possible to predict to what extent any undetected recombinant viral RNAs were produced, or what the biological properties of the recombinants could be.

More recently, Turturo *et al.* (2008) have taken a potentially more powerful comparative approach. Using a highly sensitive RT-PCR assay that specifically amplifies recombinant viruses, they showed that equivalent populations of recombinant viruses were present in plants expressing a CMV CP transgene infected with a related virus (either CMV or *Tomato aspermy virus*—TAV) and in nontransgenic plants infected simultaneously with both viruses. These results show that, since equivalent populations were observed, there was no evidence for novel recombination mechanisms in the transgenic plants. This approach of course also has its limits, since undetected novel recombinant viruses could possibly have been produced at levels below the threshold of detection.

The perception of potential risks has had a strong effect on decisions regarding which resistance strategies to develop. As soon as it was clear that effective virus resistance could be obtained with VRTPs in which only short virus-derived RNA fragments accumulate, it was proposed that one of the advantages of these resistance strategies was to avoid the issues of potential risk associated with accumulation of viral proteins or even cistron-length viral RNA sequences. However, considering the little information available on the long-term behavior of VRTPs in the field, we strongly believe that it is premature to abandon any strategy for creating virus resistance, since as explained above, research does make it possible to delimit and minimize the potential risks that had first been envisaged in the 1990s.

VIII. WEIGHING THE BENEFITS AND RISKS. TAKING INTO ACCOUNT EFFICACY, DURABILITY, AND SAFETY

In Section I, we presented the state of present knowledge regarding the economic impact of viruses on crops. This was then followed by sections on strategies for dealing with plant viral diseases including plant breeding and creating VRTPs. We then presented briefly the potential risks associated with virus-resistant transgenic crops. In closing, we would like to reflect on the difficulty in making choices regarding the best strategies for managing a viral disease problem. Overall, the choices include: continuing as at present, introgressing natural resistance genes, and creating resistance through genetic transformation.

A. Time required

One of the real advantages of genetic transformation is that the process is relatively rapid, and at least for RNA viruses, strategies for resistance transgene design are becoming relatively standard, although there is of course always room for improvement. In contrast, introducing resistance by classical plant breeding requires starting by seeking the source germplasm, which can be quite long and arduous, and then introgressing the corresponding gene(s), something that can take up to 20 generations if transferring from a wild species, and which can be potentially complicated by close genetic linkage between the desired resistance gene(s) and ones that confer undesirable features (Conner and Jacobs, 1999). But in practice, the total duration of the process for development of a VRTP may not be so much shorter, since for VRTPs one must add on a number of years for backcrossing the transgene from the initial transformant into the interesting elite lines, and for going through the regulatory process for authorization of commercial release.

B. Cost

Clearly, for cases where the impact of a virus on a given crop and continent is on the order of billions of dollars per year, as has been reported for cassava mosaic disease in Africa (Taylor *et al.*, 2004), then even relatively costly resistance strategies can be considered for implementation. Considering the scale of this particular disease problem, we would argue that the only reasonable decision would be to attempt to advance all potentially valid strategies, which in this case include both classical breeding and transgene-mediated resistance (Patil and Fauquet, 2009; Thresh and Cooter, 2005; Vanderschuren *et al.*, 2007). But for other circumstances, there can be real uncertainties about cost effectiveness. Although in the simplest cases, VRTPs can be produced with remarkably modest technical investment, and thus presumably at relatively low cost, the hidden expenses are far from negligible. For many crops, the elite breeding lines are not the ones that can be easily transformed, and thus, as mentioned above, it may require a fairly lengthy series of backcrosses to in fact introduce the transgene into the most desirable plant genotypes. In addition, the time and effort required to go through the regulatory process is also in effect an additional cost, and it is in fact one that many consider to extremely dissuasive of developing GM crops of any sort.

C. Breadth and efficacy of resistance

Different sources of natural virus resistance can vary greatly in the breadth of resistance they confer; in some cases this can be as narrow as strain-specific gene-for-gene resistance. We believe that one of the great unknowns regarding different strategies for creating VRTPs concerns the breadth and efficacy of the resistance they confer in the field. Inoculation of VRTPs with a range of strains in the greenhouse provides some indication regarding the breadth of resistance, but is not always predictive of what will occur in the field. For instance, a resistance to *African cassava mosaic virus* (ACMV) that was promising in the greenhouse (Chellappan *et al.*, 2004), then later proved ineffective (Anon, 2006). There is also a serious lack of long-term field experience. The PRSV-resistant papaya in Hawaii has been effective for a number of years, but it has probably not been challenged by the full range of PRSV strains, and indeed it is known to not be resistant to Southeast Asian strains (Bau *et al.*, 2003). To the best of our knowledge, it is unclear whether the susceptibility to the latter strains is due to sequence divergence, or rather to differences in the virus's silencing suppressors, or both. Further, some viral diseases are caused by several related virus species. This is the case for grapevine fanleaf disease, cotton leaf curl disease, and cassava mosaic disease; such cases obviously pose remarkably difficult challenges for VRTP development in terms of breadth and efficacy of resistance.

D. Durability

Durability in the field of any resistance that is not yet fully characterized may be closely related to the issues of efficacy and breadth of resistance; in particular, if a resistance is narrowly strain-specific, one may express doubts concerning its durability. In the case of VRTPs, we have only two in which there have been enough years of large-scale field release to have some idea of the durability of the resistance, and in the case of the PRSV-resistant papaya, since it has only been grown under conditions of extreme geographic isolation (Hawaii), one might well wonder if this serves as a predictor of durability of PRSV resistance in other circumstances. Clearly, this remains one of the great unknowns regarding the ultimate usefulness of VRTPs.

IX. CONCLUSIONS

Considering the remarkable diversity of the transgenes that have been created, the development of VRTPs should be regarded as one of the great early success stories of plant biotechnology. However, in spite of the hundreds of field trials, it is quite remarkable that until now only a very small number of VRTPs have in fact reached the market, and this was only during the early years of development of GM crops. In addition, the VRTPs cultivated presently concern only crops of relatively minor global importance. Since then, skyrocketing regulatory costs, increased opposition to GMOs, particularly in certain parts of the world, and the potential negative trade impacts have certainly slowed emergence of new commercial releases. However, this seems to be in the process of changing, since according to a recent survey, several VRTPs are now clearly in the pipeline for commercial release in different countries (Stein and Rodríguez-Cerezo, 2009). These include PPV-resistant plum in the United States, PVY-resistant potato in Argentina, rice resistant to *Rice tungro bacilliform virus* (RTBV) in India, and bean resistant to *Bean golden mosaic virus* (BGMV) in Brazil.

Although the technology for developing VRTPs—at least ones that are resistant to RNA viruses—is relatively mature, significant challenges remain. There is an obvious need for further development of strategies for resistance to plant DNA viruses, several of which are particularly important in developing countries. Some of these, which cause important diseases such as cassava mosaic, cotton leaf curl or banana streak may be particularly difficult to manage because of the remarkable diversity of the viruses involved. Particularly in the former two cases, the etiological agent might best be described as a cloud of interacting, recombining viral species.

Even in terms of the advancement of our understanding of VRTPs, the future prospects of large-scale commercial release are important; we believe that only through more field studies, both large and small scale, will we accumulate sufficient knowledge to determine which of the many strategies available is best able to combine efficacy, durability, and safety.

REFERENCES

Abel, P. P., Nelson, R. S., De, B., Hoffmann, N., Rogers, S. G., Fraley, R. T., and Beachy, R. N. (1986). Delay of disease development in transgenic plants that express the tobacco mosaic virus coat protein gene. *Science* **232:**738–743.

Acord, B. D. (1996). Availability of determination of nonregulated status for a squash line genetically engineered for virus resistance. *Fed. Regist.* **61:**33484–33485.

Adair, T. L., and Kearney, C. M. (2000). Recombination between a 3-kilobase tobacco mosaic virus transgene and a homologous viral construct in the restoration of viral and nonviral genes. *Arch. Virol.* **145:**1867–1883.

Agrios, G. N. (2005). *Plant Pathology*, 5th Edn. Elsevier, London.

Alexandratos, N. (2005). Countries with rapid population growth and resource constraints: Issues of food, agriculture and development. *Popul. Dev. Rev.* **31:**237–258.

Anon(1995). *International standards for phytosanitary measures.* Secretariat of the International Plant Protection Convention, FAO, Rome.

Anon(2003). *Evaluation of allergenicity of genetically modified foods.* FAO, Rome.

Anon, A. (2006). *Danforth Centre Cassava Update, May 26, 2006.* Donald Danforth Plant Science Center, St. Louis, MO. http://www.danforthcenter.org/NEWSMEDIA/NewsDetail.asp?nid=119.

Asurmendi, S., Berg, R. H., Smith, T. J., Bendahmane, M., and Beachy, R. N. (2007). Aggregation of TMV CP plays a role in CP functions and in coat-protein-mediated resistance. *Virology* **366:**98–106.

Atherton, K. T. (2002). Safety assessment of genetically modified crops. *Toxicology* **181–182:**421–426.

Barton, K. A., Binns, A. N., Matzke, A. J., and Chilton, M. D. (1983). Regeneration of intact tobacco plants containing full length copies of genetically engineered T-DNA, and transmission of T-DNA to R1 progeny. *Cell* **32:**1033–1043.

Bartsch, D., Schmidt, M., Pohl-Orf, M., Haag, C., and Schuphan, I. (1996). Competitiveness of transgenic sugar beet resistant to beet necrotic yellow vein virus and potential impact on wild beet populations. *Mol. Ecol.* **5:**199–205.

Bau, H. J., Cheng, Y. H., Yu, T. A., Yang, J. S., and Yeh, S. D. (2003). Broad-spectrum resistance to different geographic strains of *Papaya ringspot virus* in coat protein gene transgenic papaya. *Phytopathology* **93:**112–120.

Baulcombe, D. C. (1996). Mechanisms of pathogen-derived resistance to viruses in transgenic plants. *Plant Cell* **8:**1833–1844.

Bendahmane, M., Fitchen, J. H., Zhang, G., and Beachy, R. N. (1997). Studies of coat protein-mediated resistance to tobacco mosaic tobamovirus: Correlation between assembly of mutant coat proteins and resistance. *J. Virol.* **71:**7942–7950.

Bendahmane, M., Chen, I., Asurmendi, S., Bazzini, A. A., Szecsi, J., and Beachy, R. N. (2007). Coat protein-mediated resistance to TMV infection of *Nicotiana tabacum* involves multiple modes of interference by coat protein. *Virology* **366:**107–116.

Bonfim, K., Faria, J. C., Nogueira, E. O. P. L., Mendes, E. A., and Aragao, F. J. L. (2007). RNAi-mediated resistance to *Bean golden mosaic virus* in genetically engineered common bean (*Phaseolus vulgaris*). *Mol. Plant Microbe Interact.* **20:**717–726.

Boonrod, K., Galetzka, D., Nagy, P. D., Conrad, U., and Krczal, G. (2004). Single-chain antibodies against a plant viral RNA-dependent RNA polymerase confer virus resistance. *Nat. Biotechnol.* **22:**856–862.

Borja, M., Rubio, T., Scholthof, H. B., and Jackson, A. O. (1999). Restoration of wild-type virus by double recombination of tombusvirus mutants with a host transgene. *Mol. Plant Microbe Interact.* **12:**153–162.

Bos, L. (1982). Crop losses caused by viruses. *Crop Prot.* **1:**263–282.

Brown, L. R. (2005). Pushing beyond the Earth's limits, Chapter 1. *Outgrowing the Earth: The Food Security Challenge in an Age of Falling Water Tables and Rising Temperatures.* Earth Policy Institute, Washington, DC.

Bucher, E., Lohuis, D., van Poppel, P. M., Geerts-Dimitriadou, C., Goldbach, R., and Prins, M. (2006). Multiple virus resistance at a high frequency using a single transgene construct. *J. Gen. Virol.* **87:**3697–3701.

Capodici, J., Kariko, K., and Weissman, D. (2002). Inhibition of HIV-1 infection by small interfering RNA-mediated RNA interference. *J. Immunol.* **169:**5196–5201.

Castle, S., Palumbo, J., and Prabhaker, N. (2009). Newer insecticides for plant virus disease management. *Virus Res.* **141:**131–139.

Chellappan, P., Masona, M. V., Vanitharani, R., Taylor, N. J., and Fauquet, C. M. (2004). Broad spectrum resistance to ssDNA viruses associated with transgene-induced gene silencing in cassava. *Plant Mol. Biol.* **56:**601–611.

Cline, W. R. (2007). *Global Warming and Agriculture: Impact Estimates by Country.* Peter G. Peterson Institute for International Economics, Washington, DC.

Cockburn, A. (2002). Assuring the safety of genetically modified (GM) foods: The importance of an holistic, integrative approach. *J. Biotechnol.* **98:**79–106.

Conner, A. J., and Jacobs, J. M. (1999). Genetic engineering of crops as potential source of genetic hazard in the human diet. *Mutat. Res.* **443:**223–234.

Costa, A. S., and Müller, G. W. (1980). Tristeza control by cross protection: A US–Brazil cooperative success. *Plant Dis.* **64:**538–541.

Das, A. T., Brummelkamp, T. R., Westerhout, E. M., Vink, M., Madiredjo, M., Bernards, R., and Berkhout, B. (2004). Human immunodeficiency virus type 1 escapes from RNA interference-mediated inhibition. *J. Virol.* **78:**2601–2605.

de Haan, P., Gielen, J. J. L., Prins, M., Wijkamp, I. G., van Schepen, A., Peters, D., van Grinsven, M. Q. J. M., and Goldbach, R. (1992). Characterization of RNA-mediated resistance to *Tomato spotted wilt virus* in transgenic tobacco plants. *BioTechnology* **10:**1133–1137.

Donini, M., Morea, V., Desiderio, A., Pashkoulov, D., Villani, M. E., Tramontano, A., and Benvenuto, E. (2003). Engineering stable cytoplasmic intrabodies with designed specificity. *J. Mol. Biol.* **330:**323–332.

Duan, C. G., Wang, C. H., Fang, R. X., and Guo, H. S. (2008). Artificial MicroRNAs highly accessible to targets confer efficient virus resistance in plants. *J. Virol.* **82:**11084–11095.

Ellstrand, N. C. (2003). Current knowledge of gene flow in plants: Implications for transgene flow. *Philos. Trans. R. Soc. Lond. B Biol. Sci.* **358:**1163–1170.

Ewen, S. W., and Pusztai, A. (1999). Effect of diets containing genetically modified potatoes expressing *Galanthus nivalis* lectin on rat small intestine. *Lancet* **354:**1353–1354.

Fitchen, J. H., and Beachy, R. N. (1993). Genetically engineered protection against viruses in transgenic plants. *Annu. Rev. Microbiol.* **47:**739–763.

Friess, N., and Maillet, J. (1996). Influence of *Cucumber mosaic virus* infection on the intraspecific competitive ability and fitness of purslane (*Portulaca oleracea*). *New Phytol.* **132:**103–111.

Friess, N., and Maillet, J. (1997). Influence of *Cucumber mosaic virus* infection on the competitive ability and reproduction of chickweed (*Stellaria media*). *New Phytol.* **135:**667–674.

Fuchs, M., and Gonsalves, D. (2007). Safety of virus-resistant transgenic plants two decades after their introduction: Lessons from realistic field risk assessment studies. *Annu. Rev. Phytopathol.* **45:**173–202.

Fuchs, M., Chirco, E. M., McFerson, J. R., and Gonsalves, D. (2004). Comparative fitness of a wild squash species and three generations of hybrids between wild × virus-resistant transgenic squash. *Environ. Biosafety Res.* **3:**17–28.

Gal, S., Pisan, B., Hohn, T., Grimsley, N., and Hohn, B. (1992). Agroinfection of transgenic plants leads to viable cauliflower mosaic virus by intermolecular recombination. *Virology* **187:**525–533.

Garcia-Arenal, F., and McDonald, B. A. (2003). An analysis of the durability of resistance to plant viruses. *Phytopathology* **93:**941–952.

Garrett, K. A., Dendy, S. P., Frank, E. E., Rouse, M. N., and Travers, S. E. (2006). Climate change effects on plant disease: Genomes to ecosystems. *Annu. Rev. Phytopathol.* **44:**489–509.

Genoves, A., Navarro, J. A., and Pallas, V. (2006). Functional analysis of the five melon necrotic spot virus genome-encoded proteins. *J. Gen. Virol.* **87:**2371–2380.

Goldbach, R., and Peters, D. (1996). Molecular and biological aspects of tospoviruses. *In* "The Bunyaviridae" (R. M. Elliott, ed.), pp. 129–157. Plenum Press, New York, NY.

Gonsalves, D. (1998). Control of *Papaya ringspot virus* in papaya: A case study. *Annu. Rev. Phytopathol.* **36:**415–437.

Goodman, R. E., Vieths, S., Sampson, H. A., Hill, D., Ebisawa, M., Taylor, S. L., and van Ree, R. (2008). Allergenicity assessment of genetically modified crops—What makes sense? *Nat. Biotechnol.* **26:**73–81.

Goodwin, J., Chapman, K., Swaney, S., Parks, T. D., Wernsman, E. A., and Dougherty, W. G. (1996). Genetic and biochemical dissection of transgenic RNA-mediated virus resistance. *Plant Cell* **8:**95–105.

Gottula, J., and Fuchs, M. (2009). Toward a quarter century of pathogen-derived resistance and practical approaches to plant virus disease control. *Adv. Virus Res.* **75:**161–183.

Greene, A. E., and Allison, R. F. (1994). Recombination between viral RNA and transgenic plant transcripts. *Science* **263:**1423–1425.

Hamilton, A. J., and Baulcombe, D. C. (1999). A species of small antisense RNA in posttranscriptional gene silencing in plants. *Science* **286:**950–952.

Helliwell, C., and Waterhouse, P. (2003). Constructs and methods for high-throughput gene silencing in plants. *Methods* **30:**289–295.

Hull, R. (1990). The use and misuse of viruses in cloning and expression in plants. *In* "Recognition and Response in Plant–Virus Interactions" (R. S. S. Fraser, ed.), pp. 443–457. Springer, Berlin.

Hull, R. (2002). *Matthew's Plant Virology*, 4th Edn. Academic Press, San Diego, CA.

Irwin, M. E., Ruesink, W. G., Isard, S. A., and Kampmeier, G. E. (2000). Mitigating epidemics caused by non-persistently transmitted aphid-borne viruses: The role of the pliant environment. *Virus Res.* **71:**185–211.

Jackson, A. L., Bartz, S. R., Schelter, J., Kobayashi, S. V., Burchard, J., Mao, M., Li, B., Cavet, G., and Linsley, P. S. (2003). Expression profiling reveals off-target gene regulation by RNAi. *Nat. Biotechnol.* **21:**635–637.

Jacquemond, M., Teycheney, P. Y., Carrère, I., Navas-Castillo, J., and Tepfer, M. (2001). Resistance phenotypes of transgenic tobacco plants expressing different *Cucumber mosaic virus* (CMV) coat protein genes. *Mol. Breed.* **8:**85–94.

Johnson, K. L., Raybould, A. F., Hudson, M. D., and Poppy, G. M. (2007). How does scientific risk assessment of GM crops fit within the wider risk analysis? *Trends Plant Sci.* **12:**1–5.

Jones, R. A. (2004). Using epidemiological information to develop effective integrated virus disease management strategies. *Virus Res.* **100:**5–30.

Kajihara, H., Kameya-Iwaki, M., Oonaga, M., Kimura, I., Sumida, Y., Ooi, Y., and Ito, S. (2008). Field studies on cross-protection against *Japanese yam mosaic virus* in Chinese yam (*Dioscorea opposita*) with an attenuated strain of the virus. *J. Phytopathol.* **156:**75–78.

Kalantidis, K., Denti, M. A., Tzortzakaki, S., Marinou, E., Tabler, M., and Tsagris, M. (2007). Virp1 is a host protein with a major role in *Potato spindle tuber viroid* infection in *Nicotiana* plants. *J. Virol.* **81:**12872–12880.

Kamachi, S., Mochizuki, A., Nishiguchi, M., and Tabei, Y. (2007). Transgenic *Nicotiana benthamiana* plants resistant to cucumber green mottle mosaic virus based on RNA silencing. *Plant Cell Rep.* **26:**1283–1288.

Kang, B. C., Yeam, I., and Jahn, M. M. (2005). Genetics of plant virus resistance. *Annu. Rev. Phytopathol.* **43:**581–621.

Kapadia, S. B., Brideau-Andersen, A., and Chisari, F. V. (2003). Interference of hepatitis C virus RNA replication by short interfering RNAs. *Proc. Natl. Acad. Sci. USA* **100:**2014–2018.

Karimi, M., Depicker, A., and Hilson, P. (2007). Recombinational cloning with plant gateway vectors. *Plant Physiol.* **145:**1144–1154.

Komar, V., Vigne, E., Demangeat, G., Lemaire, O., and Fuchs, M. (2008). Cross-protection as control strategy against *Grapevine fanleaf virus* in naturally infected vineyards. *Plant Dis.* **92:**1689–1694.

König, A., Cockburn, A., Crevel, R. W. R., Debruyne, E., Grafstroem, R., Hammerling, U., Kimber, I., Knudsen, I., Kuiper, H. A., Peijnenburg, A. A. C. M., Penninks, A. H., Poulsen, M., *et al.* (2004). Assessment of the safety of foods derived from genetically modified (GM) crops. *Food Chem. Toxicol.* **42:**1047–1088.

Kosaka, Y., Ryang, B.-S., Kobori, T., Shiomi, H., Yasuhara, H., and Kataoka, M. (2006). Effectiveness of an attenuated *Zucchini yellow mosaic virus* isolate for cross-protecting cucumber. *Plant Dis.* **90:**67–72.

Kreuze, J. F., Klein, I. S., Lazaro, M. U., Chuquiyuri, W. J. C., Morgan, G. L., Mejía, P. G. C., Ghislain, M., and Valkonen, J. P. T. (2008). RNA silencing-mediated resistance to a crinivirus (*Closteroviridae*) in cultivated sweet potato (*Ipomoea batatas* L.) and development of sweet potato virus disease following co-infection with a potyvirus. *Mol. Plant Pathol.* **9:**589–598.

Krubphachaya, P., Juricek, M., and Kertbundit, S. (2007). Induction of RNA-mediated resistance to *Papaya ringspot virus* type W. *J. Biochem. Mol. Biol.* **40:**404–411.

Kuiper, H. A., Kleter, G. A., Noteborn, H. P., and Kok, E. J. (2002). Substantial equivalence—An appropriate paradigm for the safety assessment of genetically modified foods? *Toxicology* **181–182:**427–431.

Lacomme, C., Hrubikova, K., and Hein, I. (2003). Enhancement of virus-induced gene silencing through viral-based production of inverted-repeats. *Plant J.* **34:**543–553.

Levy, L. (2006). *Plum pox virus* in the United States of America. *OEPP/EPPO Bull.* **36:**217–218.

Lin, S. S., Wu, H. W., Elena, S. F., Chen, K. C., Niu, Q. W., Yeh, S. D., Chen, C. C., and Chua, N. H. (2009). Molecular evolution of a viral non-coding sequence under the selective pressure of amiRNA-mediated silencing. *PLoS Pathog.* **5:**e1000312.

Lindbo, J. A., Siva-Rosales, L., Proebsting, W. M., and Dougherty, W. G. (1993). Induction of a highly specific antiviral state in transgenic plants: Implications for regulation of gene expression and virus resistance. *Plant Cell* **5:**1749–1759.

Lopez, M. M., Llop, P., Olmos, A., Marco-Noales, E., Cambra, M., and Bertolini, E. (2009). Are molecular tools solving the challenges posed by detection of plant pathogenic bacteria and viruses? *Curr. Issues Mol. Biol.* **11:**13–46.

Losey, J. E., Rayor, L. S., and Carter, M. E. (1999). Transgenic pollen harms monarch larvae. *Nature* **399:**214.

Lucioli, A., Sallustio, D. E., Barboni, D., Berardi, A., Papacchioli, V., Tavazza, R., and Tavazza, M. (2008). A cautionary note on pathogen-derived sequences. *Nat. Biotechnol.* **26:**617–619.

Magaña-Gomez, J. A., and Calderón de la Barca, A. M. (2009). Risk assessment of genetically modified crops for nutrition and health. *Nutr. Rev.* **67:**1–16.

Maskell, L. C., Raybould, A. F., Cooper, J. I., Edwards, M.-L., and Gray, A. J. (1999). Effects of *Turnip mosaic virus* and *Turnip yellow mosaic virus* on the survival, growth and reproduction of wild cabbage (*Brassica oleracea*). *Ann. Appl. Biol.* **135:**401–407.

Matthews, R. F. (1991). *Plant Virology*, 3rd Edn. Academic Press, San Diego, CA.

Medley, T. L. (1994). Availability of determination of nonregulated status for virus resistant squash. *Fed. Regist.* **59:**64187–64189.

Morroni, M., Thompson, J. R., and Tepfer, M. (2008). Twenty years of transgenic plants resistant to *Cucumber mosaic virus*. *Mol. Plant Microbe Interact.* **21:**675–684.

Mueller, E., Gilbert, J., Davenport, G., Brigneti, G., and Baulcombe, D. C. (1995). Homology-dependent resistance: Transgenic virus resistance in plants related to homology-dependent gene silencing. *Plant J.* **7:**1001–1013.

Niu, Q.-W., Lin, S.-S., Reyes, J. L., Chen, K.-C., Wu, H.-W., Yeh, S.-D., and Chua, N.-H. (2006). Expression of artificial microRNAs in transgenic *Arabidopsis thaliana* confers virus resistance. *Nat. Biotechnol.* **24:**1420–1428.

Noris, E., Lucioli, A., Tavazza, R., Caciagli, P., Accotto, G. P., and Tavazza, M. (2004). *Tomato yellow leaf curl Sardinia virus* can overcome transgene-mediated RNA silencing of two essential viral genes. *J. Gen. Virol.* **85:**1745–1749.

Oerke, E.-C., Dehne, H.-W., Schönbeck, F., and Weber, A. (1994). *Crop production and crop protection: Estimated losses in major food and cash crops*. Elsevier, Amsterdam.

Palukaitis, P. (1991). Virus-mediated genetic transfer in plants. *In* ''Risk Assessment in Genetic Engineering'' (M. Levin and H. Strauss, eds.), pp. 140–162. McGraw-Hill, New York, NY.

Patil, B. L., and Fauquet, C. M. (2009). Cassava mosaic geminiviruses: Actual knowledge and perspectives. *Mol. Plant Pathol.* **10:**1–17.

Pawloski, L. C., Deal, R. B., McKinney, E. C., Burgos-Rivera, B., and Meagher, R. B. (2005). Inverted repeat PCR for the rapid assembly of constructs to induce RNA interference. *Plant Cell Physiol.* **46:**1872–1878.

Pennazio, S., Roggero, P., and Conti, M. (1996). Yield losses in virus-infected crops. *Arch. Phytopathol. Plant Prot.* **30:**283–296.

Pflieger, S., Blanchet, S., Camborde, L., Drugeon, G., Rousseau, A., Noizet, M., Planchais, S., and Jupin, I. (2008). Efficient virus-induced gene silencing in *Arabidopsis* using a 'one-step' TYMV-derived vector. *Plant J.* **56:**678–690.

Poppy, G. (2000). GM crops: Environmental risks and non-target effects. *Trends Plant Sci.* **5:**4–6.

Poppy, G. M. (2004). Geneflow from GM plants—Towards a more quantitative risk assessment. *Trends Biotechnol.* **22:**436–438.

Prins, M., Laimer, M., Noris, E., Schubert, J., Wassenegger, M., and Tepfer, M. (2008). Strategies for antiviral resistance in transgenic plants. *Mol. Plant Pathol.* **9:**73–83.

Qi, A., Dewar, A. M., and Harrington, R. (2004). Decision making in controlling virus yellows of sugar beet in the UK. *Pest Manage. Sci.* **60:**727–732.

Qu, F., Ren, T., and Morris, T. J. (2003). The coat protein of turnip crinkle virus suppresses posttranscriptional gene silencing at an early initiation step. *J. Virol.* **77:**511–522.

Qu, J., Ye, J., and Fang, R. (2007). Artificial microRNA-mediated virus resistance in plants. *J. Virol.* **81:**6690–6699.

Quemada, H., Strehlow, L., Decker-Walters, D. S., and Staub, J. E. (2008). Population size and incidence of virus infection in free-living populations of *Cucurbita pepo*. *Environ. Biosafety Res.* **7:**185–196.

Raybould, A., and Cooper, I. (2005). Tiered tests to assess the environmental risk of fitness changes in hybrids between transgenic crops and wild relatives: The example of virus resistant *Brassica napus*. *Environ. Biosafety Res.* **4:**127–140.

Reddy, D. V. R., Sudarshana, M. R., Fuchs, M., Rao, N. C., and Thottappilly, G. (2009). Genetically engineered virus-resistant plants in developing countries: current status and future prospects. *Adv. Virus Res.* **75:**185–220.

Ribeiro, S. G., Lohuis, H., Goldbach, R., and Prins, M. (2007). *Tomato chlorotic mottle virus* is a target of RNA silencing but the presence of specific short interfering RNAs does not guarantee resistance in transgenic plants. *J. Virol.* **81:**1563–1573.

Ritzenthaler, C. (2005). Resistance to plant viruses: Old issue, news answers? *Curr. Opin. Biotechnol.* **16:**118–122.

Rodoni, B. (2009). The role of plant biosecurity in preventing and controlling emerging plant virus disease epidemics. *Virus Res.* **141:**150–157.

Safarnejad, M. R., Fischer, R., and Commandeur, U. (2009). Recombinant-antibody-mediated resistance against *Tomato yellow leaf curl virus* in *Nicotiana benthamiana*. *Arch. Virol.* **154:**457–467.

Sanford, J. C., and Johnston, S. A. (1985). The concept of parasite-derived resistance—Deriving resistance genes from the parasite's own genome. *J. Theor. Biol.* **113:**395–405.

Shin, D., Lee, H., Kim, S. I., Yoon, Y., and Kim, M. (2009). Optimization of linear double-stranded RNA for the production of multiple siRNAs targeting hepatitis C virus. *RNA* **15:**898–910.

Spence, N., and Garcia-Arenal, F. (2005). Overview of key virus problems in Europe. *ResistVir. Co-ordination of Research on Genetic Resistance to Plant Pathogenic Virus and Their Vectors in European Crops.* http://www.resistvir-db.org/key_reports.htm.

Stein, A. J., and Rodríguez-Cerezo, E. (2009). The global pipeline of new GM crops: Implications of asynchronous approval for international trade. *JRC Technical Report.* JRC, Luxembourg.

Sudarshana, M. R., Roy, G., and Falk, B. W. (2007). Methods for engineering resistance to plant viruses. *Methods Mol. Biol.* **354:**183–195.

Tavladoraki, P., Benvenuto, E., Trinca, S., De Martinis, D., Cattaneo, A., and Galeffi, P. (1993). Transgenic plants expressing a functional single-chain Fv antibody are specifically protected from virus attack. *Nature* **366:**469–472.

Taylor, N., Chavarriaga, P., Raemakers, K., Siritunga, D., and Zhang, P. (2004). Development and application of transgenic technologies in cassava. *Plant Mol. Biol.* **56:**671–688.

Tepfer, M. (1993). Viral genes and transgenic plants: What are the potential environmental risks? *BioTechnology* **11:**1125–1132.

Tepfer, M. (2002). Risk assessment of virus-resistant transgenic plants. *Annu. Rev. Phytopathol.* **40:**467–491.

Tester, M. (2001). Depolarizing the GM debate. *New Phytol.* **149:**9–16.

Teycheney, P.-Y., and Tepfer, M. (1999). Gene flow from virus-resistant transgenic crops to wild relatives or to infecting viruses. *In* "Gene Flow and Agriculture: Relevance for Transgenic Crops" (P. J. W. Lutman, ed.), British Crop Protection Council Symposium Proceedings No. 72, pp. 191–196.

Thackray, D. J., Diggle, A. J., Berlandier, F. A., and Jones, R. A. (2004). Forecasting aphid outbreaks and epidemics of *Cucumber mosaic virus* in lupin crops in a Mediterranean-type environment. *Virus Res.* **100:**67–82.

Thomas, C. L., Leh, V., Lederer, C., and Maule, A. J. (2003). Turnip crinkle virus coat protein mediates suppression of RNA silencing in *Nicotiana benthamiana. Virology* **306:**33–41.

Thompson, D. (2006). *Plum pox virus* (PPV) in Canada. *OEPP/EPPO Bull.* **36:**206.

Thresh, J. M., and Cooter, R. J. (2005). Strategies for controlling cassava mosaic virus disease in Africa. *Plant Pathol.* **54:**587–614.

Tougou, M., Furutani, N., Yamagishi, N., Shizukawa, Y., Takahata, Y., and Hidaka, S. (2006). Development of resistant transgenic soybeans with inverted repeat-coat protein genes of soybean dwarf virus. *Plant Cell Rep.* **25:**1213–1218.

Turturo, C., Friscina, A., Gaubert, S., Jacquemond, M., Thompson, J. R., and Tepfer, M. (2008). Evaluation of potential risks associated with recombination in transgenic plants expressing viral sequences. *J. Gen. Virol.* **89:**327–335.

van der Vlugt, R. (2006). Plant viruses in European agriculture: Current problems and future aspects. *In* "Virus Diseases and Crop Biosecurity" (J. I. Cooper, T. Kuhne, and V. P. Polishuk, eds.), pp. 33–44. Springer, Dordrecht.

Vanderschuren, H., Stupak, M., Fütterer, J., Gruissem, W., and Zhang, P. (2007). Engineering resistance to geminiviruses—Review and perspectives. *Plant Biotechnol. J.* **5:**207–220.

Vanderschuren, H., Alder, A., Zhang, P., and Gruissem, W. (2009). Dose-dependent RNAi-mediated geminivirus resistance in the tropical root crop cassava. *Plant Mol. Biol.* **70:**265–272.

Varrelmann, M., Palkovics, L., and Maiss, E. (2000). Transgenic or plant expression vector-mediated recombination of *Plum pox virus. J. Virol.* **74:**7462–7469.

Vigne, E., Komar, V., and Fuchs, M. (2004). Field safety assessment of recombination in transgenic grapevines expressing the coat protein gene of *Grapevine fanleaf virus. Transgenic Res.* **13:**165–179.

Villani, M. E., Roggero, P., Bitti, O., Benvenuto, E., and Franconi, R. (2005). Immunomodulation of *Cucumber mosaic virus* infection by intrabodies selected *in vitro* from a stable single-framework phage display library. *Plant Mol. Biol.* **58:**305–316.

Voinnet, O. (2008). Post-transcriptional RNA silencing in plant–microbe interactions: A touch of robustness and versatility. *Curr. Opin. Plant. Biol.* **11:**464–470.

Wang, L., Luo, Y. Z., Zhang, L., Jiao, X. M., Wang, M. B., and Fan, Y. L. (2008). Rolling circle amplification-mediated hairpin RNA (RMHR) library construction in plants. *Nucleic Acids Res.* **36:**e149.

Wang, R. Y., Stork, J., and Nagy, P. D. (2009). A key role for heat shock protein 70 in the localization and insertion of tombusvirus replication proteins to intracellular membranes. *J. Virol.* **83:**3276–3287.

Waterworth, H. E., and Hadidi, A. (1998). Economic losses due to plant viruses. *In* ''Plant Virus Disease Control'' (A. Hadidi, R. K. Khetarpal, and H. Koganezawa, eds.). APS Press, St. Paul, MN.

Wielopolska, A., Townley, H., Moore, I., Waterhouse, P., and Helliwell, C. (2005). A high-throughput inducible RNAi vector for plants. *Plant Biotechnol. J.* **3:**583–590.

Wilkinson, M., and Tepfer, M. (2009). Fitness and beyond: Preparing for the arrival of GM crops with ecologically important novel characters. *Environ. Biosafety Res.* **8:**1–14.

Wilkinson, M. J., Sweet, J., and Poppy, G. M. (2003). Risk assessment of GM plants: Avoiding gridlock? *Trends Plant Sci.* **8:**208–212.

Wilson, T. M. (1993). Strategies to protect crop plants against viruses: Pathogen-derived resistance blossoms. *Proc. Natl. Acad. Sci. USA* **90:**3134–3141.

Wilson, J. A., and Richardson, C. D. (2005). Hepatitis C virus replicons escape RNA interference induced by a short interfering RNA directed against the NS5b coding region. *J. Virol.* **79:**7050–7058.

Yamamura, K., and Kiritani, K. (1998). A simple method to estimate the potential increase in the number of generations under global warming in temperate zones. *Appl. Entomol. Zool.* **33:**289–298.

Zrachya, A., Kumar, P. P., Ramakrishnan, U., Levy, Y., Loyter, A., Arazi, T., Lapidot, M., and Gafni, Y. (2007). Production of siRNA targeted against TYLCV coat protein transcripts leads to silencing of its expression and resistance to the virus. *Transgenic Res.* **16:**385–398.

CHAPTER 3

Signaling in Induced Resistance

John P. Carr,* Mathew G. Lewsey,* and **Peter Palukaitis**[†,‡]

Contents

* Department of Plant Sciences, University of Cambridge, Cambridge CB2 3EA, United Kingdom, E-mail: jpc1005@cam.ac.uk
† Scottish Crop Research Institute, Invergowrie, Dundee DD2 5DA, United Kingdom
‡ Division of Environmental and Life Sciences, Seoul Women's University, Seoul 139-774, Korea

Advances in Virus Research, Volume 76
ISSN 0065-3527, DOI: 10.1016/S0065-3527(10)76003-6

Abstract

Induced mechanisms are by definition imperceptible or less active in uninfected, unstressed, or untreated plants, but can be activated by pathogen infection, stress, or chemical treatment to inhibit the replication and movement of virus in the host. In contrast, defenses that are pre-existing or serve to limit virus propagation and spread in otherwise susceptible hosts are considered to be "basal" in nature. Both forms of resistance can be genetically determined. Most recessive resistance genes that control resistance to viruses appear not to depend upon inducible mechanisms but rather maintain basal resistance by producing nonfunctional variants of factors, specifically translation initiation factors, required by the virus for successful exploitation of the host cell protein synthetic machinery. In contrast, most dominant resistance genes condition the induction of broad-scale changes in plant biochemistry and physiology that are activated and regulated by various signal transduction pathways, particularly those regulated by salicylic acid, jasmonic acid, and ethylene. These induced changes include localized plant cell death (associated with the hypersensitive response, HR) and the upregulation of resistance against many types of pathogen throughout the plant (systemic acquired resistance, SAR). Unfortunately, it is still poorly understood how virus infection is inhibited and restricted during the HR and in plants exhibiting SAR. Resistance to viruses is not always genetically predetermined and can be highly adaptive in nature. This is exemplified by resistance based on RNA silencing, which appears to play roles in both induced and basal resistance to viruses. To counter inducible resistance mechanisms, viruses have acquired counter-defense factors to subvert RNA silencing. Some of these factors may affect signal transduction pathways controlled by salicylic acid and jasmonic acid. In this chapter, we review current knowledge of defensive signaling in resistance to viruses including the nature and roles of low molecular weight, proteinaceous, and small RNA components of defensive signaling. We discuss the differences and similarities of defenses and defensive signaling directed against viral versus nonviral pathogens, the potential role of RNA silencing as an effector in resistance and possible regulator of defensive signaling, crosstalk and overlap between antiviral systems, and interference with and manipulation of host defensive systems by the viruses themselves.

I. WHAT IS INDUCED RESISTANCE?

Historically, the term "induced resistance" has been applied to the activation, by either pathogen attack, abiotic stress, or chemical treatment, of novel defense mechanisms. By "novel," it is usually meant that these mechanisms are imperceptible in uninfected, unstressed, or untreated plants. Induced resistance mechanisms are most apparent in, and often most easily studied in, noninvaded tissue after other parts of the plant have been infected and they often contribute to the phenomena of local and systemic acquired resistance (LAR and SAR, respectively; see Loebenstein, 2009). A typical feature of induced resistance is that it is often triggered in a highly specific fashion by one particular pathogen but results in the activation of LAR and SAR against a large variety of diverse microbes and viruses (Hammerschmidt, 2009). For example, a gene-for-gene type interaction, such as that involving the attempted infection by *Tobacco mosaic virus* (TMV) of a resistant tobacco plant (i.e., carrying the *N* resistance gene), not only results in the restriction of TMV to few cells in the vicinity of the inoculation site, but also enhances the plant's ability to inhibit infection by unrelated viruses, as well as bacteria, fungi, and oomycetes (reviewed by Gilliland *et al.*, 2003; Lewsey *et al.*, 2009; Loebenstein, 2009; Palukaitis and Carr, 2008).

As our knowledge of plant–virus interactions has improved, these definitions have become less distinct. Most notably, recent and rapid developments in the field of RNA silencing have revealed that this highly adaptive and specific mechanism of gene regulation participates in the maintenance of basal resistance and several induced resistance mechanisms, as well as affecting defensive signaling (see Section IV). RNA silencing also represents an inducible resistance mechanism that shows high specificity with respect to both its induction and its effects on the invading virus.

In this chapter, we review the current status of the field of induced resistance and while we will adhere to many of the conventional classifications of resistance into induced, genetically determined, adaptive, etc., we would encourage the reader to remember that, in Nature, these various mechanisms intersect and overlap in ways that we still do not fully understand.

II. SIGNALING IN GENETICALLY DETERMINED RESISTANCE

Genetically based resistance is often viewed as the ideal form of in-field protection for crops against plant disease. When genetic resistance works it is economical, nonpolluting and it is often the only option when seeking to protect a crop against a viral disease (Maule *et al.*, 2007). Another advantage that genetic resistance has for protection against viral diseases is that the evolution of resistance-breaking virus strains occurs more slowly than the evolution of resistance-breaking bacterial, fungal, or oomycete pathogens (García-Arenal and MacDonald, 2003; García-Arenal *et al.*, 2003; Lecoq *et al.*, 2004). This might be considered surprising, considering the error-prone nature of viral replication and the consequent tendency toward a high mutation rate for viral genomes (for more on this important topic, see Chapter 1).

Single dominant, single recessive, or multiple genes can confer resistance to viruses (Maule *et al.*, 2007). But the greatest amount of information on the signaling processes regulating resistance to viruses has been obtained from studies of interactions involving plant hosts in possession of single dominant resistance (*R*) genes. Nevertheless, the investigation of other forms of genetic resistance to viruses has also informed our understanding of plant–virus interaction mechanisms in susceptible and resistant hosts.

A. Resistance conditioned by recessive, semidominant, and multiple genes

Studies of *r* genes controlling resistance to viruses have revealed profoundly important insights into mechanisms of resistance and the identity of a class of host factors required for successful viral infection (reviewed by Robaglia and Caranta, 2006; Truniger and Aranda, 2009). A combination of mapping of natural *r* genes and their creation by mutagenesis of *Arabidopsis thaliana* (hereafter referred to as Arabidopsis) has led to the conclusion that most of these genes encode variants of eukaryotic initiation factor (eIF) 4 subcomponents (Robaglia and Caranta, 2006; Truniger and Aranda, 2009). This has been best characterized for *r* gene-carrying cultivars or mutant Arabidopsis lines showing resistance to potyviruses and bymoviruses. The 5′-termini of the RNAs of these viruses are covalently linked to VPg (virus protein-genome linked) proteins. VPg proteins perform some of the functions of a methyl-G cap that would be found at the 5′-terminus of most cellular mRNA, namely allowing or enhancing interactions with the cellular translation machinery. VPg proteins can have very specific requirements for binding to factors eIF4E, eIF(iso)4E, or eIF4G and genes encoding variants or mutant versions of these factors will confer resistance to these viruses (Borgstrom and Johansen, 2001; Moury *et al.*, 2004). Interestingly, infection by certain viruses that do not possess VPg proteins, including *Cucumber mosaic virus* (CMV) and *Turnip crinkle virus* (TCV), is inhibited in plants harboring naturally occurring *r* genes or carrying mutant genes encoding eIF4E or G variants (Yoshii *et al.*, 2004).

However, these resistance mechanisms are "passive," in this case meaning that they depend on the absence of a factor necessary for successful infection by the virus, and they have not yielded data on either defensive signaling or any of the inducible antiviral defenses. Nevertheless, there is no reason to rule out the possibility that future studies of antiviral *r* genes may reveal active, rather than passive, modes of action. Indeed, a recent study of the recessive *tcm-1* locus, which conditions resistance to tomato against the begomovirus *Tomato yellow leaf curl virus*, suggested that the resistance conditioned by *tcm-1* resembles recovery and is therefore likely to involve induction of RNA silencing against the virus (see Section IV) (García-Cano *et al.*, 2008). *Tm-1*, probably the best-characterized semidominant antiviral resistance gene, can block infection of tomato plants by the tobamovirus *Tomato mosaic virus* (ToMV) in a gene dosage-dependent manner by inhibiting the replication of the virus (Fraser and Loughlin, 1980; Meshi *et al.*, 1988; Motoyoshi and Oshima, 1977; Watanabe *et al.*, 1987). This will be described in detail in Section V.D.1.

Most agronomic traits in crop plants, including virus resistance, do not segregate as simple monogenic characters. Rather, they are quantitative in nature and controlled by multiple genes (Maule *et al.*, 2007). RNA silencing, which underlies several types of resistance, is itself an example of multigenic or polygenic resistance, as first noted by Dougherty and colleagues (Goodwin *et al.*, 1996) (Section IV). The activity of both *r* genes and *R* genes (Section II.B) can be affected and complicated by the presence or absence of additional genes. For example, the effectiveness of the resistance mediated by the gene *pvr23* against *Potato virus Y* (PVY) varies depending upon the background into which it is introgressed (Palloix *et al.*, 2009). The product of the *HRT* gene enables plants of the Dijon line of Arabidopsis to recognize and resist TCV. HRT protein function is affected by an additional protein ("regulates resistance to TCV," RRT), which negatively regulates levels of the HRT protein. Thus, only plants homozygous for the recessive *rrt* allele are normally able to express *HRT*-mediated resistance to TCV (Chandra-Shekara *et al.*, 2004; Kachroo *et al.*, 2000).

B. Resistance conditioned by dominant resistance (*R*) genes

The study of dominant *R* gene-regulated responses has proved to be the richest seam for extracting information on induced resistance to viruses and the signals responsible for its establishment. However, not all single dominant genes trigger induced resistance mechanisms; some condition various forms of basal, passive resistance in which there is a constitutively produced barrier to one or more stages in infection. Examples of "passive" *R* genes would include: *RTM1–3* (*restriction of tobacco etch virus movement*) and related genes that control potyvirus movement in

Arabidopsis (see Section V.A.3); a locus in Arabidopsis, *DSTM1*, which slows but does not prevent TMV systemic movement (Serrano *et al.*, 2008); or the gene in cowpea (cv. Arlington) encoding an inhibitor of proteolytic processing of the *Cowpea mosaic virus* (CPMV) polyproteins (Bruening, 2006; Ponz and Bruening, 1986; Ponz *et al.*, 1988). In contrast, the majority of "antiviral" *R* genes trigger active defense responses.

1. R proteins and resistance to viruses: General features

Most *R* genes characterized to date control resistance mechanisms characterized by a local lesion reaction; either a hypersensitive response (HR) or extreme resistance, where the virus is localized to a few cells (in the HR) or one cell (in extreme resistance) (Maule *et al.*, 2007; Palukaitis and Carr, 2008). Loebenstein (2009) has reviewed the characteristics of the various types of local lesion response.

All of the R proteins encoded by "antiviral" resistance genes that have been isolated from diverse crop and model plants belong to the nucleotide-binding site–leucine-rich repeat (NBS–LRR) class and are located intracellularly. These proteins can be further subdivided depending on whether they have a Toll/interleukin receptor-1 (TIR) like domain (e.g., the N protein; Whitham *et al.*, 1994) or a coiled-coil (CC) like domain within the N-terminus (e.g., the Rx protein; Moffett *et al.*, 2002). However, there are no structural or other features of "antiviral" R proteins that sets them apart from the NBS–LRR proteins that confer resistance to bacteria, fungi, oomycetes, or invertebrates (Jones and Dangl, 2001), and the signaling processes that they trigger are, as far as we know, identical and lead to the activation of defenses against a broad spectrum of pathogens or pests, not just viruses (Maule *et al.*, 2007; Moffett, 2009; Murphy *et al.*, 1999; Palukaitis and Carr, 2008). For more details on R protein structure and mode of action, the reader is referred to Section V. A.2 in this review and to the recent review by Moffett (2009).

2. The effector-triggered immunity paradigm and the role of R proteins in defense against viruses

Resistance genes allow plants to recognize specific pathogens in a so-called gene-for-gene manner, that is, a direct or indirect interaction between the R protein and an elicitor, the product of a corresponding avirulence (*Avr*) gene in the pathogen (Kachroo, 2006; Schoelz, 2006). It is now becoming more common to refer to this type of interaction as effector-triggered immunity (ETI), particularly with regard to the mode of action of *R* genes conferring resistance to bacterial pathogens such as *Pseudomonas syringae* (Boller and He, 2009). This redefinition has come about because in many cases, NBS–LRR proteins are responding to attacks on components of plant defense networks regulating the recognition and response to MAMPS (microbe-associated molecular patterns) also known as PAMPS (pathogen-associated molecular patterns).

Such recognition, often referred to as PAMP-triggered immunity (PTI), underlies certain basal and nonhost resistance phenomena.

When virulent bacteria mount attacks on PTI, they inject "effector" molecules into the plant cell using the type III secretion system (reviewed by Lindeberg *et al.*, 2009). Once in the cell, the effector molecules modify or destroy components of the basal defense network, for example the proteolytic cleavage by the AvrPphB effector of PBS1 or of RIN4 by the AvrRPT2 effector in Arabidopsis plants infected with appropriate strains of *P. syringae* (Axtell and Staskawicz, 2003; Shao *et al.*, 2003). Plants have evolved the ability to recognize effector molecules using NBS–LRR genes and the coevolution of plants and their cellular pathogens (not only bacteria but also fungi and oomycetes) has been driven by the appearance of new effectors and of corresponding R proteins to "guard" effector targets in the PTI network (Chisholm *et al.*, 2006), and possibly of host molecules that mimic PTI components to "decoy" the effectors (van der Hoorn and Kamoun, 2008). However, MAMPS such as conserved amino acid sequences within flagellin or the prokaryotic elongation factor Tu (Boller and Felix, 2009) are not produced by viruses and these bacterial MAMPS are perceived by PAMP recognition receptors (PRRs) that possess extracellular recognition/binding domains (Boller and He, 2009; Chisholm *et al.*, 2006). It would be unlikely that the known PRRs can detect viral gene products, which in plants remain intracellularly localized.

The possibility remains, however, that plants may possess another, undiscovered type of PRR tailored to the detection of viruses or to common virus-induced effects in the initially infected cell. There is no direct evidence for their existence but a number of observations hint at the ability of plants to respond biochemically to virus infection even in the absence of an effective induced resistance response. These include the detection of a transient burst of reactive oxygen species (ROS), a common early indicator of defense-related biochemical changes (see Section III.D), either in TMV-susceptible (*nn* genotype) tobacco following treatment with the virus or purified TMV coat protein (CP) (Allan *et al.*, 2001), or in response to potexvirus CP sequences (Baurès *et al.*, 2008), and the observation of long-distance induction of ROS and nitric oxide (NO) production in advance of virus spread in susceptible Arabidopsis and tomato plants systemically infected with *Cauliflower mosaic virus* (CaMV) and ToMV, respectively (Fu *et al.*, 2010; Love *et al.*, 2005).

Is it appropriate to use the term ETI to refer to gene-for-gene relationships between plants and viruses? While it may be pointless for viruses to produce effectors that defeat PTI against bacteria or other cellular pathogens, there are other mechanisms that contribute to basal resistance, which would make more appropriate viral effector targets. Probably the most obvious of these is RNA silencing, which has roles not only in the destruction of viral RNAs (Section IV) but also, as it happens, in the regulation of PTI against bacterial pathogens and so may be an important mechanism in basal resistance to many pathogen types (Agorio and Vera, 2007; Navarro *et al.*, 2006; Ruiz-Ferrer and Voinnet, 2009). Moffett (2009) offers a model for the evolution of ETI against viruses based on the interplay between viruses and the RNA-silencing system.

However, the evidence for a viral equivalent of the ETI paradigm, in which components of the host's RNA-silencing pathways are targeted by viral "effectors" and "guarded" by NBS–LRR proteins, is equivocal. Viruses encode proteins that function as suppressors of RNA silencing (see Section IV.A.2). However, several viral gene products that act as elicitors in gene-for-gene type interactions have no known silencing suppressor activity. Examples of viral silencing suppressors that do act as gene-for-gene elicitors include the TMV 126 kDa replicase protein (elicitor of *N*-mediated HR in *Nicotiana* spp.); the TCV CP (also known as P38 and the elicitor of *HRT*-mediated HR in Arabidopsis), the *Potato virus X* (PVX) p25 movement protein (elicitor of the *Nb*-mediated HR in potato) (Malcuit *et al.*, 1999), or the P6 transactivator protein of CaMV (elicitor of HR in *N. edwardsonii*) (Cooley *et al.*, 2000; Deleris *et al.*, 2006; Harries *et al.*, 2008; Kiraly *et al.*, 1999; Kobayashi and Hohn, 2004; Love *et al.*, 2007a; Whitham *et al.*, 1994). In addition, expression of the 2b silencing suppressor protein of *Tomato aspermy virus* (TAV, a cucumovirus), using a TMV-based expression vector, triggered an HR-like response in tobacco (Li *et al.*, 1999) (see Section IV.A.2). Counterexamples, where viral elicitors are not viral silencing suppressors include the TMV CP, variants of which can elicit the HR in *Nicotiana* plants carrying the *N′* gene; the PVX CP, variants of which elicit extreme resistance in potato plants carrying the *Rx* gene; and the ToMV 30 kDa movement protein, which elicits the HR in tomato plants carrying the *R* genes *Tm-2* or *Tm-2*^{2} (Bendahmane *et al.*, 1995; Culver *et al.*, 1994; Moffett *et al.*, 2002; Pfitzner, 2006).

In conclusion, it is evident that our current understanding of the place of virus-triggered, R protein-mediated resistance induction within the wider network of signaling that coordinates basal (or nonhost) resistance with inducible defenses currently lags behind that for interactions involving phytopathogenic bacteria.

III. LOW MOLECULAR WEIGHT CHEMICAL SIGNALS IN INDUCED RESISTANCE

A. The biosynthesis and occurrence of salicylic acid and its derivatives, during incompatible and compatible interactions with viruses

Salicylic acid (SA; 2-hydroxybenzoic acid) plays a central role in the signal transduction pathway that results in the induction of SAR and it is required for localization of viral and other pathogens during the HR (reviewed by Alvarez, 2000; Hammerschmidt, 2009). SA is required for the expression of a group of proteins that collectively are referred to as pathogenesis-related (PR) proteins, many of which have antimicrobial properties, but which are thought to be unlikely to include directly antiviral proteins (see Section V).

The effects of SA on viruses are still probably not as well understood as SA-mediated resistance to fungi, oomycetes, or bacteria. SA can induce inhibition of virus replication, cell-to-cell movement, and systemic movement but the precise effects of SA-induced resistance on the life cycle of a virus can differ between hosts and between viruses (Chivasa and Carr, 1998; Chivasa *et al.*, 1997; Love *et al.*, 2005, 2007b; Mayers *et al.*, 2005; Murphy and Carr, 2002; Wong *et al.*, 2002). In tobacco and Arabidopsis some of these effects may depend on the action of a component of the host RNA-silencing pathway, RNA-dependent RNA polymerase 1 (RdRp1) (Gilliland *et al.*, 2003; Xie *et al.*, 2001; Yang *et al.*, 2004; Yu *et al.*, 2003) (see Sections IV and V). However, signaling, downstream of SA, transduced through changes in redox, or mitochondrially generated ROS also appears to play a role in resistance (Gilliland *et al.*, 2003; Love *et al.*, 2005, 2007b; Mayers *et al.*, 2005; Murphy *et al.*, 2004; Singh *et al.*, 2004; Wang *et al.*, 2008) (see Sections IV and V).

SA biosynthesis is induced most strongly during HR lesion development. During the TMV-induced HR in *NN* genotype tobacco, biosynthesis occurs initially in and around the developing lesions, but later throughout the entire plant concomitant with the onset of SAR (Huang *et al.*, 2006; Malamy *et al.*, 1990; Métraux *et al.*, 1990; Nobuta *et al.*, 2007; Strawn *et al.*, 2007). SA is rapidly conjugated to glucose to form SA-2-*O*-β-D-glucoside (SAG), which accumulates in cell vacuoles (Dean and Mills, 2004; Enyedi *et al.*, 1992; Hennig *et al.*, 1993). Klessig and colleagues suggested that SAG is a storage form of SA from which free, biologically active SA can be released in response to secondary infection in SAR-expressing plants (Hennig *et al.*, 1993).

The induction of SAR and, to some extent, maintenance of basal resistance to pathogen infection depends on SA-mediated signaling. Consequently, if SA accumulation is inhibited by engineering plants to express the salicylate-degrading enzyme SA hydroxylase (*nahG*-transgenic plants; Gaffney *et al.*, 1993; Mur *et al.*, 1997), or if SA production is decreased by mutation of SA biosynthetic genes (Nawrath and Métraux, 1999; Wildermuth *et al.*, 2001), plants are not able to express SAR and are supersusceptible to both virulent and avirulent pathogens.

Although SA is needed for the maintenance of basal resistance, successful pathogen localization during the HR and the establishment of SAR, an increase in its biosynthesis can occur during a compatible interaction with certain virulent pathogens. With respect to virus infection, systemic, non-necrotizing infections do not normally trigger biosynthesis of SA and the consequent induction of SA-responsive genes (*PR* genes, etc.) (Malamy *et al.*, 1990). However, SA-responsive gene expression has been observed in Arabidopsis (Huang *et al.*, 2005; Whitham *et al.*, 2003) and increased SA accumulation was observed in potato (Krečič-Stres *et al.*, 2005) and tobacco (M.G. Lewsey and A.M. Murphy, unpublished data) during systemic infections with CMV and potyviruses (*Turnip mosaic virus* in Arabidopsis and PVY in potato). These results appear to be paradoxical. For example, in the case of CMV, elevation in SA levels in the host, due to treatment with exogenous SA or induction of endogenous SA biosynthesis by an avirulent pathogen prior to infection with CMV, will inhibit the systemic movement of the virus (Ji and Ding, 2001; Mayers *et al.*, 2005; Naylor *et al.*, 1998). However, elicitation of SA biosynthesis by CMV infection is a slow process that is only apparent after systemic movement of the virus to all parts of the plant has already occurred (M.G. Lewsey and A.M. Murphy, unpublished data). Because of the counterdefensive action of the CMV 2b protein, continued replication or local movement of CMV in systemically infected tissues of the plant will be unaffected by increased SA levels (Ji and Ding, 2001; Murphy and Carr, 2002; Naylor *et al.*, 1998). It remains to be seen how the induction of what is nominally a defensive response can benefit CMV.

B. Signaling mediated by ethylene and jasmonates

Jasmonic acid (JA), an oxygenated fatty acid (oxylipin), and several of its derivatives (methyl-JA and certain JA-amino acid conjugates, particularly JA-Ile) are signals in resistance to certain bacterial and fungal pathogens and against insect and nematode pests (Chini *et al.*, 2007; Devoto, and Turner, 2003; Farmer *et al.*, 2003; Lorenzo and Solano, 2005; Rojo *et al.*, 2003; Staswick and Tiryaki, 2004). Together with the gaseous plant hormone ethylene (ethene), JA regulates a SA-independent, systemic resistance pathway inducible by nonpathogenic microbes ("induced systemic resistance," ISR), which primes resistance to pathogenic fungi and bacteria in Arabidopsis (Ton *et al.*, 2002). However, based on studies with TCV, it appears that ISR is not effective against infection by most viruses (Loebenstein, 2009; Pieterse *et al.*, 2009; Ton *et al.*, 2002), although there may be exceptions (see Section IV.C).

The interplay, or "crosstalk," between the pathways and the "choices" made between expression of SA-induced antimicrobial gene products (e.g., PR proteins) and JA-induced antimicrobial gene products (e.g., defensins) is regulated at several levels. In Arabidopsis, the products of the *PAD4* and *EDS1* genes promote SA-mediated gene induction and repress gene expression induced by JA. These regulatory proteins are themselves negatively regulated through phosphorylation catalyzed by mitogen-activated protein kinase (MAPK) systems (see Section V.B.1). Although the JA and SA pathways are viewed predominantly as mutually antagonistic, transcriptomic studies have revealed significant positive as well as negative crosstalk between the two pathways (Schenk *et al.*, 2000). Other work has shown that in the earliest stages of a TMV-induced HR in *NN* genotype tobacco, there is a complex interplay between the synthesis of JA and SA with early MAPK activity, and complex changes in auxin and abscisic acid levels (Kenton *et al.*, 1999; Kovač *et al.*, 2009; Liu *et al.*, 2004b). One might also speculate, based on findings with JA-insensitive mutants suggesting that JA regulates methyl-SA production (Attaran *et al.*, 2009), that transient changes in JA level in or around the developing lesion may affect the production of this highly mobile SA derivative (see Section III.C). Recent work has also revealed crosstalk between JA-regulated signaling and gene expression with the pathways of RNA silencing (see Section IV.C).

Ethylene is produced copiously during the HR and is a strong inducer of certain *PR* genes. Ethylene-dependent signaling is indispensable for the maintenance of basal resistance to fungi and bacteria (Geraats *et al.*, 2007; van Loon and van Strien, 1999), but it is not clear if this is also true for basal resistance to viruses. However, since ethylene biosynthesis from 1-aminocyclopropane-1-carboxylic acid results in the release of cyanide (Siegien and Bogatek, 2006), which can stimulate resistance to viruses (Chivasa and Carr, 1998), it is conceivable that this may contribute to the induction of resistance to viruses via the mitochondrial signaling pathway (see Section V.B.3). It is also possible that ethylene plays a role as a negative regulator of SA-mediated resistance to viruses. It has been proposed that this may explain a decrease in susceptibility to infection with CaMV observed by Milner and colleagues in the ethylene signaling mutant *etr1* in Arabidopsis (Love *et al.*, 2007b).

C. Long-distance signaling in SAR induction: A perennial conundrum

It is still not clear what signals are needed for the establishment of SAR in uninfected tissues of plants following an HR. In cucumber plants infected with necrotizing pathogens, and in *Ricinus communis* SA can be detected in phloem sap, consistent with a role in the spread of SAR induction to all parts of the plant (Métraux *et al.*, 1990; Rasmussen *et al.*, 1991; Rocher *et al.*, 2009). However, results, sometimes conflicting with each other, obtained from grafting experiments with independently produced lines of *NahG*-transgenic tobacco (Darby *et al.*, 2000; Vernooij *et al.*, 1994), cast doubt on the idea that SA is necessarily an essential translocated signal in SAR induction.

Subsequently, the isolation of the Arabidopsis *dir1* mutant, defective in a gene encoding a lipid-transfer protein, suggested that a lipid or lipid-derived substance could be the translocated SAR-inducing signal (Maldonado *et al.*, 2002). JA or one of its derivatives was put forward as a possible candidate for this role (Truman *et al.*, 2007), although there are other candidates (see below). In contrast, work from the Klessig laboratory showed that transgenic tobacco or Arabidopsis plants silenced for the gene encoding a SA-binding protein (SABP2) were compromised in gene expression of PR proteins and SAR induction (Park *et al.*, 2007; Vlot *et al.*, 2008a). SABP2 is an esterase that can convert methyl-SA to SA (Kumar *et al.*, 2006). Methyl-SA is produced in tissues undergoing an HR (Shulaev *et al.*, 1997). It is a volatile chemical that can diffuse into the air and even induce SAR in nearby, noninfected plants (Shulaev *et al.*, 1997). Taken together these studies appear to provide evidence suggesting that methyl-SA is an important long-distance signal in the induction of SAR following HR induction.

The roles of JA and in SAR induction have been questioned. Shah and colleagues (Chaturvedi *et al.*, 2008) isolated phloem extracts from Arabidopsis plants expressing SAR that could induce systemic resistance to pathogens in healthy Arabidopsis and wheat plants. This suggests that the phloem-mobile chemical(s) that induce SA signals are highly conserved between monocots and dicots. However, while phloem exudates from plants harboring mutations in genes controlling fatty acid biosynthesis did not have the same resistance-inducing effect, addition of JA or methyl-JA did not restore SAR induction. Furthermore, JA was not detectable by GC–MS in the resistance-inducing exudates from wild-type plants (Chaturvedi *et al.*, 2008). Work from Zeier's group indicates that SAR induction may be enhanced in plants carrying mutations in genes controlling JA biosynthesis and perception, further bringing into question the role of JA in SAR signaling (Attaran *et al.*, 2009).

Attaran *et al.* (2009) have also further investigated the role methyl-SA in SAR induction. While they did detect methyl-SA and SAG (see Section III.A) in phloem exudates, which is consistent with a role for these chemicals in SAR signaling, they contend that in tissues undergoing an HR most of the methyl-SA produced will be lost to the atmosphere and will never reach uninfected parts of the plant in sufficient amounts to stimulate SAR. Furthermore, they suggest that certain pathogens, in the case of their study, a coronatine-producing strain of *P. syringae*, will enhance conversion of SA to methyl-SA to dissipate SA levels and counter LAR. Their most compelling evidence is that mutant plants that are unable to synthesize methyl-SA could still exhibit SAR induction (accompanied by SA biosynthesis in uninfected leaves) following inoculation with an avirulent pathogen (Attaran *et al.*, 2009). However, we are faced with a puzzling conflict between two sets of apparently definitive experiments based on knockout of synthesis (Attaran *et al.*, 2009) or perception (Park *et al.*, 2007) of methyl-SA. If the conclusions of Attaran *et al.* (2009), namely that methyl-SA is not indispensible for SAR induction, are vindicated this may still not completely rule out a role for methyl-SA in SAR induction, if this process can be induced by multiple chemical factors (see below).

Recently, attention has returned to other potential lipid-based SAR signals. The dicarboxylic fatty acid azelaic acid (1,7-heptane dicarboxylic acid) is a natural product that until recently was best known as an ingredient in topical treatments for acne (Fitton and Goa, 1991). Azelaic acid shows increased accumulation in Arabidopsis plants following an HR and is phloem-mobile (Jung *et al.*, 2009). Although treatment with exogenous azelaic acid did not induce SA biosynthesis and it was proposed that this compound may prime synthesis (Jung *et al.*, 2009). Exogenous azelaic acid did cause a slight induction of a gene encoding a lipid-transfer protein (AZA1). Although AZA1 and DIR1 are not particularly similar, *aza1* mutant plants had similar characteristics to *dir1*, in that they are defective in SAR induction, and both *aza1* and *dir1* mutant plants were insensitive to azelaic acid treatment, suggesting that both may participate in long-distance signaling (Jung *et al.*, 2009).

If an interim conclusion can be reached concerning signaling in SAR it is that the system is unlikely to be dependent solely on one signal or even one chemical class of signals. Recent commentaries by Parker (2009) and Vlot *et al.* (2008b) include models involving more than one phloem-mobile signal. We should not be surprised by the possibility that SAR induction involves multiple and possibly redundant signals. Modeling of other biological signaling networks (e.g., the circadian regulation of plant gene expression) indicates that increased complexity, such as the inclusion of feedback loops and backup or alternate pathways, in a network will enhance its robustness against perturbations resulting, for instances from mutation, pathogen attack, or abiotic stress (Harmer, 2009). Thus, in functional and evolutionary terms it makes sense for plants not to depend upon a single phloem-mobile chemical to act as the SAR signal since the factors required for the biosynthesis and perception of such a substance would make ideal targets for a pathogen effector molecule.

D. Signaling by reactive oxygen, calcium, and nitric oxide

ROS have long been recognized as signals in defense, most notably during the oxidative burst or bursts that occur very early in the HR during a gene-for-gene (ETI) response (Heath, 2000) and during PTI (see Section II.B.2). ROS have several roles in the HR, but from the signaling point of view, perhaps two are the most important. Firstly, the oxidative burst activates Ca^{2+} ion influx across the plasma membrane via cyclic nucleotide-gated channels, in addition to mobilization of Ca^{2+} ions from intracellular stores (Ma and Berkowitz, 2007; Torres and Dangl, 2005). The cytoplasmic domains of the NADPH oxidase proteins have "EF-hand" motifs, characteristic of proteins regulated by Ca^{2+} ion levels (Keller *et al.*, 1998). This permits changes in Ca^{2+} flux to function both upstream and downstream of ROS production, resulting in a positive feedback on ROS production and, in concert with NO, helping to drive cell death in the HR. A second effect of changes in Ca^{2+} ion flux in the cytoplasm is triggering of the activity of calcium-dependent protein kinases, as well as highly complex MAPK cascades (see Section V.B.1).

ROS generated in the mitochondrion may also play roles in defensive signaling, particularly with respect to the induction by SA, of resistance to viruses (Singh *et al.*, 2004). It has been proposed that alternative oxidase (AOX) is one of the factors that influence this form of defensive signaling (see Section V.B.3).

NO is an important signal in plant defense. For example, the relative levels of NO and H_2O_2 appear to regulate programmed cell death during an HR (Delledonne *et al.*, 2001), and regulate defense gene expression both at the point of infection and in distal tissues, in part by inducing the biosynthesis of SA (Song and Goodman, 2001). NO also appears, together with SA, to coregulate ethylene biosynthesis during the HR (Mur *et al.*, 2008). NO may exert some of its effects via modulation of cyclic nucleotide-based signaling (Wendehenne *et al.*, 2004) and by the S-nitrosylation of protein targets (Hong *et al.*, 2008). NO also may stimulate changes in nuclear gene expression and defensive signaling indirectly through inhibition of cytochrome oxidase (Huang *et al.*, 2002). This process can lead to changes in mitochondrial redox and the induction of specific sets of defense-related genes (Maxwell *et al.*, 2002). This may offer another mechanism by which NO can influence resistance to viruses via the AOX-regulated mitochondrial signaling pathway (W.-S. Liang and J.P. Carr, unpublished data; see Section V.B.3). At present, the available data on the mechanism(s) of NO production in plants are confusing (reviewed by Hong *et al.*, 2008).

E. Novel signals in defense against viruses and other pathogens

The best-studied low molecular weight chemical signals in plant defense are SA, JA, NO, ethylene, ROS, and to a lesser extent Ca^{2+}. Potentially, however, these may represent only a minority of signals that exist and that are required to maintain basal resistance or trigger SAR and other induced resistance mechanisms. Novel signals continue to be discovered; for example, azelaic acid is a recently proposed SAR signal (see Section III. C). Future metabolomic analyses of plants during the induction phase of resistance phenomena may reveal many more, as may continuing efforts to isolate defense signaling mutant plant lines. The following two examples describe defensive signals that were unknown until recently, yet appear to play key roles in, respectively, SA-dependent and SA-independent resistance to viruses and other phytopathogens.

1. Extracellular ATP: A negative regulator of SA-mediated defense

Adenosine-5′-triphosphate (ATP) is an energy-rich metabolite of fundamental importance in all organisms. It is a key substrate and cofactor in a wide range of intracellular biochemical processes. Surprisingly, given its value as a biochemical intermediate, ATP is secreted by plant, animal, and microbial cells (see Chivasa *et al.*, 2005 and references therein). This extracellular ATP (eATP) can act as a signal in response to abiotic and biotic stresses. For example, studies with root hair cells show that eATP stimulates ROS production and the influx of Ca^{2+} ions across the plasma membrane (Demidchik *et al.*, 2003, 2009; Jeter *et al.*, 2004; Lew and Dearnaley, 2000) and may stimulate NO production (Foresi *et al.*, 2007; Wu and Wu, 2008). In leaves, a decrease in eATP can lead to the induction of light-induced necrosis indicating that it is a key signal molecule that modulates programmed cell death in plants (Chivasa *et al.*, 2005, 2009b). A complex relationship appears to exist between eATP, SA, and the signaling pathways that they control (Chivasa *et al.*, 2009a,b, 2010). Under appropriate illumination conditions, in which eATP depletion does not induce necrosis, decreasing the levels of eATP in tobacco leaves by injecting a mix of glucose and hexokinase, or injecting a nonhydrolyzable ATP analog induced *PR*-gene expression and increased resistance to TMV and *P. syringae* (Chivasa *et al.*, 2009a). Conversely, injecting exogenous ATP into leaves caused a decrease in basal SA levels, while SA treatment of tobacco cell cultures caused a collapse in the concentration of ATP in the medium (Chivasa *et al.*, 2009a). Thus, eATP appears to be a negative regulator of SA-dependent defense and potentially of SA biosynthesis.

2. Phytate: A signal in the maintenance of basal resistance to viruses and other phytopathogens

Phytic acid (*myo*-inositol hexakisphosphate, $InsP_6$) is an important phosphate store and signal molecule in animals and a key compound for phosphate storage in plants (Irvine and Schell, 2001; Raboy, 2003). In plants $InsP_6$ is also a signal molecule. This was shown during studies of the response of guard cells to the drought stress hormone abscisic acid in which it was found that $InsP_6$ is two orders of magnitude more potent as a trigger of the release of endomembrane-stored Ca^{2+} than the better known inositol phosphate signal, $InsP_3$ (Lemtiri-Chlieh *et al.*, 2000, 2003). Unfortunately, the phytic acid present in plants can have undesirable effects. In humans, phytate can inhibit the uptake of dietary iron and zinc (Raboy, 2001). Also, when it enters the diets of monogastric farm animals, such as pigs, it is poorly absorbed and its presence in farm runoff can enhance phosphate pollution in streams and rivers (Raboy, 2001). For these reasons low phytate plant lines have been developed.

While investigating the properties of plants of transgenic potato lines in which $InsP_6$ accumulation had been decreased, it was noted that these plants were more susceptible to infection by PVY, TMV (Murphy *et al.*, 2008), and a bacterial pathogen (B. Otto, A.M. Murphy, D.E. Hanke, and J.P. Carr, unpublished data). Similarly, Arabidopsis mutant plants carrying a defective gene for $InsP_5$ 2-kinase, which catalyzes the final step of $InsP_6$ biosynthesis, had decreased $InsP_6$ levels and were also hypersusceptible to infection by TMV and the bacterial pathogen *P. syringae*. The first step in $InsP_6$ synthesis is the conversion of glucose 6-phosphate to *myo*-inositol 3-monophosphate catalyzed by *myo*-inositol 3-phosphate synthase (IPS) and in Arabidopsis there is a family of three *IPS* genes. Plants carrying mutations in the genes *AtIPS1* and *2* were both found to have lower levels of $InsP_6$. Curiously, however, only the *Atips2* mutant plants showed hypersusceptibility to infection with viruses (TMV, CMV, *Turnip mosaic virus*, and CaMV), a fungus (*Botrytis cinerea*) and *P. syringae* (Murphy *et al.*, 2008). The results suggest that only a specific pool of $InsP_6$ is involved in defensive signaling.

The effect of $InsP_6$ on susceptibility to infection is independent of SA signaling and it is not yet clear how $InsP_6$ regulates defense against such a broad range of pathogens. Kazan and Manners (2009) have suggested that since TIR-1, the auxin receptor, requires $InsP_6$ as a structural cofactor, altering $InsP_6$ levels in plants may disrupt auxin-mediated defensive signaling. A similar argument might be made with regard to COI1, the jasmonate receptor, which is a similar ubiquitin ligase/F-box type protein (Chini *et al.*, 2007; Thines *et al.*, 2007) that might also require $InsP_6$ as a cofactor. However, neither auxin-mediated nor JA-mediated resistance mechanisms affect such a broad range of pathogens (Kazan and Manners, 2009; see Section III.B), and there is no evidence of antagonism between $InsP_6$ and SA, which is a characteristic of auxin-mediated resistance (B. Otto, A.M. Murphy, D.E. Hanke, and J.P. Carr, unpublished data; Wang *et al.*, 2007). These early findings regarding $InsP_6$ hint at the existence of a completely novel defensive signaling pathway.

IV. RNA SILENCING AND INDUCED RESISTANCE

A. RNA silencing

1. Mechanisms of RNA silencing

RNA silencing consists of a number of interconnected pathways that limit the synthesis, stability, and translatability of RNA molecules. The common feature of these pathways is that they are sequence-specific and this specificity is provided by small (s)RNAs, in the 20–25 nucleotide (nt) size range, that are complementary to sequences in the target molecule

(Baulcombe, 2006; Csorba *et al.*, 2009; Hamilton and Baulcombe, 1999; Llave *et al.*, 2002). RNA-silencing pathways occur in a diverse range of eukaryotic organisms including algae, higher plants, fungi, and animals (Cogoni and Macino, 1998; Fire *et al.*, 1999; Molnár *et al.*, 2007). In plants RNA silencing decreases gene expression in three ways: by degradation of transcripts; inhibiting translation of mRNAs, or by promoting targeted methylation of DNA to effect transcriptional gene silencing (Bäurle *et al.*, 2007; Brodersen *et al.*, 2008; Jones *et al.*, 1999). Although outside the scope of this chapter, which is focused on resistance, it has been found that the triggering of RNA-silencing pathways can also contribute to pathogenicity, particularly during viroid infections (Gómez *et al.*, 2008; Itaya *et al.*, 2007).

a. RNA silencing directed against viruses RNA silencing aids in defense against both RNA and DNA viruses (Blevins *et al.*, 2006; Bouché *et al.*, 2006; Deleris *et al.*, 2006; Moissiard and Voinnet, 2006; Qu *et al.*, 2008). The cleavage of double-stranded (ds)RNA molecules into short 20–25 nt duplexes by ribonuclease III-like DICER enzymes (in plants "DICER-like": DCL) can be regarded as the initiation in the direction of silencing against a specific RNA sequence (Kurihara and Watanabe, 2004; Schauer *et al.*, 2002) (Fig. 1). Production of dsRNA is characteristic of RNA virus infections and both DNA and RNA virus-specific transcripts may contain structural features (e.g., hairpin loops, pseudoknots, etc.) containing double-stranded domains. However, for a number of reasons it is thought that dsRNA replication "intermediates" may not be major targets for DCL-mediated cleavage (reviewed in detail by Csorba *et al.*, 2009). However, dsRNA molecules generated from single-stranded RNA templates by cellular RdRp activity may act as DCL substrates in antiviral silencing (Qi *et al.*, 2009). Nevertheless, the precise mechanisms governing selection of target sequences by DCLs, for example within folded regions of single-stranded RNAs (Molnár *et al.*, 2005), are still incompletely understood, and remain the focus of ongoing research (Csorba *et al.*, 2009).

Four DCL enzymes, DCLs 1–4, have been identified in Arabidopsis, which generate dsRNA products of 20–25, 22, 24, and 21 nt, respectively. In addition to any cellular RNA targets they may have (Section IV.A.1.b), DCLs 2–4 (but predominantly DCLs 4 and 2) also target transcripts from a range of RNA viruses, while DCL1 has been implicated in the cleavage of RNA molecules derived from DNA viruses (Blevins *et al.*, 2006; Bouché *et al.*, 2006; Csorba *et al.*, 2009; Deleris *et al.*, 2006; Kurihara and Watanabe, 2004; Moissiard and Voinnet, 2006; Qu *et al.*, 2008; Schauer *et al.*, 2002). The dsRNA products of DCL activity are unwound, yielding short-interfering (si)RNAs that are incorporated into an RNA-induced silencing complex (RISC), the protein component of which is a member of the ARGONAUTE (AGO) family.

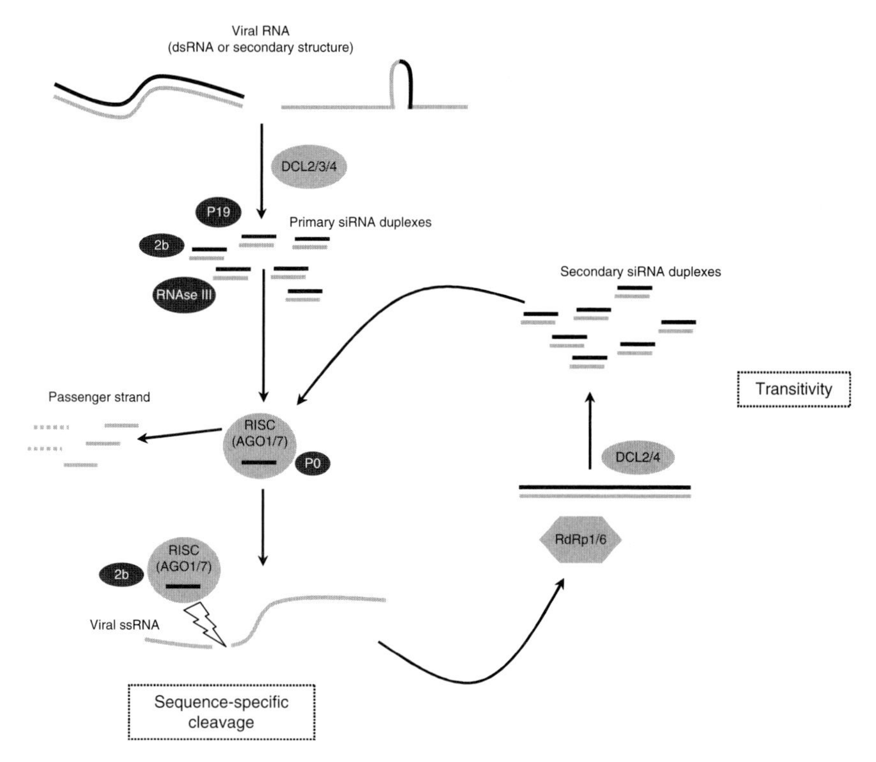

FIGURE 1 RNA silencing directed against RNA viruses in plants. Double-stranded (ds) regions within viral RNA molecules are cleaved by the host's dicer-like (DCL) ribonucleases (predominantly DCLs 2 and 4 and to some extent by DCL3), producing primary ds small-interfering (si)RNA duplexes. These are unwound to yield a passenger strand (i.e., degraded) and a primary siRNA, which is loaded into a RISC (RNA-induced silencing complex): the only known protein components of which are AGO (Argonaute) proteins. So far, only AGOs 1 and 7 have been demonstrated to have direct roles in antiviral silencing. The bound siRNA directs the endoribonuclease activity of the RISC against viral RNAs containing sequences complementary to the primary siRNA. The activity of host-encoded RNA-dependent RNA polymerases (RdRps), for example RdRp1 or 6, is thought to generate further dsRNA, which is cleaved by DCLs 2 and 4 to yield secondary siRNA duplexes. These siRNAs may then direct additional AGO-mediated cleavage against viral-derived RNAs. Host-encoded proteins are shown in light gray. Silencing suppressors (dark gray) encoded by cucumoviruses (2b protein), tombusviruses (P19), poleroviruses (P0), and the silencing-inhibiting RNAse III of the crinivirus *Sweet potato chlorotic stunt virus* are depicted at proposed sites of action (see main text for additional details and references). Adapted from Lewsey and Carr (2009b) with permission from Elsevier.

AGO proteins are ribonucleases, with structural and mechanistic affinities to RNAse H enzymes, a class of enzymes that cleave RNA/DNA duplexes (Parker *et al.*, 2004). The incorporation of an sRNA endows AGO proteins with a sequence-specific ''slicing'' activity against target RNAs (Baumberger and Baulcombe, 2005). Ten AGO family proteins have been identified in Arabidopsis and, thus far, two members, AGOs 1 and 7, have confirmed roles in antiviral slicing, with AGO4 having a role in the inhibition of viral mRNA translation (Bhattacharjee *et al.*, 2009; Qu *et al.*, 2008; Zhang *et al.*, 2006b).

b. RNA silencing in regulation of cellular transcripts The steady state levels and translatability of certain cellular mRNAs are regulated by RNA silencing. In this case, specificity for AGO slicing or translational inhibitor activity is provided by two classes of sRNAs known as micro (mi)RNAs and *trans*-acting short-interfering (tasi)RNAs (Baumberger and Baulcombe, 2005; Brodersen *et al.*, 2008; Lanet *et al.*, 2009; Llave *et al.*, 2002; Vaucheret, 2005) (Fig. 2). Biogenesis and activity of these sRNAs is tightly linked. *MIR* genes encode partly ds miRNA precursor transcripts that are processed by DCL1 into 20–25 nt RNA duplexes that are unwound to yield single-stranded miRNAs that incorporate into RISCs (Aukerman and Sakai, 2003; Baumberger and Baulcombe, 2005; Reinhart *et al.*, 2002). AGOs 1, 7, and 10 have known roles in miRNA-directed RNA cleavage and translational inhibition (Baumberger and Baulcombe, 2005; Brodersen *et al.*, 2008; Montgomery *et al.*, 2008). Among their other targets, miRNAs direct cleavage of the primary transcripts of *TAS* genes, which encode tasiRNAs (Allen *et al.*, 2005). Cleaved *TAS* transcripts are templates for dsRNA synthesis by cellular RdRps (Allen *et al.*, 2005). These dsRNA products are processed into sRNA duplexes by the activity of DCL3, separated into single-stranded tasiRNAs, and bind AGO proteins to direct sequence-specific RNA slicing (Allen *et al.*, 2005).

c. Epigenetic silencing and resistance to geminiviruses Small RNAs direct *de novo* DNA methylation via the RNA-directed DNA methylation (RdDM) pathway, which can result in transcriptional silencing of nuclear genes, transposons, and repetitive elements (Chan *et al.*, 2004; Lippman *et al.*, 2004; Lister *et al.*, 2008; Zhang *et al.*, 2006a,b) (Fig. 3). Interestingly, in susceptible plants infected with a virus (TMV) there is an increased rate of somatic recombination accompanied by an increase in DNA methylation, including in regions of the plant genome containing copies of potential NBS–LRR genes. This was maintained into subsequent generations, suggesting that this epigenetic response may provide an early stage in the adaptation of plants to attack by ''new'' pathogens (Boyko *et al.*, 2007; Kovalchuk *et al.*, 2003).

Recent work by Raja *et al.* (2008) indicates that the RdDM pathway can also target viral DNA. RdDM is guided by siRNAs of 21–24 nt, but predominantly so by those of 24 nt (Lister *et al.*, 2008). These siRNAs are generated by the combined action of DNA-dependent RNA polymerase

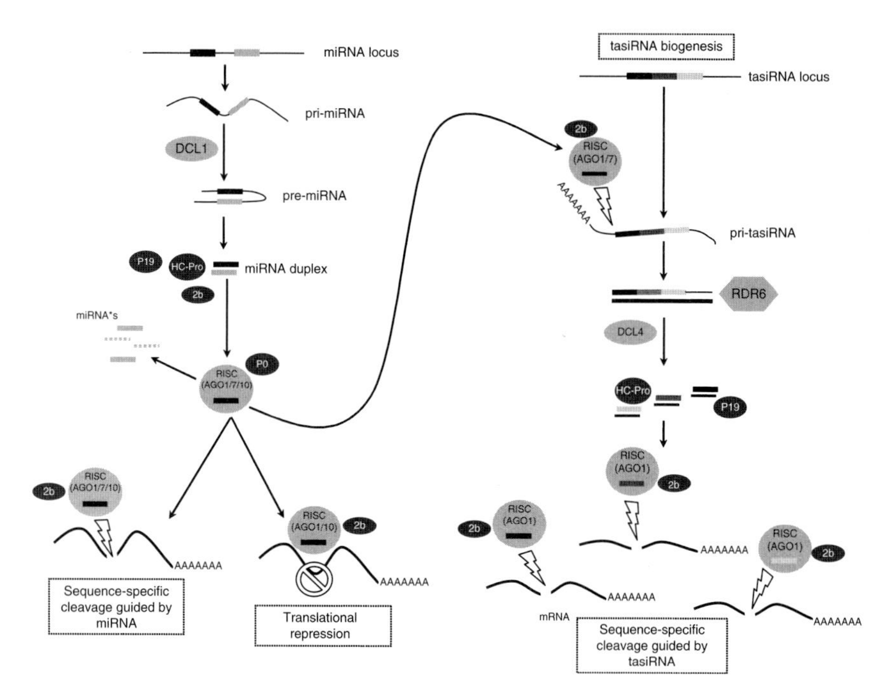

FIGURE 2 A cartoon depicting RNA silencing directed against endogenous transcripts by micro (mi)RNAs and *trans*-activating (ta)siRNAs. The miRNA precursor pri-miRNA is transcribed from a genomic miRNA locus. It forms a foldback structure and is cleaved by DCL1, yielding a miRNA duplex. The miRNA* strand (equivalent to the passenger siRNA strand: see Fig. 1) is released from the miRNA duplex, and the mature miRNA is loaded into a RISC that may cleave a target transcript or inhibit its translation into protein. Currently, AGOs 1 and 10 are known to be involved in miRNA-directed cleavage and translational repression, while AGO7 has a demonstrated role solely in cleavage. miRNAs may also bear sequence homology with, and direct the cleavage of, pri-tasiRNAs. These are the transcripts of endogenous tasiRNA loci. The cleaved transcripts are used as templates for synthesis of double-stranded RNA by the host-encoded RdRp6, then processed into tasiRNA duplexes (~21 nt) by DCL4. The duplexes are separated and loaded into RISC, to direct the cleavage activity of AGO1 against endogenous transcripts with which they have sequence homology. Host-encoded proteins are depicted in light gray. Proposed sites of disruption of the miRNA pathway by silencing suppressors (dark gray) encoded by cucumoviruses (2b protein), tombusviruses (P19), poleroviruses (P0), and *Sweet potato chlorotic stunt virus* (RNAse III) are shown. Adapted from Lewsey and Carr (2009b) with permission from Elsevier.

PolIV, RDR2, and DCL3 and subsequently bind to AGOs 4 and 6 (Chan *et al.*, 2004; Matzke *et al.*, 2009; Mosher *et al.*, 2008; Zheng *et al.*, 2007). These AGO-associated siRNAs direct the sequence-specific activity of the DNA methyltransferase, DRM2 (Chan *et al.*, 2004). RdDM also requires the activity of DNA-directed RNA polymerase Pol V (previously

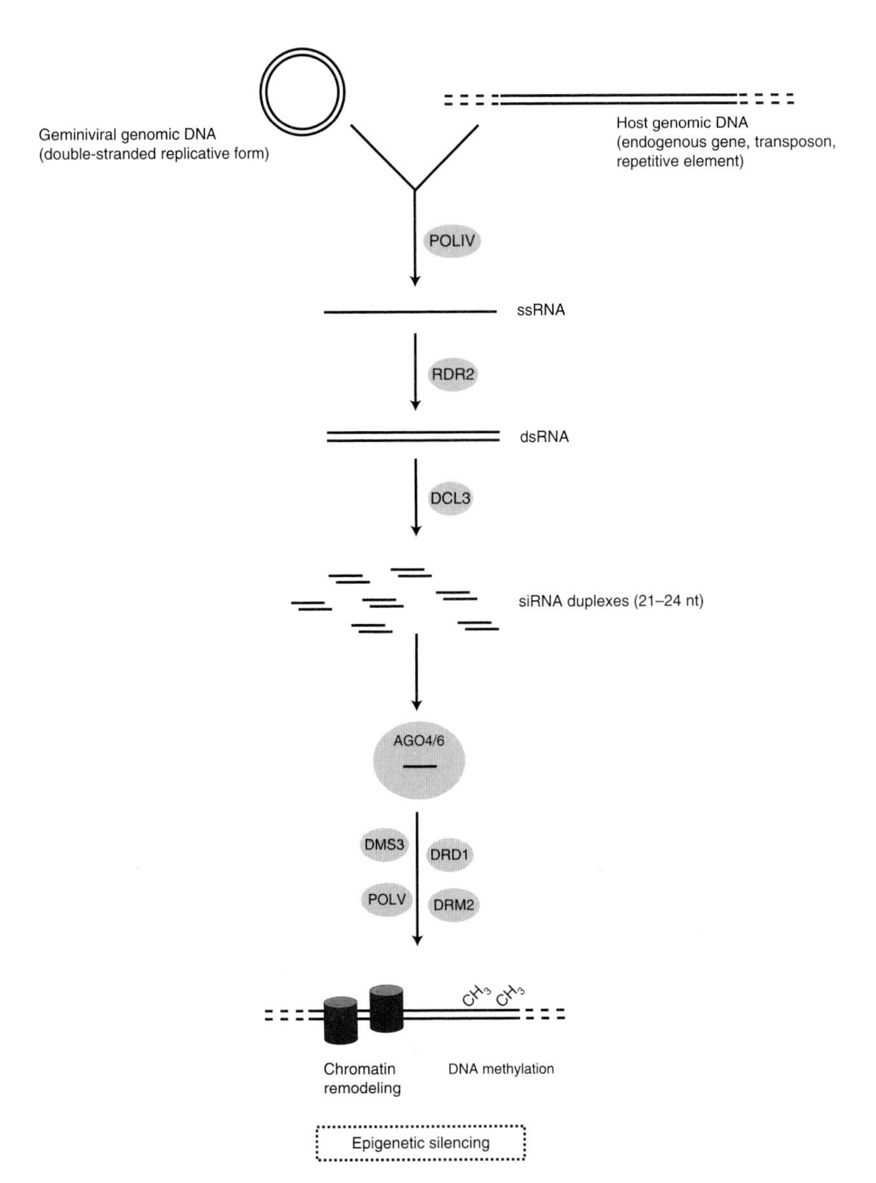

FIGURE 3 A cartoon depicting epigenetic silencing of geminiviruses and genomic loci (which include endogenous genes, transposons, and repetitive elements). Single-stranded (ss)RNA complementary to the target locus is generated by the DNA-directed RNA polymerase PolIV, and dsRNA is generated from this by the action of the host-encoded RdRp, RDR2. The dsRNA is cleaved by DCL3, yielding siRNAs bearing sequence homology to the target locus. The siRNAs are bound by AGOs 4 and 6 and direct epigenetic silencing in a sequence-specific manner. This process involves the proteins DMS3, DRD1, POLV, and DRM2. It may result in DNA methylation and/or chromatin remodeling at genomic loci as well as affecting the genomes of DNA viruses.

called PolIVb), the chromatin remodeling protein DRD1, and a putative chromosome architectural protein, DMS3 (Kanno *et al.*, 2004, 2008; Matzke *et al.*, 2009; Mosher *et al.*, 2008). While the mechanistic details of RdDM await further refinement, it is increasingly evident that RNA silencing is tightly linked to DNA methylation and chromatin remodeling (Matzke *et al.*, 2009).

RdDM and resistance to DNA viruses appear to be linked. Raja *et al.* (2008) found that Arabidopsis plants belonging to methylation-deficient mutant lines were hypersusceptible to infection by two geminiviruses (plant-infecting circular ssDNA viruses). Hypersusceptibility of mutant plants correlated with decreased methylation of viral DNA, whereas in wild-type plants undergoing recovery from infection viral DNA is highly cytosine-methylated (Raja *et al.*, 2008). Differing methylated forms of histone H3 were associated with double-stranded replicative forms of viral DNA, although it is less clear from this study how these modifications correlated with resistance or susceptibility to geminivirus infection. Importantly, Raja *et al.* (2008) demonstrated that host recovery from geminivirus infection requires that RdDM pathway component AGO4 and correlates with hypermethylation of the geminiviral DNA. This establishes a likely role for RdDM in resistance to DNA viruses. However, siRNA-directed silencing of geminivirus-encoded RNAs also occurs during infection (Blevins *et al.*, 2006). Further understanding is needed to assess the relative importance of RdDM versus RNA silencing in plant resistance to DNA viruses.

2. Disruption and exploitation of RNA silencing by viruses

Virtually all plant viruses encode proteins that inhibit RNA silencing (reviewed by Csorba *et al.*, 2009; Diaz-Pendon and Ding, 2008; Lewsey *et al.*, 2009). Viroids, which do not encode proteins, employ features of their own RNA sequences to subvert and possibly exploit RNA silencing (Gómez *et al.*, 2009). Among viral silencing suppressor proteins, the most common mode of action is sequestration of siRNAs to prevent their loading into RISC, as exemplified by the tombusvirus P19 protein (Csorba *et al.*, 2009). However, other modes of action include the targeting of proteinaceous silencing components for ubiquitin/proteasome-mediated degradation, which is the case for AGO1 and the P0 protein encoded by poleroviruses (Bortolamiol *et al.*, 2007), or the degradation of ds siRNAs, by the RNase III encoded by *Sweet potato chlorotic stunt virus*, into 14 bp fragments that are too small to program RISC (Cuellar *et al.*, 2009). Other suppressor proteins appear to target protein and RNA components of silencing. The CMV 2b protein was shown to bind to AGO1 and inhibit slicer activity, leading to an initial conclusion that this was solely responsible for its silencing suppressor activity (Zhang *et al.*, 2006a). However, the 2b proteins of CMV and another cucumovirus

(TAV) also bind sRNAs (Chen *et al.*, 2008; González *et al.*, 2010; Goto *et al.*, 2007; Rashid *et al.*, 2008). Subsequent study of the interactions of mutant variants of the CMV 2b protein with AGOs 1 and 4, and siRNA indicate that the latter activity is indispensible for silencing suppression activity (González *et al.*, 2010).

Many silencing suppressors also inhibit miRNA/tasiRNA-mediated regulation of RNA turnover and mRNA translation, causing extensive disruption of host transcript abundance (Chapman *et al.*, 2004; Chellappan *et al.*, 2005; Chen *et al.*, 2004; Diaz-Pendon and Ding, 2007; Dunoyer *et al.*, 2004; Lanet *et al.*, 2009; Lewsey *et al.*, 2007; Zhang *et al.*, 2006a,b) (Section IV.A.1.b). This results in developmental abnormalities and underlies many disease symptoms typically seen in virus-infected plants. Thus, symptom induction due to silencing suppressor proteins has been ascribed to coincidence, resulting from off-target effects of suppressors on components conserved or shared between pathways involved in silencing mediated by miRNA/tasiRNAs and virus-specific siRNAs (Chapman *et al.*, 2004; Chellappan *et al.*, 2005; Chen *et al.*, 2004; Dunoyer *et al.*, 2004). However, a comparison of the effects of expression of 2b proteins from two CMV strains in transgenic Arabidopsis showed that inhibition of siRNA-mediated silencing does not lead inevitably to effects on miRNA-regulated host gene regulation (Lewsey *et al.*, 2007). This study implies that, rather than being incidental, symptom induction may be in some way advantageous to certain virus strains.

There is further evidence that viruses have evolved means to alter host transcript abundance and signaling by manipulation of the sRNA pathways, which implies that symptom induction may provide some selective advantage for at least some viruses. The caulimovirus CaMV is a pararetrovirus, that is, a virus with a DNA genome, which is replicated via an RNA intermediate (Hull, 2009). The RNA intermediate for CaMV is the 35S RNA that also acts as a polycistronic mRNA, and which has extensive secondary structure within its 5′-translational leader sequence (Hemmings-Mieszczak *et al.*, 1997). This sequence is a major template for generation of virus-derived siRNAs that accumulate extensively in CaMV-infected turnip and Arabidopsis plants (Moissiard and Voinnet, 2006). Many of these leader-derived siRNAs bear near-perfect sequence homology with Arabidopsis transcripts (Moissiard and Voinnet, 2006). Using a GFP-sensor transgene in CaMV-infected plants, Moissiard and Voinnet (2006) determined that leader-derived siRNAs are biologically active *in planta*. Further bioinformatic analysis identified more than 100 host transcripts potentially targeted by CaMV-encoded siRNAs, many of which were downregulated during virus infection.

Similarly, 4874 siRNAs derived from the crucifer-infecting strain of TMV (TMV-Cg) found in infected Arabidopsis had potential targets within the host's own transcript population (Qi *et al.*, 2009). The predicted target mRNAs encoded proteins with a range of functions, including defense responses, RNA processing, and transcription factors. However, changed abundance could only be detected for two of 16 predicted target transcripts during experimental validation (Qi *et al.*, 2009). Potentially, this indicates another layer of resistance whereby virus-encoded siRNAs are prevented by the host plant from having biological effects. Alternatively, virus-encoded siRNAs may cause translational inhibition of their targets (instead of degradation), as has been shown for some miRNAs (Lanet *et al.*, 2009). Qi *et al.* (2009) also identified 2978 virus-derived siRNAs with predicted host transcript targets in TCV-infected Arabidopsis (Qi *et al.*, 2009). Taken together, these data support the hypothesis that some (as yet unclear) benefits accrue to viruses (and viroids, see Section IV.A.1) by allowing certain of their nucleic acid sequences to act as templates for siRNA generation. We may speculate that this is a strategy for turning the host's own RNA-silencing machinery against cellular mRNAs encoding defensive factors.

B. Connections between RNA silencing and induced resistance

RNA silencing appears to play roles in SA-induced resistance to viruses and the inhibition of virus gene expression and replication during the HR. In this section, we describe several lines of evidence for crosstalk and overlap between RNA-silencing pathways and signal transduction pathways governed by SA.

1. RdRp1: An SA-inducible RNA-silencing component

Cellular RdRps have important roles in plant RNA-silencing pathways, providing amplification of silencing through generation of siRNA-primed dsRNA synthesis and initiation of antiviral silencing through *de novo* synthesis of dsRNA (Section IV.A.1.a) (Csorba *et al.*, 2009). RdRp1 is necessary for the maintenance of basal resistance to several RNA viruses including, for example, TMV and PVY but not CMV (Section V.D.5). RdRp1 has also been implicated in the regulation of host mRNAs encoding proteins involved in JA- and SA-regulated defensive pathways (see Sections IV.C and V.D.5).

The connection between SA-induced resistance and RdRp1 was first discovered in Z. Chen's laboratory, where it was shown that *NtRDR1* gene expression and RdRp activity was increased in tobacco following SA treatment (Xie *et al.*, 2001). However, increased gene expression and activity also were found following systemic infection with a compatible strain of TMV (Xie *et al.*, 2001); a separate induction process that is unlikely to be SA-induced, since this infection does not trigger increased SA biosynthesis (Malamy *et al.*, 1990). *NtRDR1* orthologs have been characterized in *Nicotiana glutinosa* (*NgRDR1*), *N. benthamiana* (*NbRDR1m*),

A. thaliana (*AtRDR1*), and *Medicago truncatula* (*MtRDR1*), and these genes are inducible by virus infection or SA treatment (Liu *et al.*, 2009; Yang *et al.*, 2004; Yu *et al.*, 2003). In rice (*Oryza sativa*), *OsRDR1*, also is induced by SA treatment and virus infection. Interestingly, the transcriptional response of the *OsRDR1* gene to rice yellow mottle virus infection is altered in transgenic rice plants expressing the Arabidopsis *NPR1* gene (Quilis *et al.*, 2008). This result might suggest a link between RNA silencing and defensive signaling mediated by NPR1. However, it should be interpreted with caution because the role of this NPR1 in antiviral defense is currently unclear (see Section V.B.2).

It is highly likely that RdRp1 activity contributes to SA-induced resistance to certain viruses. However, it is a dispensable component, as tobacco plants transformed with an antisense transgene to downregulate *NtRDR1* transcript levels still exhibited SA-induced resistance to TMV (Xie *et al.*, 2001). Similarly, *N. benthamiana*, which expresses a nonfunctional RdRp1 enzyme (Yang *et al.*, 2004), is still able to exhibit SA-induced resistance to TMV (Fu, 2008; Lee, 2009). Since certain resistance-inducing chemicals, such as antimycin A, trigger AOX-regulated resistance (Section V.B.3) without triggering increased *NtRDR1* expression, it has been suggested that induction of RdRp1 activity is only one of several parallel antiviral resistance mechanisms that are inducible by SA (Gilliland *et al.*, 2003; Singh *et al.*, 2004).

2. Viral suppressors of RNA silencing and SA signaling

The second line of evidence suggesting that RNA silencing and SA-induced resistance are connected came from experiments showing that certain viral silencing suppressor proteins also suppress SA-mediated defense. In a seminal study done in S.W. Ding's laboratory, it was shown that the CMV 2b protein enables the virus to evade SA-induced resistance to replication and local movement, and also represses the transcriptional response of *AOX1a* to SA treatment (Ji and Ding, 2001). The significance of this is that AOX is a likely regulatory component of a virus-specific SA-responsive signaling pathway (see Section V.B.3). The P1/HCPro protein from *Tobacco etch virus* (TEV; a potyvirus) also perturbs SA-mediated signaling, but the precise nature of these effects remains unresolved (Alamillo *et al.*, 2006; Pruss *et al.*, 2004). In a study by the Vance laboratory (Pruss *et al.*, 2004), P1/HCPro was found to enhance the SA defense pathway, while a study by the García group (Alamillo *et al.*, 2006) demonstrated that P1/HCPro altered the temporal dynamics of SA-responsive gene expression. It is possible that some suppressors of silencing encoded by viruses may also act as elicitors of ETI, a response that is dependent upon SA signaling for pathogen localization (for discussion of this see Section II.B.2 and Moffett, 2009).

3. SA-mediated defense: A balancing act between RNA-silencing and other resistance mechanisms

Can SA-induced resistance still occur in plants when the antiviral RNA-silencing pathway is no longer able to function? Recent experiments show that the three DCLs required for RNA silencing against RNA viruses, DCLs 2–4, are dispensable for SA-induced resistance to TMV and CMV (Lewsey and Carr, 2009a). In *dcl2/3/4* triple mutant plants, SA treatment was still able to induce effective resistance against both viruses, indicating that antiviral cleavage mediated by DCLs 2–4 is not absolutely required for effective SA-mediated defense. However, this does not exclude a role for RNA silencing in a pathway that is parallel or redundant with other forms of SA-mediated defense, nor does it rule out an indispensible role for SA-induced resistance to other viruses. We also cannot rule out a contribution from the other six non-DCL RNase III enzymes encoded by Arabidopsis (Bouché *et al.*, 2006).

On balance, the available data indicate that SA stimulates more than one antiviral defense pathway (in addition to its several effects on nonviral pathogens). While investigating the effects of HCPro on plant defenses, Alamillo *et al.* (2006) demonstrated that SA might potentiate RNA silencing against the *Potyvirus Plum pox virus* (PPV): indicated by a decrease in PPV-derived siRNA in SA-depleted *NahG*-transgenic tobacco plants. These workers also determined that the transcriptional responses of *PR-1*, *PR-2* (gene markers for SAR induction: Section V.D.2), and *AOX1a* (associated with SA-induced virus resistance: Section V.B.3) following SA treatment were altered in plants expressing a TEV-derived *HCPro* transgene. These results are consistent with other experiments where perturbation of silencing, either through modification of expression of a silencing component, such as RdRp1 (Rakhshandehroo *et al.*, 2009), or by transgenic expression of a viral silencing suppressor, such as the CMV 2b protein (Ji and Ding, 2001), affected the expression of plant genes associated with induced resistance. Consequently, it should be considered that links between RNA silencing and SA-mediated defense need not necessarily be direct; that is, through cleavage of viral RNA. Rather, some of the links between RNA silencing and defensive signaling may be regulatory in nature, involving the control of expression of defense-related cellular mRNAs (discussed in Section V.D.5).

A recent, groundbreaking study by the Moffett laboratory (Bhattacharjee *et al.*, 2009) demonstrated that translational inhibition by RNA silencing plays a role in ETI (Section II.B.2). This was achieved by coagroinfiltration into *N. benthamiana* of expression constructs harboring (1) the tobacco *N* gene, (2) a cDNA encoding the p50 fragment of the TMV replicase, and (3) a cDNA encoding a PVX vector expressing green fluorescent protein (PVX–GFP). Upon coinfiltration of these three constructs, no GFP fluorescence was observed, no PVX CP could be detected and no significant cell death was induced, indicating effective antiviral resistance in the absence of hypersensitive cell death. The region of the PVX genome encoding the viral CP was identified to be the PVX target or elicitor of this antiviral resistance, and upon its deletion accumulation of GFP was restored. However, in the presence or absence of the CP coding region, accumulation of the subgenomic RNA encoding the GFP was unchanged, irrespective of GFP accumulation. Furthermore, in the presence of the CP coding region, the subgenomic RNA did not associate with ribosomes. Taken together, these data indicate that GFP accumulation was being inhibited at a translational level. This translational repression was dependent upon AGO4, a component of the RNA-silencing machinery (Bhattacharjee *et al.*, 2009). Interestingly, PVX is a positive-sense RNA virus that replicates in the cytoplasm (Yoshinari and Hemenway, 2003) but the only previously known role for AGO4 was in RdDM and epigenetic silencing of DNA, which occurs in the nucleus (see Section IV.A.1.c).

C. Jasmonic acid and RNA silencing: Implications for virus transmission

JA and its derivatives are important signals in plant anti-insect defenses (Section III.B). Recently, it was discovered that in *Nicotiana attenuata*, silencing of *NaRDR1* altered accumulation of JA, its derivatives, and the hormone ethylene, which exhibits extensive overlap with JA with respect to its effects on plant defense (Section III.B) (Pandey and Baldwin, 2007; Pandey *et al.*, 2008). In *NaRDR1*-silenced transgenic plants it was found that biosynthesis of the secondary metabolite nicotine was decreased and that, consequently, the plants exhibited enhanced susceptibility to feeding by tobacco hornworm caterpillars (*Manduca sexta*) and another insect, the cell content feeder *Tupiocoris notatus* (Pandey and Baldwin, 2007). Further analysis by this group showed that JA-mediated signaling and resistance to insect herbivores in *N. attenuata*, and presumably in other plants, are highly dependent upon regulation by RNA-silencing pathways (Pandey *et al.*, 2008).

This work may have important implications for plant–virus transmission since many viruses are spread by insects, the majority by aphids and whiteflies (Hull, 2009). JA-mediated resistance to insects does affect aphids and whiteflies (Ellis *et al.*, 2002; Zarate *et al.*, 2007) and since RNA silencing regulates JA signaling (Pandey and Baldwin, 2007; Pandey *et al.*, 2008), could viral RNA-silencing suppressors indirectly aid virus transmission by inhibiting JA-mediated signaling? Work in our laboratory with the CMV 2b silencing suppressor suggests that this is a possibility (M.G. Lewsey *et al.*, unpublished data; H. Ziebell *et al.*, unpublished data).

However, there is growing evidence that viral gene products may also disrupt JA-mediated signaling in other ways as well.

For example, the C1 pathogenicity factor encoded by a DNA β-satellite molecule of the geminivirus *Tomato yellow leaf curl China virus* was shown to inhibit expression of five JA-regulated genes (Yang *et al.*, 2008). It does so by interacting directly with the Arabidopsis protein ASYMMETRIC LEAVES 1 (AS1), which is a negative regulator of JA defenses (Nurmberg *et al.*, 2007; Yang *et al.*, 2008). Interestingly, it was found that on tobacco and cotton plants infected with this virus or another whitefly transmitted geminivirus, *Tobacco curly shoot virus*, population growth of the whitefly *Bemisia tabaci* was accelerated (Jiu *et al.*, 2007).

Another geminivirus protein, the multifunctional C2/L2, disrupts a range of hormone-response pathways, including JA signaling (R. Lozano-Durán and E.R. Bejarano, unpublished data). It does so by disrupting the activity of the Skp, Cullin, F-box containing (SCF) complexes involved in hormone perception. This likely is achieved by impairing the rubylation/derubylation (i.e., conjugation to the "RELATED TO UBIQUITIN" regulatory protein; Hotton and Callis, 2008) function of the Arabidopsis COP9 signalosome, which regulates SCF function. Interestingly, the C2/L2 protein is also known to be a silencing suppressor encoded by certain geminivirus species (Bisaro, 2006), although in this case its effects on JA signaling do not appear to be related to its silencing activity. Taken together, these studies indicate that a number of plant viruses disrupt JA signaling and possibly anti-insect defense and that this may help to favor transmission in some way. Alternatively, interference with JA-mediated signaling may have a more direct benefit for the pathogen, since at least one virus (CMV) is inhibited by a JA-dependent mechanism induced in Arabidopsis plants following their colonization by beneficial bacteria (Ryu *et al.*, 2004), which contrasts for findings with other viruses, such as TCV (see Section III.B).

V. PROTEIN FACTORS IN SIGNALING OR RESISTANCE RESPONSES

In the defense response to virus infection, there are a number of protein factors that respond to the initial activation of resistance genes and subsequent signaling. These include the resistance gene products, signal transduction factors that interact directly with the resistance gene products, kinases, transcription factors, various regulators that influence the resistance response and are affected by specific low molecular weight signaling molecules (e.g., phytohormones), and various effector proteins (Fig. 4). These have a direct or indirect effect on the virus, a virus-encoded protein, or on host factors that need to interact with either virus-encoded proteins or

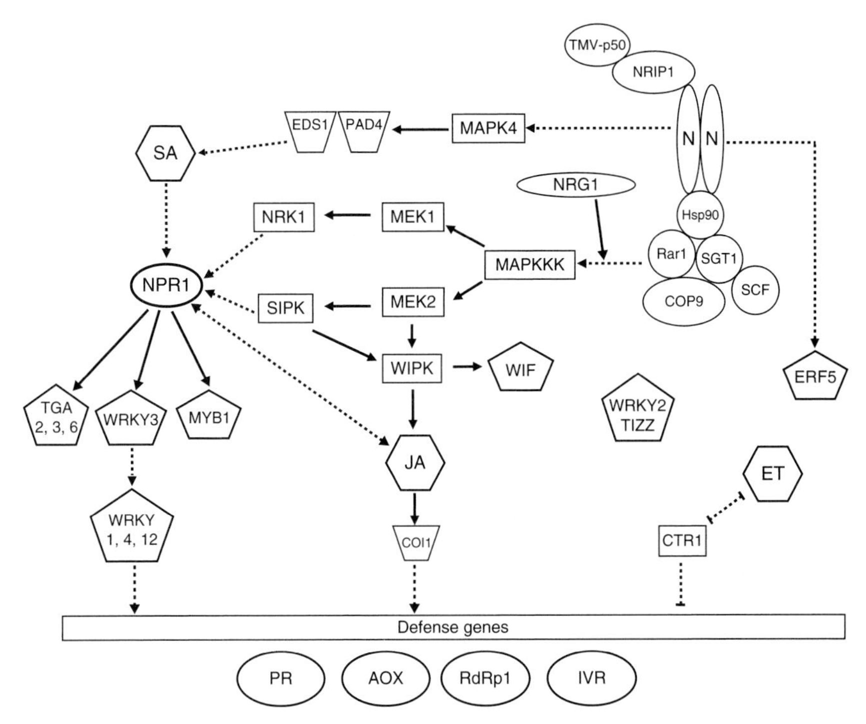

FIGURE 4 An overview of the signaling pathways involved in virus resistance. This diagram is based primarily on published data for the triggering of the *N* gene-mediated resistance response in tobacco by the TMV p50 elicitor, but also incorporates some data on the interactions of TCV with Arabidopsis and tobamoviruses with resistant capsicum. The various proteins are discussed and defined in the main text. The upper part of the figure shows the relationship of various signaling pathway factors to each other, with proteins that are known to interact indicated by their clustering and proteins known to act directly on other proteins indicated by solid arrows. Factors that have some further-ranging effect on another protein or factor are indicated by dotted arrows or lines with T-shaped termini, indicating that these factors repress other factors. The symbols for the different classes of signaling factors (above the "Defense genes") denote upstream signaling proteins (spheres and ellipses in the upper right), kinases (rectangles), transcription factors (TFs; pentagons), phytohormones (hexagons), and regulatory genes (trapezoids). Actual proteins at the end of the pathways known to be involved in defense are indicated by ellipses under "Defense genes," although the position of AOX may more properly be in the upper, signaling group. In many cases, the direct connections are not known, or are assumed (dotted arrows), and in the case of the defense genes, it is only known that PRs, AOX, and RdRp1 are induced by SA. However, with the exception of the PRs, it is not clear how SA does this or which TFs are involved. Many additional far-ranging interactions (so-called crosstalk) between some factors are not depicted in this two-dimensional diagram.

viral RNA to facilitate infection. Different viruses also trigger different signal pathways that induce resistance. For example, in *N* gene-mediated resistance to TMV, SA acts as the primary signal for most of the defense responses, but JA and ethylene may also function as independent signal molecules (Caplan and Dinesh-Kumar, 2006; Guo *et al.*, 2000; Ohtsubo *et al.*, 1999; Seo *et al.*, 2001). By contrast, CMV-induced resistance in Arabidopsis, mediated by the *RCY1* gene, is SA and ethylene dependent, but JA independent (Takahashi *et al.*, 2002, 2004), while TCV resistance in the same species, mediated by the *HRT* allele, is SA dependent, but JA and ethylene independent (Kachroo *et al.*, 2000).

In this section, we consider two examples of dominant resistance gene products that operate through signaling pathways to confer resistance to specific viruses (Sections V.A.1 and V.A.2), as well as dominant resistance and recessive resistance gene products that do not operate through signaling pathways (Sections V.A.3 and V.A.4). We also will consider the known phytohormone-responsive, signaling factors downstream of various resistance genes, including kinases, NPR1, AOX, COI, and EDS1/PAD4 (Sections V.B.1–V.B.5), various transcription factors (Section V.C), and various host effector proteins (Section V.D). Although many of these components and their upstream or downstream effectors or regulators are shown in Fig. 4, it is not possible to illustrate all of the known interactions, since various components are involved in feedback loops with other components inside out outside specific pathways.

A. Resistance gene products

While many other dominant resistance genes have been isolated and characterized (see Section II.B and also reviews by Maule *et al.*, 2007; Moffett, 2009; Palukaitis and Carr, 2008), little is known about their subsequent interactions or functions in resistance beyond what is described in various sections of this review.

1. Tobacco *N*-gene product

The *N*-gene product is a 131.4 kDa protein, but through alternate splicing of the *N*-gene transcript there is also a larger mRNA produced that encodes a truncated protein of 75.3 kDa (Dinesh-Kumar and Baker, 2000; Whitham *et al.*, 1994). The full-length N protein contains three designated functional domains: amino acids 8–150 contains a TIR domain (i.e., with sequence similarity to a *Drosophila* *T*oll/human *i*nterleukin-1 *r*eceptor); amino acids 216–325 contains an ATP/GTP-binding motif (amino acids 216–223), which is called the P loop, as well as two kinase domains (amino acids 297–301 and 320–325), referred to as a NBS domain; and amino acids 590–928 contains a LRR domain (Whitham *et al.*, 1994).

The TIR domain of N protein interacts with the chloroplast protein NRIP1 (for *N r*eceptor-*i*nteracting *p*rotein 1, a chloroplast rhodanese sulfurtransferase; Caplan *et al.*, 2008) in the cytoplasm, but only after NRIP1 has interacted with the TMV-encoded elicitor of the *N*-gene response (Burch-Smith *et al.*, 2007; Caplan *et al.*, 2008). (In various studies, a 50-kDa fragment derived from the TMV 126-kDa replication-associated protein was used as the elicitor, as this was the minimum-size active elicitor and was easier to work with than the intact 126-kDa protein; Abbink *et al.*, 1998; Erickson *et al.*, 1999.) Presumably after these interactions have occurred, the N protein oligomerizes through its TIR domain (Mestre and Baulcombe, 2006) and this complex is then transported to the nucleus (Caplan *et al.*, 2008). These steps occur independently and therefore upstream of NRG1 (for *N r*equirement *g*ene 1, encoding a protein with a CC–NBS–LRR structure) that is required for activation of the N-mediated response (Peart *et al.*, 2005). The LRR domain of N protein also interacts with HSP90 (for *h*eat-*s*hock *p*rotein 90, an ATP-dependent protein chaperone that facilitates protein folding; Liu *et al.*, 2004a; Lu *et al.*, 2003), which interacts with the signaling factors RAR1 (for "required for *Mla*12 resistance") and SGT1 [for Suppressor of G2 allele of skp1, a ubiquitin ligase-associated protein; both SGT1 and SKP1 are components of SCF (for *S*kp1/*C*ullin/*F*-box protein)-type E3 ubiquitin ligase complex]; both RAR1 and SGT1 also interact with each other (Botër *et al.*, 2007; Liu *et al.*, 2002a, 2004a; Lu *et al.*, 2003; Peart *et al.*, 2002; Takahashi *et al.*, 2003). It is thought that all three proteins (HSP90, RAR1, and SGT1) act as cochaperones in the folding of resistance proteins such as N (Schulze-Lefert, 2004).

NRG1-mediated resistance is also SGT1 dependent (Peart *et al.*, 2005). Both RAR1 and SGT1 also interact with the COP9 signalosome (for *co*nstitutive *p*hotomorphogenic phenotype, locus 9; Wei and Deng, 1999), which is a multisubunit complex that is involved in protein degradation mediated by the ubiquitin–26S proteasome pathway (Wei and Ding, 1992). SGT1 and RAR1 also interact with the SKP1 and Cullin1 components of the SCF complex (Liu *et al.*, 2002a, 2004a), which is involved in protein degradation. (These various interactions leading to one or more complexes are depicted in Fig. 4 as a single complex.) Silencing expression of these components in *N* gene-transgenic, *N. benthamiana* compromised *N*-mediated resistance to TMV (Liu *et al.*, 2002a,b). This complex is then believed to signal various responses that occur, including kinase-activated cell death responses (Liu *et al.*, 2004b; Takabatake *et al.*, 2007), although the next key steps starting from this complex are not known.

2. Potato *Rx*-gene product

The potato *Rx1* gene, located on chromosome XII, encodes a protein of 107.5 kDa, which contains the various domains typical of other (leucine zipper) CC–NBS–LRR resistance proteins, including three conserved kinase motifs in the N-terminal NBS domain (Bendahmane *et al.*, 1999). The C-terminus of the NBS domain also contains two ARC subdomains (for *a*poptosis, *R*-gene products and CED-4) conserved in a number of proteins involved in innate immunity and apoptosis in plants and animals (see Moffett, 2009). Despite this, in potato the resistance to PVX conditioned by Rx is extreme and does not involve programmed cell death. The functionally similar *Rx2* gene, originating from a different breeding source than *Rx1* and located on chromosome V, is 95% identical to *Rx1* in sequence, with 40 amino acid differences: 35 of which are in the CC and NBS domains and only five of which are in the LRR domain (Bendahmane *et al.*, 2000). Mutation of the LRR domain altered the *R*-gene recognition specificity from some strains of PVX to include other strains of PVX as well as another distinct potexvirus, *Poplar mosaic virus* (Farnham and Baulcombe, 2006). The CC–NBS domain was shown to interact with the LRR domain *in planta*, as did the CC and NBS–LRR domains, but not if the P-loop motif of the NBS domain was missing (Moffett *et al.*, 2002). The ARC1 subdomain was shown to be required for binding of the LRR domain to the N-terminal domain (Rairdan and Moffett, 2006). The ARC2 domain also was required to maintain an autoinactive state for the Rx protein, in the absence of the *Rx* elicitor (Rairdan and Moffett, 2006).

The elicitor of the *Rx* genes is the CP of PVX (Bendahmane *et al.*, 1995), although the CP did not interact directly with the Rx protein (Moffett *et al.*, 2002). Nevertheless, the CP was shown to interfere with the interaction *in planta* between either CC–NBS and LRR or CC and NBS–LRR, while coexpression of all either pair of Rx subdomain proteins and the PVX CP resulted in an HR, suggesting that activation of the Rx function required the sequential disruption of at last two intramolecular interactions (Moffett *et al.*, 2002). The C-terminal part of the LRR domain was shown to be the recognition domain for the PVX CP (Rairdan and Moffett, 2006). Disruption of the interaction between LRR and ARC was not required for signal initiation (Rairdan and Moffett, 2006). A highly conserved motif within the CC domain mediated an intermolecular interaction with the rest of the Rx protein, which was dependent on several other domains in Rx (Rairdan *et al.*, 2008). On the other hand, The Rx NB domain was shown to be sufficient for inducing an HR in an agroinfiltration assay in *N. benthamiana* (Rairdan *et al.*, 2008).

As was described above for the N protein (Liu *et al.*, 2004a), the Rx protein also required and interacted with the RAR1–SGT1–HSP90 complex (Botër *et al.*, 2007; Lu *et al.*, 2003). Rx also requires HSP90 for stability (Lu *et al.*, 2003) and hence accumulation (Botër *et al.*, 2007). Silencing SGT1 and HSP90 expression also led to *Rx1*-mediated resistance to PVX being compromised (Lu *et al.*, 2003; Peart *et al.*, 2002) as well as the Rx NB domain-mediated HR (Rairdan *et al.*, 2008). The Ran (for *Ra*s-like *n*uclear)

GTPase activating protein, RanGAP2, bound to the Rx protein, through the N-terminal CC domain and RanGAP2 was shown to be required for the Rx-mediated resistance responses; however, RanGAP2 did not interact with the PVX CP *in planta* (Sacco *et al.*, 2007; Tameling and Baulcombe, 2007). When overexpressed in tobacco with the CC-NB domains of Rx, RanGAP2 elicited a CP-independent activation of the Rx-mediated HR, although this was much weaker when RanGAP2 was overexpressed with the full-length Rx protein (Sacco *et al.*, 2007). More than one region of the CC domain of Rx is involved in the interaction with RanGAP2 (Rairdan *et al.*, 2008).

Rx1 expressed in transgenic *N. benthamiana* was able to engender resistance to PVX, with no visible HR (Bendahmane *et al.*, 1999). On the other hand, such transgenic plants also showed a resistance to other PVX strains, as well as to more distantly related potexviruses, which is not seen in potato plants possessing the *Rx* gene (Baurès *et al.*, 2008). These data indicated that the elicitor corresponding to the Rx protein is a structural feature of potexvirus CPs rather than a specific sequence, and that additional factors account for the higher specificity of recognition in potato.

3. Arabidopsis *RTM1*, *2*, and *3*

Three genes in Arabidopsis are required for the *RTM* resistance to TEV, which have been designated *RTM1*, *RTM2*, and *RTM3*. The *RTM1* gene encodes a protein similar to the lectin jacalin (Chisholm *et al.*, 2000), while the *RTM2* gene encodes a 41-kDa protein similar to small heat-shock proteins, although it was not responsive to temperature (Whitham *et al.*, 2000). The *RTM3* gene was not characterized (Chisholm *et al.*, 2000). Both *RTM1* and *RTM2* were expressed only in cells of the vasculature and the expressed proteins localized to sieve elements (Chisholm *et al.*, 2001). As the phenotype of the RTM-mediated resistance is a block in long-distance movement, these proteins must somehow function in this process by a yet unknown mechanism.

4. eIF4E

Recessive resistance against specific potyviruses in various plant species is conditioned by mutants of the translation factors eIF4E and its isomer, eIF(iso)4E. Similar resistance against a carmovirus in melon, and two bymoviruses in barley is conferred only by mutants of eIF4E, while resistance to a sobemovirus in rice is mediated by the translation factor eIF4G variants and resistance to CMV in Arabidopsis is conferred by variants of eIF4E and eIF4G (reviewed by Maule *et al.*, 2007; Palukaitis and Carr, 2008; Robaglia and Caranta, 2006; Truniger and Aranda, 2009). The translation factors encoded by these resistance genes are all involved in the formation of the 40S ribosome complex and are required for the initiation of the translation of cap-dependent mRNAs, as well as viral

RNAs possessing a VPg (Robaglia and Caranta, 2006). The eIF4E translation factor interacted with the VPg of several potyviruses, while variants of eIF4E encoded by the resistance allele generally (but not always) did not show such interactions (Beauchemin *et al.*, 2007; Charron *et al.*, 2008; Kang *et al.*, 2005; Michon *et al.*, 2006; Yeam *et al.*, 2007). The exact mechanism of the eIF4E-mediated resistance is not known, but it affects replication and apparently also cell-to-cell movement (Gao *et al.*, 2004).

B. Early signaling factors

1. Kinases

A number of protein kinases have been implicated in the defense response against viruses as well as other pathogens, in both Arabidopsis and tobacco. Some of those that have been identified are in the SA-mediated response pathway, while others occur in pathways involving other signal chemicals (see Fig. 4). With regard to virus resistance signaling, most of the work has been done with TMV in either *N*-gene tobacco or *N. benthamiana* plants expressing an *N* transgene, although much work also has been done in Arabidopsis with respect to infection by TCV or CMV.

a. CTR1 An Arabidopsis mutant (*ctr1*) that constitutively exhibits phenotypes of ethylene application was selected by a constitutive triple-response (CTR) assay (Kieber *et al.*, 1993). CTR1 is an ETR1-associated protein kinase (a MAPKKK) that negatively regulates ethylene signaling (see Fig. 4). ETR1 is a membrane-anchored receptor involved in ethylene signaling. Silencing expression of *NbCTR1* enhanced the defense response against TMV leading to a more rapid HR (Liu *et al.*, 2004b). The constitutive expression of ethylene led to upregulation of basic *PR* genes such as basic *PR-1*, *plant defensin 1.2* (*PDF1.2*) and *chitinase B* (*CHINB*), which may enhance the HR (Liu *et al.*, 2004b; Ohtsubo *et al.*, 1999). The nature of the connection of the pathway between the N-associated signaling complex and ethylene is not known.

b. SIPK An SA-induced protein kinase (SIPK) is activated by SA, NO, fungal elicitors, and viral infection (Kumar and Klessig, 2000). Moreover, SA is required for NO-mediated activation of SIPK. Neither JA nor ethylene activated SIPK (Kumar and Klessig, 2000). SIPK interacted with an upstream MAPKK, *Nt*MEK2 (Jin *et al.*, 2003) (see Fig. 4). As the generation of ROS requires the activation of MAPKs, this response probably occurs downstream of SIPK (Ren *et al.*, 2002).

c. WIPK A wound-induced protein kinase (WIPK) also functions upstream of JA during the TMV-induced HR in *N*-gene tobacco (Seo *et al.*, 2001). WIPK is *N* gene-dependent and is probably upstream of SA (Zhang and Klessig, 1998), but was not activated by SA, JA, ethylene, or NO (Kumar and Klessig, 2000). WIPK itself was phosphorylated and also has been shown to phosphorylate a transcription factor, *Nt*WIF, the expression of which was induced during a TMV-induced HR (Yap *et al.*, 2005). Overexpression of *Nt*WIF led to increased SA accumulation, constitutive PR protein expression, and enhanced resistance to TMV (Waller *et al.*, 2006) (see Fig. 4).

d. Other kinases and kinase–kinase interactions WIPK and SIPK function cooperatively to control the production of SA, and regulate the production of JA in response to wounding (Seo *et al.*, 2007). Both WIPK and SIPK formed a complex with *Nt*MEK2 and suppressing expression of all three kinases affected the defense response against TMV (Jin *et al.*, 2003), as did suppression of the upstream MAPKKK, NPK1 (Jin *et al.*, 2002). *Nt*MEK2 controlled the expression of SIPK and WIPK, as well as other defense responses (Yang *et al.*, 2001). Suppressing expression of the MAPKK, *Nt*MEK1, and the MAPK, NRK1, attenuated the defense response against TMV in *N*-transgenic *N. benthamiana*, such that the virus was not localized by the HR response (Liu *et al.*, 2004b). A proposed model for the relationship of the various kinases in a signaling cascade, based on that proposed by Liu *et al.* (2004b) is shown in Fig. 4.

2. Nonexpressor of PR-1

NPR1, named after a mutant, *npr1* [also referred to noninducible immunity 1 (*nim1*) and salicylic acid-insensitive 1 (*sai1*)], in which application of SA did not lead to expression of the *PR* genes, is a 65-kDa ankyrin repeat protein (reviewed by Durrant and Dong, 2004). Although NPR1 is involved in SAR against bacteria and fungi (Després *et al.*, 2000; Zhou *et al.*, 2000), it does not seem to be required for SA-induced resistance to viruses in Arabidopsis (Kachroo *et al.*, 2000; Wong *et al.*, 2002); however, it appears to be required for HR-type resistance to TMV in *N* gene-expressing transgenic *N. benthamiana* (Liu *et al.*, 2002b). Prior to treatment with SA or its analogues, NPR1 was located in the cytoplasm as an oligomer, but in the presence of SA, NPR1 was reduced and the monomer units translocated to the nucleus and stimulated various bZIP, TGA transcription factors (see Section V.C.3) to bind to DNA promoter elements of the *PR-1* gene (Fan and Dong, 2002; Johnson *et al.*, 2003). In the absence of SA, any reduced NPR1 that enters the nucleus was degraded rapidly by the proteasome, after being targeted there by CUL3-based, E3 ligase-mediated ubiquitinylation. After SA-mediated reduction of NPR1 oligomers to monomers, and entry of the monomer units into the nucleus, NPR1 was phosphorylated and activated transcription, after which it also was turned over by interaction with a CUL3-based ubiquitinylation complex that recognized the phosphorylated form of the NPR1. This then allowed further phosphorylated NPR1 to cycle into the transcription initiation—NPR1 degradation cycle (Spoel *et al.*, 2009). NPR1 is also a mediator of SA-induced plant cell death associated with autophagy (an intracellular degradation process of cytoplasmic components by the vacuole; Yoshimoto *et al.*, 2009). Three other Arabidopsis proteins have been found to interact with NPR1. These were designated NIMIN-1, -2, and -3, and they were able to modify the interaction of the TGA transcriptions factors with NPR1 in yeast. Transcription of the three *AtNIMIN* genes was stimulated by SA treatment. An NPR1 mutant that could not activate SAR was unable to bind the three NIMIN proteins, suggesting that the latter proteins also are involved in this SA-induced defense signaling (Weigel *et al.*, 2001). NRR (for negative regulator of resistance) also interacted with NPR1. Overexpressing NRR in rice led to suppression of induction of defense-related genes and enhanced susceptibility to a bacterial pathogen (Chern *et al.*, 2005). Thus, NPR1 occupies a central role in signaling the activation of PR-mediated defense against pathogens, but its role in resistance to viruses remains obscure.

3. Alternative oxidase

AOX is an inner mitochondrial membrane protein involved primarily in regulating carbon metabolism through bypassing the normal passage of electrons, from NADH/NAD^+, generated in the TCA cycle, through ubiquinone/ubiquinol to Complexes III and IV of the mitochondrial electron transport chain. Passage of electrons from ubiquinol to AOX results in the conversion of oxygen to water with no generation of ATP, and also results in the generation of heat (Vanlerberghe and McIntosh, 1997). This latter effect (thermogenesis) is used in some plant species to volatilize chemicals to attract pollinators. However, AOX is also important in preventing or reducing the flow of electrons from some reduced components of the electron transport chain (Complexes I and III) directly to molecular oxygen and generating ROS (Maxwell *et al.*, 1999; Yip and Vanlerberghe, 2001), which also are generated during pathogen infection (Heath, 2000). Therefore, the role of AOX in pathogen response has been assessed and found to be important in the signaling of defense against various viruses, but not of bacteria or fungi (Chivasa *et al.*, 1997; Murphy *et al.*, 1999; Singh *et al.*, 2004; Wong *et al.*, 2002).

There are two *AOX* gene families. The *AOX1* gene family was induced by stress, while the *AOX2* gene family was either constitutive or developmentally regulated (Maxwell *et al.*, 2002; Vanlerberghe and McIntosh, 1997). The *AOX1a* gene could be induced by chemicals such as cyanide and antimycin A, which inhibited the flow of electrons through the cytochrome pathway (Complexes III and IV), resulting in an increase in mitochondrial ROS levels (Singh *et al.*, 2004; Wong *et al.*, 2002). Expression of the *AOX1a* gene, as well as AOX activity, was induced by SA and NO

(Chivasa *et al.*, 1997; Huang *et al.*, 2002). The expression of the *AOX1a* gene also increased in Arabidopsis (Lacomme and Roby, 1999; Simons *et al.*, 1999) and in tobacco (Chivasa and Carr, 1998; Lennon *et al.*, 1997) in tissues undergoing a HR induced by viruses as well as by other pathogens. Although SA is involved in signaling resistance to viruses as well as in the induction of AOX, the evidence that AOX is involved in the virus resistance came from various studies showing that stimulation of AOX activity by antimycin A or cyanide led to increased resistance to replication and/or movement of several RNA viruses (Chivasa and Carr, 1998; Wong *et al.*, 2002) as well as of CaMV (Love *et al.*, 2005, 2007b), while inhibition of AOX by application of salicylhydroxamic acid also led to a decrease in SA-induced resistance (Chivasa *et al.*, 1997). This indicates that AOX functions downstream of SA to inhibit virus accumulation, although this activation pathway is independent of NPR1 (Wong *et al.*, 2002). However, studies involving transgenic tobacco plants expressing higher or lower levels of AOX showed that AOX also can act as a negative regulator of induced resistance to viruses (Gilliland *et al.*, 2003). The specific effects of AOX that lead to a reduction of virus accumulation are not known, although AOX does not regulate the induction of RdRp1 (Gilliland *et al.*, 2003), another factor involved in the resistance response against infection by TMV (see Sections IV.B.1 and V.B.5). In fact, RdRp1 exerts an effect on *AOX1a* gene expression in tobacco (Rakhshandehroo *et al.*, 2009). Viruses also can exert effects on the transcription of AOX1a. Transgenic plants expressing the CMV 2b prevented *AOX1a* gene transcription induction by SA (Ji and Ding, 2001) and transgenic plants expressing the PPV HCPro protein inhibited SA-induced transcription of *AOX1a* (Alamillo *et al.*, 2006).

4. COI

Coronatine-insensitive 1 (COI1) is an F-box protein known to interact with components of the SCF–ubiquitin ligase complex in Arabidopsis (Devoto *et al.*, 2002; Xie *et al.*, 1998; Xu *et al.*, 2002). The mutant *coi1*, showed insensitivity to both JA and the phytotoxin coronatine (a mimic of JA) and affected both plant development and defense responses mediated by JA (Feys *et al.*, 1994). The interaction of COI1 with the SCF–ubiquitin ligase complex is needed for these JA-mediated signaling responses (Devoto *et al.*, 2002; Xu *et al.*, 2002). In Arabidopsis, *RCY1*-mediated resistance to Y-CMV (Takahashi *et al.*, 2002) and *HRT*-mediated resistance to TCV (Kachroo *et al.*, 2000) were both unaffected by the *coi1* mutant, indicating that these resistance responses did not require JA-mediated signaling. By contrast, silencing of *COI1* gene expression in *N* gene-expressing *N. benthamiana* led to the loss of resistance against TMV–GFP, indicating that JA-mediated signaling was involved in the *N* gene-mediated resistance to TMV (Liu *et al.*, 2004b), although this seems to conflict with other data (Section III.B). COI1 functioned in the inhibition of the transcription factor *At*WRKY70, which inhibited JA-response genes and activated SA-responsive genes (Li *et al.*, 2004) (see Section V.C.1).

5. EDS1 and PAD4

Some genes conferring resistance to bacterial and fungal pathogens require the *EDS1* (for enhanced disease susceptibility 1) gene for their function (Aarts *et al.*, 1998). EDS1, like PAD4 (for phytoalexin deficient), which is also required for resistance against these other pathogens, is a lipase-like protein (Falk *et al.*, 1999). Both proteins function upstream of SA-dependent *PR-1* induction and have been shown to be required for SA synthesis in Arabidopsis (reviewed by Kachroo, 2006). EDS1 functioned independently of PAD4, as well as together with PAD4, with which it interacted, in the early stages of defensive signaling (Feys *et al.*, 2001). EDS1 and PAD4 also affected SA–JA/ET signal antagonism, by functioning as activators of SA and inhibitors of the JA/ET signaling pathways. These activities were negatively regulated through MAP kinase 4, which is upstream of EDS1 and PAD4 (Brodersen *et al.*, 2006). EDS1 also could be induced by SA or SA mimics (Falk *et al.*, 1999; Peart *et al.*, 2002) and EDS1 also was expressed during a compatible interaction in *N. benthamiana* infected by *Tobacco rattle virus* or TMV–GFP, but silencing *EDS1* did not result in enhanced disease susceptibility (Peart *et al.*, 2002). On the other hand, EDS1 was required for the *N* gene-mediated resistance to TMV–GFP in *N* gene-containing *N. benthamiana* (Liu *et al.*, 2002b; Peart *et al.*, 2002), although EDS1 was not required for *Rx* gene-mediated resistance to PVX (Peart *et al.*, 2002). Both EDS1 and PAD4 were required for *HRT*-mediated resistance to TCV in Arabidopsis (Chandra-Shekara *et al.*, 2004). Thus, there is considerable variation in the requirements for these signaling components in induction of various virus resistance responses.

C. Transcription factors

A number of transcription factors have been shown to play important key roles in signaling the defense response against virus infection. Our current knowledge in this area is derived largely from work on responses to TMV in *N. tabacum* containing the *N* gene, TMV in resistant *Capsicum annum* or one of several viruses in resistant Arabidopsis.

1. WRKY transcription factors

WRKY transcription factors contain the WRKYGQK core sequence before a zinc-finger motif and bind to the W-box [(T)TGAC(C/T)] found in the promoter sequences of various genes that responded to wounding or pathogens, including the genes encoding PR proteins (Ülker and

Somssich, 2004). They have various functions and roles in the defense response. A number of WRKY transcriptions factors have been identified in the response to TMV infection in *N*-gene tobacco (see Fig. 4). *Nt*WRKY 1, 3, 4, and 12 were all stimulated by TMV infection and SA treatment, while *Nt*WRKY2 was not induced by either treatment (Chen and Chen, 2000; van Verk *et al.*, 2008; Yang *et al.*, 1999). Both *Nt*WRKY3 and *Nt*WRKY4 were induced by 3 h following infection with TMV, while *Nt*WRKY1 was induced 9 h after TMV infection, and *Nt*WRKY12 was induced transiently for 30–60 min after TMV infection and then at a high level at 48–72 h after TMV infection (Chen and Chen, 2000; van Verk *et al.*, 2008; Yang *et al.*, 1999). In chili pepper, the 60.55-kDa *Ca*WRKYa was induced by an incompatible reaction with TMV, by SA, ethylene, JA, and wounding (Park *et al.*, 2006). By contrast, the 18.8-kDa *Ca*WRKY1 was induced by SA and an avirulent strain of *Pepper mild mottle virus* (PMMoV), but not by ethylene, JA, or wounding (Oh *et al.*, 2008). *CaWRKY1* levels declined following the appearance of HR lesions. *CaWRKY1* acted as a negative regulator of SAR/LAR, as shown by expressing it constitutively in transgenic *N*-gene tobacco plants, where it engendered an increase in TMV-induced lesion size (Oh *et al.*, 2008). Both CaWRKYa and CaWRKY1 were localized to the nucleus (Oh *et al.*, 2008; Park *et al.*, 2006). The transcription factor TIZZ (for TMV-induced leucine zipper zinc-finger) bound to the TTGAC element recognized by WRKY transcription factors. TIZZ was activated by TMV in the *N*-mediated response, but was shown to be SA independent (Yoda *et al.*, 2002). Like most other factors examined by Liu and colleagues, silencing the expression of *Nt*WRKY1, 2, or 3 compromised resistance against TMV–GFP in *N*-transgenic *N. benthamiana* (Liu *et al.*, 2004b). *Nt*WRKY3 is most similar to *At*WRKY70, an Arabidopsis gene that is induced by SA and repressed by JA (Li *et al.*, 2004; Liu *et al.*, 2004b). *At*WRKY70 is downstream of NPR1. Constitutive expression of *At*WRKY70 resulted in constitutive expression of *PR* genes and enhanced pathogen resistance, while suppression of *At*WRKY70 resulted in activation of JA- and COI-dependent genes (Li *et al.*, 2004).

2. MYB1 transcription factor

Myb1 is a transcription factor, which is expressed from a tobacco ortholog of the oncogene *myb1*. Myb1 was induced by TMV infection in *NN* tobacco, but not in *nn* tobacco, which is susceptible to the virus. Myb1 also was induced in upper noninfected leaves of *NN* tobacco, indicating that it is also part of signaling in the SAR. Myb1 appeared rapidly after exogenous application of SA in both *NN* and *nn* tobacco and induced expression of the *PR* genes. Unlike expression of *PR* genes, expression of *NtMYB1* did not require protein synthesis following application of SA. Myb1 also bound preferentially to a sequence (GTTTGGT) in the promoter of the *PR-1a* gene (Yang and Klessig, 1996). Silencing expression of *NbMYB1* in *N* gene-transgenic *N. benthamiana* also attenuated the resistance response to TMV (Liu *et al.*, 2004b). Thus, Myb1 is another transcription factor that is involved in the signaling of the defense response against infection by TMV (see Fig. 4).

3. TGA transcription factors

These transcription factors have a *b*asic domain leucine *zip*per (bZIP) domain and are involved in a number of aspects of signaling during development. They are referred to as TGA transcription factors, since the first described members of this family of bZIP transcription factors bound to *TGACG*/as-1 elements in promoters; the as-1 element being a binding site for transcription factors within the CaMV 35S RNA promoter containing the TGACG motif (Katagiri *et al.*, 1989). The bZIP transcription factors TGA2, TGA3, and TGA6 (as well as TGA5, albeit to a lesser extent) all bound to NPR1, and NPR1 enhanced the binding of these transcription factors to various promoter elements of the *PR-1* gene. Furthermore, mutants in NPR1 that affected the activation of the SA-dependent signaling also affected the interaction of NPR1 with these transcription factors (Després *et al.*, 2000; Fan and Dong, 2002; Zhang *et al.*, 1999; Zhou *et al.*, 2000). Inactivation of each of these transcription factors alone did not compromise the SA-dependent signaling, while inactivation of all three did, indicating that they functioned redundantly in activating transcription mediated by NPR1 (Zhang *et al.*, 2003). TGA1 interacted with NPR1 only after a redox reaction in which the TGA1 disulfide bridges had been reduced and the NPR1 oligomer had been reduced to a monomer form; both of these changes occurred after treatment with SA (Després *et al.*, 2003; Mou *et al.*, 2003). NPR3 and 4 negatively regulate *PR*-gene expression through binding to TGA2 and its paralogs (Zhang *et al.*, 2006b). Thus, these various TGA transcription factors are located in the SA-mediated defensive signaling pathway between NPR1 and the expression of the *PR* genes and other defense genes (see Fig. 4).

4. AP2/ERF transcription factor

The *NtERF5* gene encodes a protein related in sequence to the AP2/ERF family of transcription factors. *Nt*ERF5 bound weakly to GCC-box *cis*-elements, which were known to mediate pathogen-induced transcription of several *PR* genes. *NtERF5* itself was induced by TMV infection (as well as infection with *P. syringae*). The sequence of *NtERF5* was most similar that of *AtERF1* (*Ethylene-Response Factor 1* from Arabidopsis), but since four other *ERF* genes had been identified previously in tobacco, this one was named *NtERF5*. However, this gene was not induced by application of ethylene, SA, or JA. Thus, the expression of this transcription factor is independent of these three phytohormones (Fischer and Dröge-Laser, 2004).

NtERF5 was not induced when *NN* tobacco plants were maintained at 32 °C prior to and after inoculation with TMV, but was induced within 24 h after shifting the plants to 22 °C. *NtERF5* also was induced within 3 h by wounding, although to a lesser extent and only transiently compared to after infection by TMV. Constitutive overexpression of *NtERF5* in transgenic *NN* tobacco plants led to decreases in local lesion size in the inoculated leaves of plants maintained at 22 °C, while TMV-infected, *NtERF5*-overexpressing, transgenic, tobacco plants showed an inhibition of systemic infection by TMV after a temperature shift from 32 °C down to 22 °C (Fischer and Dröge-Laser, 2004). This indicated that this defense response was independent of the other *N* gene-mediated defense responses, which are not functional at 32 °C. *NtERF5* also was induced by infection with CMV and PVX (J.-Y. Yoon and P. Palukaitis, unpublished data), as well as by PVY (Rakhshandehroo *et al.*, 2009); the latter stimulation was regulated by RdRp1 (see Section V.D.5).

5. NAC transcription factors

Several transcription factors from the NAC family [for *Petunia* NAM (no apical meristem) and Arabidopsis ATAF1, ATAF2, and CUC2 (cup-shaped cotyledons)] have been shown to have some role in regulating virus accumulation, although the roles varied with the nature of the transcription factor and the virus. The NAC transcription factors GRAB1 and 2 (for Geminivirus Rep-A binding) bound to the Rep-A protein encoded by the geminivirus wheat dwarf virus, and expression of GRAB1 and 2 in wheat cells inhibited wheat dwarf virus replication (Xie *et al.*, 1999). A replication enhancer (REn) protein of another geminivirus, *Tomato leaf curl virus*, interacted with the tomato NAC transcription factor *Sl*NAC1. The later gene was induced by virus infection (or just transient expression of REn) and this response resulted in enhanced virus accumulation (Selth *et al.*, 2005). The CP of TCV interacted with the Arabidopsis NAC transcription factor TIP (for TCV-interacting protein) in a region of the CP required for the activation of the HR in the Arabidopsis ecotype Dijon, which contains the *HRT* gene for resistance to TCV (Cooley *et al.*, 2000; Ren *et al.*, 2000). The binding of the TCV CP to TIP prevented its translocation to the nucleus (Ren *et al.*, 2005); however, the *AtTIP* gene was necessary neither for the *HRT*-mediated resistance to TCV nor for the SA-mediated induction of *HRT* gene expression, but it was necessary for basal resistance to CMV (Jeong *et al.*, 2008). The helicase domain of TMV interacted with the Arabidopsis NAC transcription factor ATAF2. Interestingly, while infection by TMV induced the expression of the *AtATAF2* gene, the interaction of the TMV helicase with ATAF2 resulted in the transcription factor being targeted for proteasome-mediated degradation. Overexpression of *AtATAF2* led to increased expression of the defense-associated marker genes *PDF1.2*, *PR-1*, and

PR-2, as did treatment of nontransgenic plants with SA, indicating that ATAF2 regulated the expression of defense-related genes. Infection by TMV prevented systemic induction of *ATAF2* and *PR-1* even when SA was applied, supporting the conclusion that the interaction between the transcription factor ATAF2 and the TMV helicase domain prevented or suppressed basal host defenses against TMV in Arabidopsis (Wang *et al.*, 2009).

D. Host effector proteins

1. *Tm-1/tm-1*

The *Tm-1*-encoded 80-kDa protein bound to sequences present in both of the tobamoviral replication proteins (126 and 183 kDa) and inhibited assembly of the viral replicase complex (Ishibashi *et al.*, 2007, 2009). *Tm-1* was introduced into cultivated tomato from a wild tomato species, but ToMV-susceptible tomato cultivars possess corresponding *tm-1* alleles; these are not null alleles. Ishibashi *et al.* (2009) have characterized one of these, $tm\text{-}1^{GCR26}$, and found that it encodes a protein that inhibited replication of the tobamoviruses *Tobacco mild green mosaic virus* (TMGMV) and PMMoV. While ToMV, TMGMV, and PMMoV could all infect tobacco, tomato was a nonhost for TMGMV and PMMoV. The results suggested that *Tm-1* and *tm-1* might represent a family of genes encoding inhibitors that control tobamovirus host range. However, the resistance was passive in nature and did not depend upon or trigger defensive signaling. Nevertheless, in the case of tomato and both PMMoV and TMGMV, this is the first known case of a resistance response against a plant virus that may be considered as due to nonhost resistance.

2. Pathogenesis-related proteins

PR proteins are induced by SA through an NPR1-dependent pathway during pathogen attack, most commonly during incompatible interactions, although they also can be induced during a non-necrotic response of plants to sequences derived from a compatible pathogen (Pruss *et al.*, 2004; Shams-Bakhsh *et al.*, 2007; Whitham *et al.*, 2003). The tobacco PR proteins induced during the HR to TMV were grouped into five families, each containing both basic and acidic isoforms, but subsequent work with PRs induced in other species and by other pathogenic agents has led to 14 families of PR proteins being described. Most of these have known functions or activities: for example, PR-2 is a β-1,3-glucanase; PR-3, -4, -8, and -11 are different types of chitinase; PR-5 is a thaumatin-like protein; PR-6 is a proteinase inhibitor; PR-7 is an endoproteinase; PR-9 is a peroxidase; PR-10 is a ribonuclease; and PR-12, -13, and -14 are a defensin, thionin, and lipid-transfer proteins, respectively (reviewed by van Loon and van Strien, 1999). Some PR proteins have antifungal and antibacterial properties (Bowles, 1990); however, as far as is known, none of them have an activity that can be clearly linked to the inhibition of virus infection. Nevertheless, as the induction of the PR proteins is a hallmark of SAR, these are used as markers for this process and for the signaling events that lead to SAR and PR protein accumulation.

3. Inhibitor of virus replication

A proteinaceous inhibitor of virus replication (IVR) induced by TMV infection in *NN* tobacco plants and protoplasts was identified by Loebenstein and colleagues (reviewed by Loebenstein, 2009; Loebenstein and Akad, 2006). Although the induction of *NtIVR* appeared to be specific to both TMV and the *N* gene, IVR was not target specific, in that IVR applied to protoplasts, leaf disks, or whole plants could inhibit accumulation of other viruses, including CMV, PVX, and PVY (Gera and Loebenstein, 1983; Loebenstein and Akad, 2006). Recently, it was shown that IVR is induced by PVY infection of *NN* genotype tobacco, but to a much lower extent than by TMV (Rakhshandehroo *et al.*, 2009). Constitutive expression of *NtIVR* in transgenic *nn* tobacco also gave resistance to TMV, as well as to *B. cinerea*, indicating that the expressed 21.6-kDa IVR protein had multiple functions. IVR accumulated in the intercellular spaces following infection of *NN* tobacco by TMV (Spiegel *et al.*, 1989), making it unlikely that IVR is a transcription factor. While this site of accumulation may have some influence on infection by *B. cinerea*, it is unlikely to affect viruses, which are intracellular pathogens. Thus, either some lower level of IVR that remains intracellular inhibits the virus, or some extracellular IVR cycles back into the cell after modification; the antiviral activity of IVR added to the external medium of protoplasts and leaf disks supports the latter possibility (Loebenstein and Gera, 1981). The inhibition of TMV accumulation in protoplasts indicates that IVR functions by inhibiting virus replication, although a further role in inhibiting translation or virus movement cannot be ruled out. IVR did not have a single-stranded RNase activity (Gera and Loebenstein, 1983), although it has never been assessed for dsRNA-specific RNase activity. Although *NtIVR*-gene expression was stimulated during the *N* gene response to infection by TMV, *NtIVR* was not induced by application of exogenous SA, and was induced in *NahG*-transgenic *NN* tobacco plants that prevent SA accumulation (M. Takeshita, J. Squires, and P. Palukaitis, unpublished data). On the other hand, inhibition of *NtIVR* mRNA accumulation in transgenic *NN* tobacco led to a reduction in SAR induced by either TMV infection of lower leaves, or the exogenous application of SA (M. Takeshita and P. Palukaitis, unpublished data), which suggests that IVR is affected by SA. It may be that the effect of SA on IVR is indirect, since SA affects *NtRDR1* gene expression and the *NtRDR1* gene regulates expression of the *NtIVR* gene (Rakhshandehroo *et al.*, 2009).

4. Antiviral factor

An antiviral factor (AVF) from *NN* tobacco infected with TMV was described that is different from the IVR, but which also was induced in TMV-infected *NN* tobacco. AVF is actually a family of phosphorylated glycoproteins that are stimulated in their synthesis by TMV infection. When mixed with TMV prior to inoculation, AVF inhibited the accumulation of TMV (Sela, 1981; Sela and Appelbaum, 1962). AVF appeared to function similar to human β-interferon in plants, since both stimulated plants to synthesize nucleotides with antiviral activities (Reichman *et al.*, 1983). However, AVF does not resemble human β-interferon in sequence, even though antibodies to human β-interferon could be used to purify AVF. Rather, the sequence of the two purified AVF glycoproteins suggested that gp35 is a β-1,3-glucanase, while gp22 is an isoform of PR-5 (Edelbaum *et al.*, 1990, 1991). The regulation of *NtAVF* gene expression has not been examined.

5. RNA-dependent RNA polymerase 1

RdRp1 is most likely the RNA-dependent RNA polymerase that was studied extensively in the 1970s and 1980s and shown to be induced by virus infection in several plant species (reviewed by Fraenkel-Conrat, 1983, 1986; Wassenegger and Krczal, 2006). The gene encoding this RdRp is designated *RDR1* (Wassenegger and Krczal, 2006), since there are six *RDRs* in Arabidopsis (Yu *et al.*, 2003), although it is not known how many there are in *Nicotiana* species. *At*RdRp1, in contrast to *At*RdRp6, is not required for RNA silencing, but may function using RNA silencing (Yu *et al.*, 2003). Alternatively, RdRp1 may function by making complementary viral RNA to the viral genome, masking structural motifs required for replication, transcription, or translation. SA induced *NtRDR1* gene expression in both *n*-gene and *N*-gene tobacco (Gilliland *et al.*, 2003; Xie *et al.*, 2001). Silencing the expression of the *NtRDR1* gene enhanced infection by TMV and PVX in tobacco plants without the *N* gene (Xie *et al.*, 2001), while it enhanced infection by PVY in *N*-gene tobacco plants (Rakhshandehroo *et al.*, 2009). Silencing the expression of the *AtRDR1* gene in Arabidopsis enhanced infection by *tobacco rattle virus* and a crucifer strain of TMV, TMV-Cg (Yu *et al.*, 2003). As mentioned above (Section IV.B.1), the *RDR1* gene in *N. benthamiana* produced a dysfunctional protein, due to the presence of an intron (Yang *et al.*, 2004), and this plant species is hypersusceptible to a large number of plant viruses. This further supports a role for RdRp1 in virus resistance. However, constitutive expression of the *MtRDR1* gene in *N. benthamiana* only inhibited disease development during infection by two tobamoviruses and not during infection by PVX or CMV (Yang *et al.*, 2004). In addition, in *N. glutinosa* while expression of the *NgRDR1* gene was stimulated by SA, methyl-JA, hydrogen peroxide and infection by CMV, PVY, and TMV, it was not stimulated by infection with PVX (Liu *et al.*, 2009). Thus, not only does the *RDR1* gene show enhanced expression after stimulation with various biotic stresses, it also shows selective enhancement by viruses, depending on the virus and the plant species. Moreover, as mentioned above (Section IV.C), *RDR1* in *N. attenuata* also is involved in regulation of JA-responsive genes, as well as the biosynthesis of JA and nicotine (Pandey and Baldwin, 2007; Pandey *et al.*, 2008). In addition, in *N. tabacum*, *RDR1* is involved in the regulation of several defense-related genes; viz, *AOX1*, *ERF5*, *IVR*, and *RDR6* (Rakhshandehroo *et al.*, 2009). This further exemplifies the complexity of the signaling between these various defense-response pathways.

6. Other effector proteins

The expression of genes encoding several proteins was induced in a specific *C. annum* cultivar carrying the L^2 resistance gene and infected by an incompatible strain of TMV (TMV-P_0). These included genes encoding a sequence-nonspecific ribonuclease (*CaPR-10*: Park *et al.*, 2004), a cell wall protein (*CaTin2*: Shin *et al.*, 2003), a lipid-transfer protein (*CaLTP1*; Park *et al.*, 2002), and an alanine aminotransferase (*CaAlaAT1*: Kim *et al.*, 2005). Constitutive, expression of the 23-kDa Tin2 cell wall protein in transgenic tobacco led to a temporary partial resistance to both TMV and CMV (Shin *et al.*, 2003). In addition, the 18-kDa, PR-10 RNase induced in leaves could degrade both viral and plant RNAs, the activity of which was enhanced by phosphorylation (Park *et al.*, 2004). *CaPR-10* gene expression was induced by SA, JA, ethylene, sodium chloride, and the herbicide methyl viologen (Paraquat), the last of which generates superoxide free radicals (Park *et al.*, 2004), while *CaTin2* gene expression was induced by SA, JA, ethylene, and more slowly by abscisic acid, sodium chloride, and methyl viologen (Shin *et al.*, 2003). How these various proteins function in the natural defense against virus infection in this incompatible response between the virus and the L^2 gene in *C. annum* is not known; although it may be considered that the RNase is released to degrade the viral genome or at least limit its replication.

Coexpression of the N protein and the TMV 50-kDa elicitor in *N. benthamiana* did not elicit an HR, unlike in tobacco (Mestre and Baulcombe, 2006). However, when coexpression of these proteins was done at the same time as infection by PVX–GFP, the accumulation of PVX CP and GFP were blocked (Bhattacharjee *et al.*, 2009). This effect was shown to be due to an inhibition of the translation of the viral RNAs, due to sequences in the PVX CP ORF and to require the factor AGO4 (Bhattacharjee *et al.*, 2009), which previously had been shown to also exert translational control on RNAs to which it becomes tethered (Pillai *et al.*, 2004, 2007; Wu and Belasco, 2008). The inhibition of PVX–GFP by *Rx*-gene variants also was compromised by silencing the expression of AGO4 (Bhattacharjee *et al.*, 2009). Thus, AGO4-mediated tethering, sequestration of RNAs, and possibly degradation of RNAs appear to be involved in the RNA inhibition stage of some *R* gene-mediated resistance responses, but not in the HR stage.

VI. CONCLUDING THOUGHTS

The diagram of various components involved in the signaling responses against viruses (Fig. 4) illustrates the limitations of what we know about these processes. There is still much to be learned about the interactions between signaling pathways mediated by various phytohormones (SA, JA, and ethylene) as well as other chemical mediators (NO, Ca^{2+}, etc.) and the effects of various feedback loops within a pathway. The steps from the early signaling complexes involving the viral inducers, the NBS–LRR resistance gene products, and the protein HSP90, SGT1, and RAR, which lead to activation of the MAP kinases are still unknown. This also is true for the role and regulation of the transcription factors ERF5, TIZZ, WRKY2, and WIF. Various transcription factors such as MYB, as well as various TGAs and WRKYs all activate transcription of the *PR-1a* gene, but as PR-1 has no role in virus infection, it is unknown what role these transcription factors have in the signaling of either resistance or the development of a HR. What plant effectors are involved in resistance to viruses mediated by JA and ethylene, and the pathways leading to activation of these effectors remain largely unknown. The position of RdRp1 as both a direct effector on virus accumulation, as well as a regulator of the expression of other defense-related genes, and its own regulation by SA as well as SA-independent pathways makes for a complex yet intriguing network in need of further analysis. The regulation of IVR production is not clear and the subtle effects observed by changes in AOX levels on the inhibition of virus accumulation and the feedback on SA-mediated defense also are all areas that need further exploration, as do the overlaps between these systems and RNA silencing. Furthermore, it is not clear if findings in a limited number of model systems (e.g., *Nicotiana* spp., Arabidopsis) fairly represent the totality of induced resistance mechanisms in plants.

Thus, in the field of induced resistance to viruses, we are very much aware that we are missing much of what is going on and that much work remains to be done on this complex system. We have no doubt the study of induced resistance signaling, most likely combined with insights from computational and systems biology, will continue to yield results of intellectual interest for many years to come. Additionally, we remain optimistic, provided that studies are extended beyond model species, that the accelerating rate of discovery in this area will aid the development of badly needed advances in crop protection and disease management.

ACKNOWLEDGMENTS

We are grateful to colleagues who allowed us to mention their work prior to publication, including R. Lozano-Durán and E.R. Bejarano (University of Malaga, Spain), H. Ziebell (Cornell University), A.M. Murphy, B. Otto, and D.E. Hanke (Cambridge University). We are also grateful to A.M. Murphy for critically reading the manuscript. Research in the J.P.C. lab is funded by grants from the Biotechnological and Biological Sciences (grants BB/D008204/1 and BB/F014376/1), The Leverhulme Trust, the European Union, and the Isaac Newton Trust.

REFERENCES

Aarts, M. G. M., Hekkert, B. T., Holub, E. B., Beynon, J. L., Stiekema, W. J., and Pereira, A. (1998). Identification of *R*-gene homologous DNA fragments genetically to disease resistance loci in *Arabidopsis thaliana*. *Molec. Plant Microbe Interact.* **11**:251–258.

Abbink, T. E. M., Tjernberg, P. A., Bol, J. F., and Linthorst, H. J. M. (1998). Tobacco mosaic virus helicase domain induces necrosis in *N* gene-carrying tobacco in the absence of virus replication. *Mol. Plant Microbe Interact.* **11**:1242–1246.

Agorio, A., and Vera, P. (2007). ARGONAUTE4 is required for resistance to *Pseudomonas syringae* in Arabidopsis. *Plant Cell* **19**:3778–3790.

Alamillo, J. M., Saénz, P., and García, J. A. (2006). Salicylic acid-mediated and RNA-silencing defense mechanisms cooperate in the restriction of systemic spread of plum pox virus in tobacco. *Plant J.* **48**:217–227.

Allan, A. C., Lapidot, M., Culver, J. N., and Fluhr, R. (2001). An early tobacco mosaic virus-induced oxidative burst in tobacco indicates extracellular perception of the virus coat protein. *Plant Physiol.* **126**:97–108.

Allen, E., Xie, Z., Gustafson, A. M., and Carrington, J. C. (2005). microRNA-directed phasing during *trans*-acting siRNA biogenesis in plants. *Cell* **121**:207–221.

Alvarez, M. E. (2000). Salicylic acid in the machinery of hypersensitive cell death and disease resistance. *Plant Mol. Biol.* **44**:429–442.

Attaran, E., Zeier, T. E., Griebel, T., and Zeier, J. (2009). Methyl salicylate production and jasmonate signaling are not essential for systemic acquired resistance in Arabidopsis. *Plant Cell* **21**:954–971.

Aukerman, M. J., and Sakai, H. (2003). Regulation of flowering time and floral organ identity by a MicroRNA and its *APETALA2*-like target genes. *Plant Cell* **15**:2730–2741.

Axtell, M. J., and Staskawicz, B. J. (2003). Initiation of RPS2-specified disease resistance in Arabidopsis is coupled to the AvrRpt2-directed elimination of RIN4. *Cell* **112**:369–377.

Baulcombe, D. C. (2006). Short silencing RNA: The dark matter of genetics? *Cold Spring Harbor Symp. Quant. Biol.* **71**:13–20.

Baumberger, N., and Baulcombe, D. C. (2005). Arabidopsis ARGONAUTE1 is an RNA slicer that selectively recruits microRNAs and short interfering RNAs. *Proc. Natl. Acad. Sci. USA* **102**:11928–11933.

Baurès, I., Candresse, T., Leveau, A., Bandahmane, A., and Sturbois, B. (2008). The *Rx* gene confers resistance to a range of *Potexviruses* in transgenic *Nicotiana* plants. *Mol. Plant Microbe Interact.* **21**:1154–1164.

Bäurle, I., Smith, L., Baulcombe, D. C., and Dean, C. (2007). Widespread role for the flowering-time regulators FCA and FPA in RNA-mediated chromatin silencing. *Science* **318:**109–112.

Beauchemin, C., Boutet, N., and Laliberté, J. F. (2007). Visualization of the interaction between the precursors of VPg, the viral protein linked to the genome of *Turnip mosaic virus*, and the translation eukaryotic initiation factor iso4E in planta. *J. Virol.* **81:**775–782.

Bendahmane, A., Köhm, B. A., Dedi, C., and Baulcombe, D. C. (1995). The coat protein of PVX is a strain-specific elicitor of Rx1-mediated virus resistance in potato. *Plant J.* **8:**933–941.

Bendahmane, A., Querci, M., Kanyuka, K., and Baulcombe, D. C. (1999). The *Rx* gene from potato controls separate virus resistance and cell death responses. *Plant Cell* **11:**781–791.

Bendahmane, A., Querci, M., Kanyuka, K., and Baulcombe, D. C. (2000). *Agrobacterium* transient expression system as a tool for the isolation of disease resistance genes: Application to the *Rx2* locus in potato. *Plant J.* **21:**73–81.

Bhattacharjee, S., Zamora, A., Azhar, M. T., Sacco, M. A., Lambert, L. H., and Moffett, P. (2009). Virus resistance induced by NB–LRR proteins involves Argonaute4-dependent translational control. *Plant J.* **58:**940–951.

Bisaro, D. M. (2006). Silencing suppression by geminivirus proteins. *Virology* **344:**158–168.

Blevins, T., Rajeswaran, R., Shivaprasad, P. V., Beknazariants, D., Si-Ammour, A., Park, H.-S., Vazquez, F., Robertson, D., Meins, F., Jr., Hohn, T., and Pooggin, M. M. (2006). Four plant Dicers mediate viral small RNA biogenesis and DNA virus induced silencing. *Nucleic Acids Res.* **34:**6233–6246.

Boller, T., and Felix, G. (2009). A renaissance of elicitors: Perception of microbe-associated molecular patterns and danger signals by pattern-recognition receptors. *Annu. Rev. Plant Biol.* **60:**379–406.

Boller, T., and He, S. Y. (2009). Innate immunity in plants: An arms race between pattern recognition receptors in plants and effectors in microbial pathogens. *Science* **324:**742–744.

Borgstrom, B., and Johansen, I. E. (2001). Mutations in pea seed-borne mosaic potyvirus genome-linked protein VPg alter pathotype specific virulence in *Pisum sativum*. *Mol. Plant Microbe Interact.* **14:**707–714.

Bortolamiol, D., Pazhouhandeh, M., Marrocco, K., Genschik, P., and Ziegler-Graff, V. (2007). The Polerovirus F box protein P0 targets ARGONAUTE1 to suppress RNA silencing. *Curr. Biol.* **17:**1615–1621.

Botër, M., Amigues, B., Peart, J., Breuer, C., Kadota, Y., Casais, C., Moore, G., Kleanthous, C., Ochsenbein, F., Shirasu, K., and Guerois, R. (2007). Structural and functional analysis of SGT1 reveals that its interaction with HSP90 is required for the accumulation of Rx, an R protein involved in plant immunity. *Plant Cell* **19:**3791–3804.

Bouché, N., Lauressergues, D., Gasciolli, V., and Vaucheret, H. (2006). An antagonistic function for Arabidopsis DCL2 in development and a new function for DCL4 in generating viral siRNAs. *EMBO J.* **25:**3347–3356.

Bowles, D. J. (1990). Defense-related proteins in higher plants. *Annu. Rev. Biochem.* **59:**873–907.

Boyko, A., Kathiria, P., Zemp, F. J., Yao, Y., Pogribny, I., and Kovalchuk, I. (2007). Transgenerational changes in the genome stability and methylation in pathogen-infected plants (virus-induced plant genome instability). *Nucleic Acids Res.* **35:**1714–1725.

Brodersen, P., Petersen, M., Nielsen, H. B., Zhu, S., Newman, M.-A., Shokat, K. M., Rietz, S., Parker, J., and Mundy, J. (2006). Arabidopsis MAP kinase 4 regulates salicylic acid- and jasmonic acid/ethylene-dependent responses via EDS1 and PAD4. *Plant J.* **47:**532–546.

Brodersen, P., Sakvarelidze-Achard, L., Bruun-Rasmussen, M., Dunoyer, P., Yamamato, Y. Y., Sieburth, L., and Voinnet, O. (2008). Widespread translational inhibition by plant miRNAs and siRNAs. *Science* **320:**1185–1190.

Bruening, G. (2006). Resistance to infection. *In* "Natural Resistance Mechanisms of Plants to Viruses" (G. Loebenstein and J. P. Carr, eds.), pp. 211–240. Springer, Dordrecht.

Burch-Smith, T. M., Schiff, M., Caplan, J. L., Tsao, J., Czymmek, K., and Dinesh-Kumar, S. P. (2007). A novel role for the TIR domain in association with pathogen-derived elicitors. *PLoS Biol.* **5:**e68.

Caplan, J., and Dinesh-Kumar, S. P. (2006). Recognition and signal transduction associated with *R* gene-mediated resistance. *In* "Natural Resistance Mechanisms of Plants to Viruses" (G. Loebenstein and J. P. Carr, eds.), pp. 73–98. Springer, Dordrecht.

Caplan, J. L., Mamaillapalli, P., Burch-Smith, T., Czymmek, K., and Dinesh-Kumar, S. P. (2008). Chloroplast protein NRIP1 mediates innate immune receptor recognition of a viral effector. *Cell* **132:**449–462.

Chan, S. W.-L., Zilberman, D., Xie, Z., Johansen, L. K., Carrington, J. C., and Jacobsen, S. E. (2004). RNA silencing genes control de novo DNA methylation. *Science* **303:**1336.

Chandra-Shekara, A. C., Navarre, D., Kachroo, A., Kang, H. G., Klessig, D. F., and Kachroo, P. (2004). Signaling requirements and role of salicylic acid in *HRT*- and *rrt*-mediated resistance against turnip crinkle virus in *Arabidopsis. Plant J.* **40:**647–659.

Chapman, E. J., Prokhnevsky, A. I., Gopinath, K., Dolja, V. V., and Carrington, J. C. (2004). Viral RNA silencing suppressors inhibit the microRNA pathway at an intermediate step. *Genes Dev.* **18:**1179–1186.

Charron, C., Nicolai, M., Gallois, J.-J., Robaglia, C., Moury, B., Palloix, A., and Caranta, C. (2008). Natural variation and functional analyses provide evidence for co-evolution between plant eIF4E and potyviral VPg. *Plant J.* **54:**56–68.

Chaturvedi, R., Krothapalli, K., Makandar, R., Nandi, A., Sparks, A. A., Roth, M. R., Welti, R., and Shah, J. (2008). Plastid omega 3-fatty acid desaturase-dependent accumulation of a systemic acquired resistance inducing activity in petiole exudates of *Arabidopsis thaliana* is independent of jasmonic acid. *Plant J.* **54:**106–117.

Chellappan, P., Vanitharani, R., and Fauquet, C. M. (2005). MicroRNA-binding viral protein interferes with Arabidopsis development. *Proc. Natl. Acad. Sci. USA* **102:**10381–10386.

Chen, C., and Chen, Z. (2000). Isolation and characterization of two pathogen- and salicylic acid-induced genes encoding WRKY DNA-binding proteins from tobacco. *Plant Mol. Biol.* **42:**387–396.

Chen, J., Li, W. X., Xie, D., Peng, J. R., and Ding, S. W. (2004). Viral virulence protein suppresses RNA silencing-mediated defense but upregulates the role of microRNA in host gene expression. *Plant Cell* **16:**1302–1313.

Chen, H. Y., Yang, J., Lin, C., and Yuan, Y. A. (2008). Structural basis for RNA-silencing suppression by *Tomato aspermy virus* protein 2b. *EMBO Rep.* **9:**754–760.

Chern, M., Canlas, P. E., Fitzgerald, H. A., and Ronald, P. C. (2005). Rice NRR, a negative regulator of disease resistance, interacts with Arabidopsis NPR1 and rice NH1. *Plant J.* **43:**623–635.

Chini, A., Fonseca, S., Fernández, G., Adie, B., Chico, J. M., Lorenzo, O., García-Casado, G., Lopez-Vidriero, I., Lozano, F. M., Ponce, M. R., Micol, J. L., and Solano, R. (2007). The JAZ family of repressors is the missing link in jasmonate signalling. *Nature* **448:**666–671.

Chisholm, S. T., Mahajan, S. K., Whitham, S. A., Yamamoto, M. L., and Carrington, J. C. (2000). Cloning of the *Arabidopsis RTM1* gene, which controls restriction of long-distance movement of tobacco etch virus. *Proc. Natl. Acad. Sci. USA* **97:**489–494.

Chisholm, S. T., Parra, M. A., Anderberg, R. J., and Carrington, J. C. (2001). Arabidopsis *RTM1* and *RTM2* genes function in phloem to restrict long-distance movement of tobacco etch virus. *Plant Physiol.* **127:**1667–1675.

Chisholm, S. T., Coaker, G., Day, B., and Staskawicz, B. J. (2006). Host–microbe interactions: Shaping the evolution of the plant immune response. *Cell* **124:**803–814.

Chivasa, S., and Carr, J. P. (1998). Cyanide restores *N* gene-mediated resistance to tobacco mosaic virus in transgenic tobacco expressing salicylic acid hydroxylase. *Plant Cell* **10:**1489–1498.

Chivasa, S., Murphy, A. M., Naylor, M., and Carr, J. P. (1997). Salicylic acid interferes with tobacco mosaic virus replication via a novel salicylhydroxamic acid-sensitive mechanism. *Plant Cell* **9:**547–557.

Chivasa, S., Ndimba, B. K., Simon, W. J., Lindsey, K., and Slabas, A. R. (2005). Extracellular ATP functions as an endogenous external metabolite regulating plant cell viability. *Plant Cell* **17:**3019–3034.

Chivasa, S., Hamilton, J. H., Murphy, A. M., Lindsey, K., Carr, J. P., and Slabas, A. (2009). Extracellular ATP: A novel regulator of pathogen defence in plants. *Plant J.* **60:**436–448.

Chivasa, S., Tomé, D. F. A., Murphy, A. M., Hamilton, J. M., Lindsey, K., Carr, J. P., and Slabas, A. R. (2009). Extracellular ATP. A modulator of cell death and pathogen defense in plants. *Plant Signal. Behav.* **4:**1078–1080.

Chivasa, S., Simon, W. J., Murphy, A. M., Carr, J. P., and Slabas, A. R. (2010). The effects of extracellular ATP on the tobacco proteome. *Proteomics* **10:**235–244.

Cogoni, C., and Macino, G. (1998). Gene silencing in *Neurospora crassa* requires a protein homologous to RNA-dependent RNA polymerase. *Nature* **399:**166–169.

Cooley, M. B., Pathirana, S., Wu, H. J., Kachroo, P., and Klessig, D. F. (2000). Members of the Arabidopsis *HRT/RPP8* family of resistance genes confer resistance to both viral and oomycete pathogens. *Plant Cell* **12:**663–667.

Csorba, T., Pantaleo, V., and Burgyán, J. (2009). RNA silencing: An antiviral mechanism. *Adv. Virus Res.* **75:**35–71.

Cuellar, W. J., Kreuze, J. F., Rajamäki, M.-L., Cruzado, K. R., Untiveros, M., and Valkonen, J. P. T. (2009). Elimination of antiviral defense by viral RNase III. *Proc. Natl. Acad. Sci. USA* **106:**10354–10358.

Culver, J. N., Stubbs, G., and Dawson, W. O. (1994). Structure–function relationship between tobacco mosaic virus coat protein and hypersensitivity in *Nicotiana sylvestris. J. Mol. Biol.* **242:**130–138.

Darby, R. M., Maddison, A., Mur, L. A. J., Bi, Y. M., and Draper, J. (2000). Cell-specific expression of salicylate hydroxylase in an attempt to separate localized HR and systemic signaling establishing SAR in tobacco. *Mol. Plant Pathol.* **1:**115–123.

Dean, J. V., and Mills, J. D. (2004). Uptake of salicylic acid 2-*O*-β-D-glucose into soybean tonoplast vesicles by an ATP-binding cassette transporter-type mechanism. *Physiol. Plant.* **120:**603–612.

Deleris, A., Gallego-Bartolome, J., Bao, J., Kasschau, K. D., Carrington, J. C., and Voinnet, O. (2006). Hierarchical action and inhibition of plant DICER-like proteins in antiviral defense. *Science* **313:**68–71.

Delledonne, M., Zeier, J., Marocco, A., and Lamb, C. (2001). Signal interactions between nitric oxide and reactive oxygen intermediates in the plant hypersensitive disease resistance response. *Proc. Natl. Acad. Sci. USA* **98:**13454–13459.

Demidchik, V., Nichols, C., Oliynyk, M., Dark, A., Glover, B. J., and Davies, J. M. (2003). Is ATP a signaling agent in plants? *Plant Physiol.* **133:**456–461.

Demidchik, V., Shang, Z., Shin, R., Thompson, E., Rubio, L., Laohavisit, A., Mortimer, J. C., Chivasa, S., Slabas, A. R., Glover, B. J., Schachtman, D. P., Shabala, S. N., *et al.* (2009). Plant extracellular ATP signalling by plasma membrane NADPH oxidase and Ca^{2+} channels. *Plant J.* **58:**903–913.

Després, C., DeLong, C., Glaze, S., Liu, E., and Fobert, P. R. (2000). The Arabidopsis NPR1/NIM1 protein enhances the DNA binding activity of a subgroup of the TGA family of bZIP transcription factors. *Plant Cell* **12:**279–290.

Després, C., Chubak, C., Rochon, A., Clark, R., Bethune, T., Desveaux, D., and Fobert, P. R. (2003). The Arabidopsis NPR1 disease protein is a novel cofactor that confers redox regulation of DNA binding activity to the basic domain/leucine zipper transcription factor TGA1. *Plant Cell* **15:**2181–2191.

Devoto, A., and Turner, J. G. (2003). Regulation of jasmonate-mediated plant responses in Arabidopsis. *Ann. Bot. (Lond.)* **92:**329–337.

Devoto, A., Nieto-Rostro, M., Xie, D., Ellis, C., Harmston, R., Patrick, E., Davis, J., Sherratt, L., Coleman, M., and Turner, J. G. (2002). COI1 links jasmonate signalling and fertility to the SCF ubiquitin-ligase complex in *Arabidopsis*. *Plant J.* **32:**457–466.

Diaz-Pendon, J. A., and Ding, S. W. (2008). Direct and indirect roles of viral suppressors of RNA silencing in pathogenesis. *Annu. Rev. Phytopathol.* **46:**303–326.

Diaz-Pendon, J. A., Li, F., Li, W. X., and Ding, S. W. (2007). Suppression of antiviral silencing by cucumber mosaic virus 2b protein in Arabidopsis is associated with drastically reduced accumulation of three classes of viral small interfering RNAs. *Plant Cell* **19:**2053–2063.

Dinesh-Kumar, S. P., and Baker, B. (2000). Alternatively spliced *N* resistance gene transcripts: Their possible role in tobacco mosaic virus resistance. *Proc. Natl. Acad. Sci. USA* **97:**1908–1913.

Dunoyer, P., Lecellier, C. H., Parizotto, E. A., Himber, C., and Voinnet, O. (2004). Probing the microRNA and small interfering RNA pathways with virus-encoded suppressors of RNA silencing. *Plant Cell* **16:**1235–1250.

Durrant, W. E., and Dong, X. (2004). Systemic acquired resistance. *Annu. Rev. Phytopathol.* **42:**185–209.

Edelbaum, O., Ilan, N., Grafi, G., Sher, N., Stram, Y., Novick, D., Tan, N., Sela, I., and Rubinstein, M. (1990). Two antiviral proteins from tobacco: Purification and characterization by monoclonal antibodies to human β-interferon. *Proc. Natl. Acad. Sci. USA* **87:**588–592.

Edelbaum, O., Sher, N., Rubinstein, M., Novick, D., Tan, N., Moyer, M., Ward, E., Ryals, J., and Sela, I. (1991). Two antiviral proteins gp35 and gp22, correspond to β-1,3-glucanase and an isoform of PR-5. *Plant Mol. Biol.* **17:**171–173.

Ellis, C., Karafyllidis, I., and Turner, J. G. (2002). Constitutive activation of jasmonate signaling in an Arabidopsis mutant correlates with enhanced resistance to *Erysiphe cichoracearum*, *Pseudomonas syringae*, and *Myzus persicae*. *Mol. Plant Microbe Interact.* **15:**1025–1030.

Enyedi, A. J., Yalpani, N., Silverman, P., and Raskin, I. (1992). Localization, conjugation, and function of salicylic acid in tobacco during the hypersensitive reaction to tobacco mosaic virus. *Proc. Natl. Acad. Sci. USA* **89:**2480–2484.

Erickson, F. L., Holzberg, S., Calderon-Urrea, A., Handley, V., Axtell, M., Corr, C., and Baker, B. (1999). The helicase domain of the TMV replicase proteins induces the *N*-mediated defence response in tobacco. *Plant J.* **18:**67–75.

Falk, A., Feys, B. J., Frost, L. N., Jones, J. D., Daniels, M. J., and Parker, J. E. (1999). *EDS1*, an essential component of *R* gene-mediated disease resistance in Arabidopsis has homology to eukaryotic lipases. *Proc. Natl. Acad. Sci. USA* **96:**3292–3297.

Fan, W., and Dong, X. (2002). In vivo interaction between NPR1 and transcription factor TGA2 leads to salicylic acid-mediated gene activation in *Arabidopsis*. *Plant Cell* **14:**1377–1389.

Farmer, E. E., Alméras, E., and Krishnamurthy, V. (2003). Jasmonates and related oxylipins in plant responses to pathogenesis and herbivory. *Curr. Opin. Plant Biol.* **6:**372–378.

Farnham, G., and Baulcombe, D. C. (2006). Artificial evolution extends the spectrum of viruses that are targeted by the disease-resistance gene from potato. *Proc. Natl. Acad. Sci. USA* **103:**18828–18833.

Feys, B. J. F., Benedetti, C. E., Penfold, C. N., and Turner, J. G. (1994). Arabidopsis mutants selected for resistance to the phytotoxin coronatine are male sterile, insensitive to methyl jasmonate, and resistant to a bacterial pathogen. *Plant Cell* **6:**751–759.

Feys, B. J., Moisan, L. J., Newman, M.-A., and Parker, J. E. (2001). Direct interaction between *Arabidopsis* disease resistance signaling proteins, EDS1 and PAD4. *EMBO J.* **20:**5400–5411.

Fire, A., Xu, S. Q., Montgomery, M. K., Kostas, S. A., Driver, S. E., and Mello, C. C. (1999). Potent and specific genetic interference by double-stranded RNA in *Caenorhabditis elegans*. *Nature* **391:**806–811.

Fischer, U., and Dröge-Laser, W. (2004). Overexpression of *NtERF5*, a new member of the tobacco ethylene response transcription factor family enhances resistance to *Tobacco mosaic virus*. *Mol. Plant Microbe Interact.* **17:**1162–1171.

Fitton, A., and Goa, K. L. (1991). Azelaic acid—A review of its pharmacological properties and therapeutic efficacy in acne and hyperpigmentary skin disorders. *Drugs* **41:**780–798.

Foresi, N. P., Laxalt, A. M., Tonón, C. V., Casalonguó, C. A., and Lamattina, L. (2007). Extracellular ATP induces nitric oxide production in tomato cell suspensions. *Plant Physiol.* **145:**589–592.

Fraenkel-Conrat, H. (1983). RNA-dependent RNA polymerases of plants. *Proc. Natl. Acad. Sci. USA* **80:**422–424.

Fraenkel-Conrat, H. (1986). RNA-directed RNA polymerases of plants. *Crit. Rev. Plant Sci.* **4:**213–226.

Fraser, R. S. S., and Loughlin, S. A. R. (1980). Resistance to tobacco mosaic virus in tomato: Effects of the *Tm-1* gene on virus multiplication. *J. Gen. Virol.* **48:**87–96.

Fu, S. F. (2008). Salicylic acid induced resistance to plant viruses. Ph.D. Thesis, University of Cambridge.

Fu, L.-J., Shi, K., Gu, M., Zhou, Y.-H., Dong, D.-K., Liang, W.-S., Song, F.-M., and Yu, J.-Q. (2010). Systemic induction and role of mitochondrial alternative oxidase and nitric oxide in a compatible tomato–*Tobacco mosaic virus* interaction. *Mol. Plant Microbe Interact.* **23:**39–48.

Gaffney, T., Friedrich, L., Vernooij, B., Negrotto, D., Nye, G., Uknes, S., Ward, E., Kessmann, H., and Ryals, J. (1993). Requirement of salicylic acid for the induction of systemic acquired resistance. *Science* **261:**754–756.

Gao, Z., Johansen, E., Eyers, S., Thomas, C. L., Ellis, T. H. N., and Maule, A. J. (2004). The potyvirus recessive resistance gene, *sbm1*, identifies a novel role for translation initiation factor eIF4E in cell-to-cell trafficking. *Plant J.* **40:**376–385.

García-Arenal, F., and MacDonald, B. A. (2003). An analysis of the durability of resistance to plant viruses. *Phytopathology* **93:**941–952.

García-Arenal, F., Fraile, A., and Malpica, J. M. (2003). Variation and evolution of plant virus populations. *Int. Microbiol.* **6:**225–232.

García-Cano, E., Resende, R. O., Boiteux, L. S., Giordano, L. B., Fernandez-Munoz, R., and Moriones, E. (2008). Phenotypic expression, stability, and inheritance of a recessive resistance to monopartite begomoviruses associated with tomato yellow leaf curl disease in tomato. *Phytopathology* **98:**618–627.

Gera, A., and Loebenstein, G. (1983). Further studies of an inhibitor of virus replication from tobacco mosaic virus-infected protoplasts of a local lesion-responding tobacco cultivar. *Phytopathology* **73:**111–115.

Geraats, B. P. J., Bakker, P. A. H. M., Linthorst, H. J. M., Hoekstra, J., and van Loon, L. C. (2007). The enhanced disease susceptibility phenotype of ethylene-insensitive tobacco cannot be counteracted by inducing resistance or application of bacterial antagonists. *Physiol. Mol. Plant Pathol.* **70:**77–87.

Gilliland, A., Singh, D. P., Hayward, J. M., Moore, C. A., Murphy, A. M., York, C. J., Slator, J., and Carr, J. P. (2003). Genetic modification of alternative respiration has differential effects on antimycin A-induced versus salicylic acid-induced resistance to *Tobacco mosaic virus*. *Plant Physiol.* **132:**1518–1528.

Gómez, G., Martínez, G., and Pallás, V. (2008). Viroid-induced symptoms in *Nicotiana benthamiana* plants are dependent on RDR6 activity. *Plant Physiol.* **148:**414–423.

Gómez, G., Martínez, G., and Pallás, V. (2009). Interplay between viroid-induced pathogenesis and RNA silencing pathways. *Trends Plant Sci.* **14:**264–269.

González, I., Martínez, L., Rakinita, D. V., Lewsey, M. G., Atienzo, F. A., Llave, C., Kalinina, N. O., Carr, J. P., Palukaitis, P., and Canto, T. (2010). Cucumber mosaic virus 2b protein subcellular targets and interactions: Their significance to RNA silencing suppressor activity. *Mol. Plant Microbe Interact.* **23:**294–303.

Goodwin, J., Chapman, K., Swaney, S., Parks, T. D., Wernsman, E. A., and Dougherty, W. G. (1996). Genetic and biochemical dissection of transgenic RNA-mediated virus resistance. *Plant Cell* **8:**95–105.

Goto, K., Kobori, T., Kosaka, Y., Natsuaki, T., and Masuta, C. (2007). Characterisation of silencing suppressor 2b of *Cucumber mosaic virus* based on examination of its small RNA binding abilities. *Plant Cell Physiol.* **48:**1050–1060.

Guo, A., Salih, G., and Klessig, D. F. (2000). Activation of a diverse set of genes during the tobacco resistance response to TMV is independent of salicylic acid; induction of a subset is also ethylene independent. *Plant J.* **21:**409–418.

Hamilton, A. J., and Baulcombe, D. C. (1999). A species of small antisense RNA in posttranscriptional gene silencing in plants. *Science* **286:**950–952.

Hammerschmidt, R. (2009). Systemic acquired resistance. *Adv. Bot. Res.* **51:**173–222.

Harmer, S. L. (2009). The circadian system in higher plants. *Annu. Rev. Plant Biol.* **60:**357–377.

Harries, P. A., Palanichelvam, K., Bhat, S., and Nelson, R. S. (2008). *Tobacco mosaic virus* 126-kDa protein increases the susceptibility of *Nicotiana tabacum* to other viruses and its dosage affects virus-induced gene silencing. *Mol. Plant Microbe Interact.* **21:**1539–1548.

Heath, M. C. (2000). Hypersensitive response-related death. *Plant Mol. Biol.* **44:**321–334.

Hemmings-Mieszczak, M., Steger, G., and Hohn, T. (1997). Alternative structures of the cauliflower mosaic virus 35S RNA leader: Implications for viral expression and replication. *J. Mol. Biol.* **267:**1075–1088.

Hennig, J., Malamy, J., Grynkiewicz, G., Indulski, J., and Klessig, D. F. (1993). Interconversion of the salicylic acid signal and its glucoside in tobacco. *Plant J.* **4:**593–600.

Hong, J. K., Yun, B.-W., Kang, J.-G., Raja, M. U., Kwon, E., Sorhagen, K., Chu, C., Wang, Y., and Loake, G. J. (2008). Nitric oxide function and signalling in plant disease resistance. *J. Exp. Bot.* **59:**147–154.

Hotton, S. K., and Callis, J. (2008). Regulation of Cullin RING ligases. *Annu. Rev. Plant Biol.* **59:**467–489.

Huang, X., von Rad, U., and Durner, J. (2002). Nitric oxide induces transcriptional activation of the nitric oxide-tolerant alternative oxidase in *Arabidopsis* suspension cells. *Planta* **215:**914–923.

Huang, Z., Yeakley, J. M., Garcia, E. W., Holdridge, J. D., Fan, J. B., and Whitham, S. A. (2005). Salicylic acid-dependent expression of host genes in compatible *Arabidopsis*–virus interactions. *Plant Physiol.* **137:**1147–1159.

Huang, W. E., Huang, L., Preston, G., Naylor, M., Carr, J. P., Li, Y., Singer, A. C., Whiteley, A. S., and Wang, H. (2006). Quantitative *in situ* assay of salicylic acid in tobacco leaves using a genetically modified biosensor strain of *Acinetobacter* sp. ADP1. *Plant J.* **46:**1073–1083.

Hull, R. (2009). Comparative Plant Virology, 2nd Edn. Academic Press, New York, NY.

Irvine, R. F., and Schell, M. J. (2001). Back in the water: The return of the inositol phosphates. *Mol. Cell. Biol.* **2:**327–338.

Ishibashi, K., Masuda, K., Naito, S., Meshi, T., and Ishikawa, M. (2007). An inhibitor of viral RNA replication is encoded by a plant resistance gene. *Proc. Natl. Acad. Sci. USA* **104:**13833–13838.

Ishibashi, K., Naito, S., Meshi, T., and Ishikawa, M. (2009). An inhibitory interaction between viral and cellular proteins underlies the resistance of tomato to non-adapted tobamoviruses. *Proc. Natl. Acad. Sci. USA* **106:**8778–8783.

Itaya, A., Zhong, X., Bundschuh, R., Qi, Y., Wang, Y., Takeda, R., Harris, A. R., Molina, C., Nelson, R. S., and Ding, B. (2007). A structured viroid RNA serves as a substrate for Dicer-like cleavage to produce biologically active small RNAs but is resistant to RNA-induced silencing complex-mediated degradation. *J. Virol.* **81:**2980–2994.

Jeong, R. D., Chandra-Shekara, A. C., Kachroo, A., Klessig, D. F., and Kachroo, P. (2008). *HRT*-mediated hypersensitive response and resistance to *Turnip crinkle virus* in Arabidopsis does not require the function of TIP, the presumed guardee protein. *Mol. Plant Microbe Interact.* **21:**1316–1324.

Jeter, C. R., Tang, W., Henaff, E., Butterfield, T., and Roux, S. J. (2004). Evidence of a novel cell signaling role for extracellular adenosine triphosphates and diphosphates in *Arabidopsis. Plant Cell* **16:**2652–2664.

Ji, L. H., and Ding, S. W. (2001). The suppressor of transgene RNA silencing encoded by *Cucumber mosaic virus* interferes with salicylic acid-mediated virus resistance. *Mol. Plant Microbe Interact.* **14:**715–724.

Jin, H., Axtell, M. J., Dahlbeck, D., Ekwenna, O., Zhang, S., Staskawicz, B., and Baker, B. (2002). NPK1, an MEKK1-like mitogen-activated protein kinase kinase kinase, regulates innate immunity and development in plants. *Dev. Cell* **3:**291–297.

Jin, H., Liu, Y., Yang, K.-Y., Kim, C. Y., Baker, B., and Zhang, S. (2003). Function of a mitogen-activated protein kinase pathway in *N* gene-mediated resistance in tobacco. *Plant J.* **33:**719–731.

Jiu, M., Zhou, X. P., Tong, L., Xu, J., Yang, X., Wan, F.-H., and Liu, S.-S. (2007). Vector–virus mutualism accelerates population increase of an invasive whitefly. *PLoS ONE* **2:**e182.

Johnson, C., Boden, E., and Arias, J. (2003). Salicylic acid and NPR1 induce the recruitment of *trans*-activating TGA factors to a defense gene promoter in Arabidopsis. *Plant Cell* **15:**1846–1858.

Jones, J. D. G., and Dangl, J. L. (2001). Plant pathogens and integrated defence responses to infection. *Nature* **411:**826–833.

Jones, L., Hamilton, A. J., Voinnet, O., Thomas, C. L., Maule, A. J., and Baulcombe, D. C. (1999). RNA–DNA interactions and DNA methylation in post-transcriptional gene silencing. *Plant Cell* **11:**2291–2301.

Jung, H. W., Tschaplinski, T. J., Wang, L., Glazebrook, J., and Greenberg, J. T. (2009). Priming in systemic plant immunity. *Science* **324:**89–91.

Kachroo, P. (2006). Host gene-mediated virus resistance mechanisms and signaling in Arabidopsis. *In* "Natural Resistance Mechanisms of Plants to Viruses" (G. Loebenstein and J. P. Carr, eds.), pp. 147–164. Springer, Dordrecht.

Kachroo, P., Yoshioka, K., Shah, J., Dooner, H. K., and Klessig, D. F. (2000). Resistance to turnip crinkle virus in Arabidopsis is regulated by two host genes, is salicylic acid dependent but *NPR1*, ethylene and jasmonate independent. *Plant Cell* **12:**677–690.

Kang, B. C., Yearn, I., Frantz, J. D., Murphy, J. F., and Jahn, M. M. (2005). The *pvr1* locus in pepper encodes a translation initiation factor eIF4E that interacts with *Tobacco etch virus* VPg. *Plant J.* **41:**392–405.

Kanno, T., Mette, M. F., Kreil, D. P., Aufsatz, W., Matzke, M., and Matzke, A. J. (2004). Involvement of putative SNF2 chromatin remodeling protein DRD1 in RNA-directed DNA methylation. *Curr. Biol.* **14:**801–805.

Kanno, T., Bucher, E., Daxinger, L., Huettel, B., Böhmdorfer, G., Gregor, W., Kreil, D. P., Matzke, M., and Matzke, A. J. (2008). A structural-maintenance-of-chromosomes hinge domain-containing protein is required for RNA-directed DNA methylation. *Nat. Genet.* **40:**670–675.

Katagiri, F., Lam, E., and Chua, N.-H. (1989). Two tobacco DNA-binding proteins with homology to the nuclear factor CREB. *Nature* **340:**727–730.

Kazan, K., and Manners, J. M. (2009). Linking development to defense: Auxin in plant–pathogen interactions. *Trends Plant Sci.* **14:**373–382.

Keller, R., Brearley, C. A., Trethewey, R. N., and Müller-Röber, B. (1998). Reduced inositol content and altered morphology in transgenic potato plants inhibited for 1D-*myo*-inositol 3-phosphate synthase. *Plant J.* **16:**403–410.

Kenton, P., Mur, L. A. J., Atzorn, R., Wasternack, C., and Draper, J. (1999). Jasmonic acid accumulation in tobacco hypersensitive response lesions. *Mol. Plant Microbe Interact.* **12:**74–78.

Kieber, J. J., Rothenberg, M., Roman, G., Feldmann, K. A., and Ecker, J. R. (1993). *CTR1*, a negative regulator of the ethylene response pathway in Arabidopsis, encodes a member of the Raf family of protein kinases. *Cell* **72:**427–441.

Kim, K. J., Park, C. J., An, J. M., Ham, B. K., Lee, B. J., and Paek, K. H. (2005). CaAlaAT1 catalyzes the alanine: 2-oxoglutarate aminotransferase reaction during the resistance response against *Tobacco mosaic virus* in hot pepper. *Planta* **221:**857–867.

Kiraly, L., Cole, A. B., Bourque, J. E., and Schoelz, J. E. (1999). Systemic cell death is elicited by the interaction of a single gene in *Nicotiana clevelandii* and gene VI of cauliflower mosaic virus. *Mol. Plant Microbe Interact.* **12:**919–925.

Kobayashi, K., and Hohn, T. (2004). The avirulence domain of *Cauliflower mosaic virus* transactivator/viroplasmin is a determinant of viral virulence in susceptible hosts. *Mol. Plant Microbe Interact.* **17:**475–483.

Kovač, M., Müller, A., Milovanovič Jarh, D., Milavec, M., Düchting, P., and Ravnikar, M. (2009). Multiple hormone analysis indicates involvement of jasmonate signalling in the early defence of potato to potato virus Y^{NTN}. *Biol. Plant.* **53:**195–199.

Kovalchuk, I., Kovalchuk, O., Kalck, V., Boyko, V., Filkowski, J., Heinlein, M., and Hohn, B. (2003). Pathogen-induced systemic plant signal triggers DNA rearrangements. *Nature* **423:**760–762.

Krečič-Stres, H., Vucak, C., Ravnikar, M., and Kovač, M. (2005). Systemic *Potato virus* Y^{NTN} infection and levels of salicylic and gentisic acids in different potato genotypes. *Plant Pathol.* **54:**441–447.

Kumar, D., and Klessig, D. F. (2000). Differential induction of tobacco MAP kinases by the defense signals nitric oxide, salicylic acid, ethylene, and jasmonic acid. *Mol. Plant Microbe Interact.* **13:**347–351.

Kumar, D., Gustafsson, C., and Klessig, D. F. (2006). Validation of RNAi silencing specificity using synthetic genes: Salicylic acid-binding protein 2 is required for innate immunity in plants. *Plant J.* **45:**863–868.

Kurihara, Y., and Watanabe, Y. (2004). Arabidopsis microRNA biogenesis through Dicer-like 1 protein functions. *Proc. Natl. Acad. Sci. USA* **101:**12753–12758.

Lacomme, C., and Roby, D. (1999). Identification of new early markers of the hypersensitive response in *Arabidopsis thaliana*. *FEBS Lett.* **459:**149–153.

Lanet, E., Delannoy, E., Sormani, R., Floris, M., Brodersen, P., Crété, P., Voinnet, O., and Robaglia, C. (2009). Biochemical evidence for translational repression by *Arabidopsis* microRNAs. *Plant Cell* **21:**1762–1768.

Lecoq, H., Moury, B., Desbiez, C., Palloix, A., and Pitrat, M. (2004). Durable virus resistance in plants through conventional approaches: A challenge. *Virus Res.* **100:**31–39.

Lee, W. S. (2009). The roles of RNA-dependent RNA polymerase 1 and alternative oxidase in anti-viral defence. Ph.D. Thesis, University of Cambridge.

Lemtiri-Chlieh, F., MacRobbie, E. A. C., and Brearley, C. A. (2000). Inositol hexakisphosphate is a physiological signal regulating the K^+-inward rectifying conductance in guard cells. *Proc. Natl. Acad. Sci. USA* **97:**8687–8692.

Lemtiri-Chlieh, F., MacRobbie, E. A. C., Webb, A. A. R., Manison, N. F., Brownlee, C., Skepper, J. N., Chen, J., Prestwich, G. D., and Brearley, C. A. (2003). Inositol hexakisphosphate mobilizes an endomembrane store of calcium in guard cells. *Proc. Natl. Acad. Sci. USA* **100:**10091–10095.

Lennon, A. M., Neuenschwander, U. H., Ribas-Carbo, M., Giles, L., Ryals, J. A., and Siedow, J. N. (1997). The effects of salicylic acid and tobacco mosaic virus infection on the alternative oxidase of tobacco. *Plant Physiol.* **115:**783–791.

Lew, R. R., and Dearnaley, J. D. W. (2000). Extracellular nucleotide effects on the electrical properties of growing *Arabidopsis thaliana* root hairs. *Plant Sci.* **153:**1–6.

Lewsey, M. G., and Carr, J. P. (2009a). DICER-LIKE proteins 2, 3 and 4 are not essential for salicylic acid-mediated resistance against two RNA viruses. *J. Gen. Virol.* **90:**3010–3014.

Lewsey, M. G., and Carr, J. P. (2009b). RNA viruses: Plant pathogenic. (D.C. Baulcombe, Section ed.)*In* ''Encyclopedia of Microbiology (Article 00351)'' (M. Schaechter, ed.), 3rd Edn., pp. 443–458. Elsevier, Oxford.

Lewsey, M., Robertson, F. C., Canto, T., Palukaitis, P., and Carr, J. P. (2007). Selective targeting of miRNA-regulated plant development by a viral counter-silencing protein. *Plant J.* **50:**240–252.

Lewsey, M., Palukaitis, P., and Carr, J. P. (2009). Plant–virus interactions: Defence and counter-defence. *In* ''Molecular Aspects of Plant Disease Resistance'' (J. Parker, ed.), pp. 134–176. Wiley-Blackwell, Oxford.

Li, H. W., Lucy, A. P., Guo, H. S., Wong, S. M., and Ding, S. W. (1999). Strong host resistance targeted against a viral suppressor of the plant gene silencing defence mechanism. *EMBO J.* **18:**2683–2691.

Li, J., Brader, G., and Palva, E. T. (2004). The WRKY70 transcription factor: A node of convergence of jasmonate-mediated and salicylate-mediated signals in plant defense. *Plant Cell* **16:**319–331.

Lindeberg, M., Cunnac, S., and Collmer, A. (2009). The evolution of *Pseudomonas syringae* host specificity and Type III effector repertoires. *Mol. Plant Pathol.* **10:**767–775.

Lippman, Z., Gendrel, A.-V. V., Black, M., Vaughn, M. W., Dedhia, N., McCombie, W. R., Lavine, K., Mittal, V., May, B., Kasschau, K. D., Carrington, J. C., Doerge, R. W., *et al.* (2004). Role of transposable elements in heterochromatin and epigenetic control. *Nature* **430:**471–476.

Lister, R., O'Malley, R. C., Tonti-Filippini, J., Gregory, B. D., Berry, C. C., Millar, A. H., and Ecker, J. R. (2008). Highly integrated single-base resolution maps of the epigenome in *Arabidopsis. Cell* **133:**523–536.

Liu, Y., Schiff, M., Serion, G., Deng, X.-W., and Dinesh-Kumar, S. P. (2002). Role of SCF ubiquitin-ligase and the COP9 signalosome in the *N* gene-mediated resistance response to *Tobacco mosaic virus. Plant Cell* **14:**1483–1496.

Liu, Y., Schiff, M., Marathe, R., and Dinesh-Kumar, S. P. (2002). Tobacco *Rar1, EDS1* and *NPR1/NIM1* like genes are required for *N*-mediated resistance to tobacco mosaic virus. *Plant J.* **30:**415–429.

Liu, Y., Burch-Smith, T., Schiff, M., Feng, S., and Dinesh-Kumar, S. P. (2004). Molecular chaperone Hsp90 associates with the resistance protein N and its signaling proteins SGT1 and Rar1 to modulate an innate immune response in plants. *J. Biol. Chem.* **279:**2101–2108.

Liu, Y., Schiff, M., and Dinesh-Kumar, S. P. (2004). Involvement of MEK1 MAPKK, NTF6 MAPK, WRKY/MYB transcription factors, COI1 and CTR1 in N-mediated resistance to tobacco mosaic virus. *Plant J.* **38:**800–809.

Liu, Y., Gao, Q. Q., Wu, B., Ai, T., and Guo, X. Q. (2009). *NgRDR1,* an RNA-dependent RNA polymerase isolated from *Nicotiana glutinosa,* was involved in biotic and abiotic stress. *Plant Physiol. Biochem.* **47:**359–368.

Llave, C., Kasschau, K. D., Rector, M. A., and Carrington, J. C. (2002). Endogenous and silencing-associated small RNAs in plants. *Plant Cell* **14:**1605–1619.

Loebenstein, G. (2009). Local lesions and induced resistance. *Adv. Virus Res.* **75:**73–117.

Loebenstein, G., and Akad, F. (2006). The local lesion response. *In* ''Natural Resistance Mechanisms of Plants to Viruses'' (G. Loebenstein and J. P. Carr, eds.), pp. 99–124. Springer, Dordrecht.

Loebenstein, G., and Gera, A. (1981). Inhibitor of virus replication released from tobacco mosaic virus-infected protoplasts of a local lesion-responding tobacco cultivar. *Virology* **114:**132–139.

Lorenzo, O., and Solano, R. (2005). Molecular players regulating the jasmonate signalling network. *Curr. Opin. Plant Biol.* **8:**532–540.

Love, A. J., Yun, B. W., Laval, V., Loake, G. J., and Milner, J. J. (2005). *Cauliflower mosaic virus*, a compatible pathogen of Arabidopsis, engages three distinct defense-signaling pathways and activates rapid systemic generation of reactive oxygen species. *Plant Physiol.* **139:**935–948.

Love, A. J., Laird, J., Holt, J., Hamilton, A. J., Sadanandom, A., and Milner, J. J. (2007). Cauliflower mosaic virus protein P6 is a suppressor of RNA silencing. *J. Gen. Virol.* **88:**3439–3444.

Love, A. J., Laval, V., Geri, C., Laird, J., Tomos, A. D., Hooks, M. A., and Milner, J. J. (2007). Components of *Arabidopsis* defense- and ethylene-signaling pathways regulate susceptibility to *Cauliflower mosaic virus* by restricting long-distance movement. *Mol. Plant Microbe Interact.* **20:**659–670.

Lu, R., Malcuit, I., Moffett, P., Ruiz, M. T., Peart, J., Wu, A.-J., Rathjen, J. P., Bendahmane, A., Day, L., and Baulcombe, D. C. (2003). High throughput virus-induced gene silencing implicates heat shock protein 90 in plant disease resistance. *EMBO J.* **22:**5690–5699.

Ma, W., and Berkowitz, G. A. (2007). The grateful dead: Calcium and cell death in plant innate immunity. *Cell. Microbiol.* **9:**2571–2585.

Malamy, J., Carr, J. P., Klessig, D. F., and Raskin, I. (1990). Salicylic acid—A likely endogenous signal in the resistance response of tobacco to viral infection. *Science* **250:**1002–1004.

Malcuit, I., Marano, M. R., Kavanagh, T. A., De Jong, W., Forsyth, A., and Baulcombe, D. C. (1999). The 25-kDa movement protein of PVX elicits *Nb*-mediated hypersensitive cell death in potato. *Mol. Plant Microbe Interact.* **12:**536–543.

Maldonado, A. M., Doerner, P., Dixon, R. A., Lamb, C. J., and Cameron, R. K. (2002). A putative lipid transfer protein involved in systemic resistance signalling in *Arabidopsis*. *Nature* **419:**399–403.

Matzke, M., Kanno, T., Daxinger, L., Huettel, B., and Matzke, A. J. (2009). RNA-mediated chromatin-based silencing in plants. *Curr. Opin. Cell Biol.* **21:**367–376.

Maule, A. J., Caranta, C., and Boulton, M. I. (2007). Sources of natural resistance to plant viruses: Status and prospects. *Mol. Plant Pathol.* **8:**223–231.

Maxwell, D. P., Wang, Y., and McIntosh, L. (1999). The alternative oxidase lowers mitochondrial reactive oxygen production in plant cells. *Proc. Natl. Acad. Sci. USA* **96:**8271–8276.

Maxwell, D. P., Nickels, R., and McIntosh, L. (2002). Evidence of mitochondrial involvement in the transduction of signals required for the induction of genes associated with pathogen attack and senescence. *Plant J.* **29:**269–279.

Mayers, C. N., Lee, K. C., Moore, C. A., Wong, S. M., and Carr, J. P. (2005). Salicylic acid-induced resistance to *Cucumber mosaic virus* in squash and *Arabidopsis thaliana*: Contrasting mechanisms of induction and antiviral action. *Mol. Plant Microbe Interact.* **18:**428–434.

Meshi, T., Motoyoshi, F., Adachi, A., Watanabe, Y., Takamatsu, N., and Okada, Y. (1988). Two concomitant base substitutions in the putative replicase genes of tobacco mosaic virus confer the ability to overcome the effects of tomato resistance gene, *Tm-1*. *EMBO J.* **7:**1575–1581.

Mestre, P., and Baulcombe, D. C. (2006). Elicitor-mediated oligomerization of the tobacco N disease resistance protein. *Plant Cell* **18:**491–501.

Métraux, J. P., Signer, H., Ryals, J., Ward, E., Wyss-Benz, M., Gaudin, J., Raschdorf, K., Schmid, E., Blum, W., and Inverardi, B. (1990). Increase in salicylic acid at the onset of systemic acquired resistance in cucumber. *Science* **250:**1004–1006.

Michon, T., Estevez, Y., Walter, J., German-Retana, S., and Le Gall, O. (2006). The potyviral virus genome-linked protein VPg forms a ternary complex with the eukaryotic initiation factors eIF4E and eIF4G and reduces eIF4E affinity for mRNA cap analogue. *FEBS Lett.* **273:**1312–1322.

Moffett, P. (2009). Mechanisms of recognition in *R* gene mediated resistance. *Adv. Virus Res.* **75:**2–33.

Moffett, P., Farnham, G., Peart, J., and Baulcombe, D. C. (2002). Interaction between domains of a plant NBS–LRR protein in disease resistance-related cell death. *EMBO J.* **21:**4511–4519.

Moissiard, G., and Voinnet, O. (2006). RNA silencing of host transcripts by cauliflower mosaic virus requires coordinated action of the four Arabidopsis Dicer-like proteins. *Proc. Natl. Acad. Sci. USA* **103:**19593–19598.

Molnár, A., Csorba, T., Lakatos, L., Várallyay, E., Lacomme, C., and Burgyán, J. (2005). Plant virus-derived small interfering RNAs originate predominantly from highly structured single-stranded viral RNAs. *J. Virol.* **79:**7812–7818.

Molnár, A., Schwach, F., Studholme, D. J., Thuenemann, E. C., and Baulcombe, D. C. (2007). miRNAs control gene expression in the single-cell alga *Chlamydomonas reinhardtii. Nature* **447:**1126–1129.

Montgomery, T. A., Howell, M. D., Cuperus, J. T., Li, D., Hansen, J. E., Alexander, A. L., Chapman, E. J., Fahlgren, N., Allen, E., and Carrington, J. C. (2008). Specificity of ARGONAUTE7-miR390 interaction and dual functionality in TAS3 trans-acting siRNA formation. *Cell* **133:**128–141.

Mosher, R. A., Schwach, F., Studholme, D., and Baulcombe, D. C. (2008). PolIVb influences RNA-directed DNA methylation independently of its role in siRNA biogenesis. *Proc. Natl. Acad. Sci. USA* **105:**3145–3150.

Motoyoshi, F., and Oshima, N. (1977). Expression of genetically controlled resistance to tobacco mosaic virus infection in isolated tomato leaf mesophyll protoplasts. *J. Gen. Virol.* **34:**499–506.

Mou, Z., Fan, W., and Dong, X. (2003). Inducers of plant systemic acquired resistance regulate NPR1 function through redox changes. *Cell* **113:**935–944.

Moury, B., Morel, C., Johansen, E., Guilbaud, L., Souche, S., Ayme, V., Caranta, C., Palloix, A., and Jacquemond, M. (2004). Mutations in potato virus Y genome-linked protein determine virulence toward recessive resistances in *Capsicum annuum* and *Lycopersicon hirsutum. Mol. Plant Microbe Interact.* **17:**322–329.

Mur, L. A. J., Bi, Y. M., Darby, R. M., Firek, S., and Draper, J. (1997). Compromising early salicylic acid accumulation delays the hypersensitive response and increases viral dispersal during lesion establishment in TMV-infected tobacco. *Plant J.* **12:**1113–1126.

Mur, L. A. J., Laarhoven, L. J. J., Harren, F. J. M., Hall, M. A., and Smith, A. R. (2008). Nitric oxide interacts with salicylate to regulate biphasic ethylene production during the hypersensitive response. *Plant Physiol.* **148:**1537–1546.

Murphy, A. M., and Carr, J. P. (2002). Salicylic acid has cell-specific effects on *Tobacco mosaic virus* replication and cell-to-cell movement. *Plant Physiol.* **128:**552–563.

Murphy, A. M., Chivasa, S., Singh, D. P., and Carr, J. P. (1999). Salicylic acid induced resistance to viruses and other pathogens: A parting of the ways? *Trends Plant Sci.* **4:**155–160.

Murphy, A. M., Gilliland, A., York, C. J., Hyman, B., and Carr, J. P. (2004). High-level expression of alternative oxidase protein sequences enhances the spread of viral vectors in resistant and susceptible plants. *J. Gen. Virol.* **85:**3777–3786.

Murphy, A. M., Otto, B., Brearley, C. A., Carr, J. P., and Hanke, D. E. (2008). A role for inositol hexakisphosphate in the maintenance of basal resistance to plant pathogens. *Plant J.* **56:**638–652.

Navarro, L., Dunoyer, P., Jay, F., Arnold, B., Dharmasiri, N., Estelle, M., Voinnet, O., and Jones, J. D. G. (2006). A plant miRNA contributes to antibacterial resistance by repressing auxin signaling. *Science* **312:**436–439.

Nawrath, C., and Métraux, J. P. (1999). Salicylic acid induction-deficient mutants of Arabidopsis express PR-2 and PR-5 and accumulate high levels of camalexin after pathogen inoculation. *Plant Cell* **11:**1393–1404.

Naylor, M., Murphy, A. M., Berry, J. O., and Carr, J. P. (1998). Salicylic acid can induce resistance to plant virus movement. *Mol. Plant Microbe Interact.* **11:**860–868.

Nobuta, K., Okrent, R. A., Stoutemyer, M., Rodibaugh, N., Kempema, L., Wildermuth, M. C., and Innes, R. W. (2007). The GH3 acyl adenylase family member PBS3 regulates salicylic acid-dependent defense responses in *Arabidopsis*. *Plant Physiol.* **144:**1144–1156.

Nurmberg, P. L., Knox, K. A., Yun, B. W., Morris, P. C., Shafiei, R., Hudson, A., and Loake, G. J. (2007). The developmental selector AS1 is an evolutionarily conserved regulator of the plant immune response. *Proc. Natl. Acad. Sci. USA* **104:**18795–18800.

Oh, S.-K., Baek, K.-H., Park, J. M., Yi, S. Y., Yu, S. H., Kamoun, S., and Choi, S. (2008). *Capsicum annuum* WRKY protein CaWRKY1 is a negative regulator of pathogen defense. *New Phytol.* **177:**977–989.

Ohtsubo, N., Mitsuhara, I., Koga, M., Seo, S., and Ohashi, Y. (1999). Ethylene promotes the necrotic lesion formation and basic PR gene expression in TMV-infected tobacco. *Plant Cell Physiol.* **40:**808–817.

Palloix, A., Ayme, V., and Moury, B. (2009). Durability of plant major resistance genes to pathogens depends on the genetic background, experimental evidence and consequences for breeding strategies. *New Phytol.* **183:**190–199.

Palukaitis, P., and Carr, J. P. (2008). Plant resistance responses to viruses. *J. Plant Pathol.* **90:**153–171.

Pandey, S. P., and Baldwin, I. T. (2007). RNA-directed RNA polymerase 1 (RdR1) mediates the resistance of *Nicotiana attenuata* to herbivore attack in nature. *Plant J.* **50:**40–53.

Pandey, S. P., Shahi, P., Gase, K., and Baldwin, I. T. (2008). Herbivory-induced changes in the small-RNA transcriptome and phytohormone signaling in *Nicotiana attenuata*. *Proc. Natl. Acad. Sci. USA* **105:**4559–4564.

Park, C. J., Shin, R., Park, J. M., Lee, G. J., You, J. S., and Paek, K. H. (2002). Induction of pepper cDNA encoding a lipid transfer protein during the resistance response to tobacco mosaic virus. *Plant Mol. Biol.* **48:**243–254.

Park, C.-J., Kim, K.-J., Shin, R., Park, J. M., Shin, Y.-C., and Paek, K.-H. (2004). Pathogenesis-related protein 10 isolated from hot pepper functions as a ribonuclease in an antiviral pathway. *Plant J.* **37:**186–198.

Park, C.-J., Shin, Y.-C., Lee, B.-J., Kim, K.-J., Kim, J.-K., and Paek, K.-H. (2006). A hot pepper gene encoding WRKY transcription factor is induced during hypersensitive response to *Tobacco mosaic virus* and *Xanthomonas campestris*. *Planta* **223:**168–179.

Park, S. W., Kaimoyo, E., Kumar, D., Mosher, S., and Klessig, D. F. (2007). Methyl salicylate is a critical mobile signal for plant systemic acquired resistance. *Science* **318:**113–116.

Parker, J. E. (2009). The quest for long-distance signals in plant systemic immunity. *Sci. Signal.* **2:**pe31.

Parker, J. S., Roe, S. M., and Barford, D. (2004). Crystal structure of a PIWI protein suggests mechanisms for siRNA recognition and slicer activity. *EMBO J.* **23:**4727–4737.

Peart, J. R., Lu, R., Sadanandom, A., Malcuit, I., Moffett, P., Brice, D. C., Schauser, L., Jaggard, D. A. W., Xiao, S. Y., Coleman, M. J., Dow, M., Jones, J. D. G., *et al.* (2002). Ubiquitin ligase-associated protein SGT1 is required for host and nonhost disease resistance in plants. *Proc. Natl. Acad. Sci. USA* **99:**10865–10869.

Peart, J., Mestre, P., Lu, R., Malcuit, I., and Baulcombe, D. C. (2005). NRG1, a CC–NB–LRR protein, mediates resistance against tobacco mosaic virus. *Curr. Biol.* **15:**968–973.

Pfitzner, A. J. P. (2006). Resistance to *Tobacco mosaic virus* and *Tomato mosaic virus* in tomato. *In* "Natural Resistance Mechanisms of Plants to Viruses" (G. Loebenstein and J. P. Carr, eds.), pp. 399–413. Springer, Dordrecht.

Pieterse, C. M., Leon-Reyes, A., Van der Ent, S., and Van Wees, S. C. (2009). Networking by small-molecule hormones in plant immunity. *Nat. Chem. Biol.* **5:**308–316.

Pillai, R. S., Artus, C. G., and Filipowicz, W. (2004). Tethering of human Ago proteins to mRNA mimics the miRNA-mediated repression of protein synthesis. *RNA* **10:**1518–1525.

Pillai, R. S., Bhattacharjee, S. N., and Filipowicz, W. (2007). Repression of protein synthesis by miRNAs: How many mechanisms? *Trends Cell Biol.* **17:**118–126.

Ponz, F., and Bruening, G. (1986). Mechanisms of resistance to plant viruses. *Annu. Rev. Phytopathol.* **24:**355–381.

Ponz, F., Glascock, C. B., and Bruening, G. (1988). An inhibitor of polyprotein processing with the characteristics of a natural virus resistance factor. *Mol. Plant Microbe Interact.* **1:**25–31.

Pruss, G. J., Lawrence, C. B., Bass, T., Li, Q. Q., Bowman, L. H., and Vance, V. (2004). The potyviral suppressor of RNA silencing confers enhanced resistance to multiple pathogens. *Virology* **320:**107–120.

Qi, X., Bao, F. S., and Xie, Z. (2009). Small RNA deep sequencing reveals role for *Arabidopsis thaliana* RNA-dependent RNA polymerases in viral siRNA biogenesis. *PLoS ONE* **4:** e4971.

Qu, F., Ye, X., and Morris, T. J. (2008). Arabidopsis DRB4, AGO1, AGO7, and RDR6 participate in a DCL4-initiated antiviral RNA silencing pathway negatively regulated by DCL1. *Proc. Natl. Acad. Sci. USA* **105:**14732–14737.

Quilis, J., Peñas, G., Messeguer, J., Brugidou, C., and San Segundo, B. (2008). The Arabidopsis AtNPR1 inversely modulates defense responses against fungal, bacterial, or viral pathogens while conferring hypersensitivity to abiotic stresses in transgenic rice. *Mol. Plant Microbe Interact.* **21:**1215–1231.

Raboy, V. (2001). Seeds for a better future: 'Low phytate' grains help to overcome malnutrition and reduce pollution. *Trends Plant Sci.* **6:**458–462.

Raboy, V. (2003). *myo*-Inositol-1,2,3,4,5,6-hexakisphosphate. *Phytochemistry* **64:**1033–1043.

Rairdan, G. J., and Moffett, P. (2006). Distinct domains in the ARC region of the potato resistance protein Rx mediate LRR binding and inhibition of activation. *Plant Cell* **18:**2082–2093.

Rairdan, G. J., Collier, S. M., Sacco, M. A., Baldwin, T. T., Boettrich, T., and Moffett, P. (2008). The coiled-coil and nucleotide binding domains of the potato Rx disease resistance protein function in pathogen and signaling. *Plant Cell* **20:**739–751.

Raja, P., Sanville, B. C., Buchmann, R. C., and Bisaro, D. M. (2008). Viral genome methylation as an epigenetic defense against geminiviruses. *J. Virol.* **82:**8997–9007.

Rakhshandehroo, F., Takeshita, M., Squires, J., and Palukaitis, P. (2009). The influence of RNA-dependent RNA polymerase 1 on *Potato virus Y* infection and on other antiviral response genes. *Mol. Plant Microbe Interact.* **22:**1312–1318.

Rashid, U. J., Hoffmann, J., Brutschy, B., Piehler, J., and Chen, J. C.-H. (2008). Multiple targets for suppression of RNA interference by *Tomato aspermy virus* protein 2B. *Biochemistry* **47:**12655–12657.

Rasmussen, J. B., Hammerschmidt, R., and Zook, M. N. (1991). Systemic induction of salicylic-acid accumulation in cucumber after inoculation with *Pseudomonas syringae* pv. *syringae*. *Plant Physiol.* **97:**1342–1347.

Reichman, M., Devash, Y., Suhadolnik, R. J., and Sela, I. (1983). Human-leukocyte interferon and the antiviral factor (AVF) from virus-infected plants stimulate plant-tissues to produce nucleotides with antiviral activity. *Virology* **128:**240–244.

Reinhart, B. J., Weinstein, E. G., Rhoades, M. W., Bartel, B., and Bartel, D. P. (2002). MicroRNAs in plants. *Genes Dev.* **16:**1616–1626.

Ren, T., Qu, F., and Morris, T. J. (2000). *HRT* gene function requires interaction between a NAC protein and viral capsid protein to confer resistance to turnip crinkle virus. *Plant Cell* **12:**1917–1925.

Ren, D., Yang, H., and Zhang, S. (2002). Cell death mediated by mitogen-activated protein kinase pathway is associated with the generation of hydrogen peroxide in *Arabidopsis. J. Biol. Chem.* **277:**559–565.

Ren, T., Qu, F., and Morris, T. J. (2005). The nuclear localization of the Arabidopsis transcription factor TIP is blocked by its interaction with the coat protein of *Turnip crinkle virus. Virology* **331:**316–324.

Robaglia, C., and Caranta, C. (2006). Translation initiation factors: A weak link in plant virus infection. *Trends Plant Sci.* **11:**40–45.

Rocher, F., Chollet, J. F., Legros, S., Jousse, C., Lemoine, R., Faucher, M., Bush, D. R., and Bonnemain, J. L. (2009). Salicylic acid transport in *Ricinus communis* involves a pH-dependent carrier system in addition to diffusion. *Plant Physiol.* **150:**2081–2091.

Rojo, E., Solano, R., and Sánchez-Serrano, J. J. (2003). Interactions between signaling compounds involved in plant defense. *J. Plant Growth Regul.* **22:**82–98.

Ruiz-Ferrer, V., and Voinnet, O. (2009). Roles of plant small RNAs in biotic stress responses. *Annu. Rev. Plant Biol.* **60:**485–510.

Ryu, C.-M., Murphy, J. F., Mysore, K. S., and Kloepper, J. W. (2004). Plant growth-promoting rhizobacteria systemically protect *Arabidopsis thaliana* against *Cucumber mosaic virus* by a salicylic acid and NPR1-independent and jasmonic acid-dependent signaling pathway. *Plant J.* **39:**381–392.

Sacco, M. A., Mansoor, S., and Moffett, P. (2007). A RanGAP protein physically interacts with the NB–LRR protein Rx, and is required for Rx-mediated viral resistance. *Plant J.* **52:**82–93.

Schauer, S. E., Jacobsen, S. E., Meinke, D. W., and Ray, A. (2002). DICER-LIKE1: Blind men and elephants in *Arabidopsis* development. *Trends Plant Sci.* **7:**487–491.

Schenk, P. M., Kazan, K., Wilson, I., Anderson, J. P., Richmond, T., Somerville, S. C., and Manners, J. M. (2000). Coordinated plant defense responses in *Arabidopsis* revealed by microarray analysis. *Proc. Natl. Acad. Sci. USA* **97:**11655–11660.

Schoelz, J. E. (2006). Viral determinants of resistance versus susceptibility. *In* "Natural Resistance Mechanisms of Plants to Viruses" (G. Loebenstein and J. P. Carr, eds.), pp. 13–43. Springer, Dordrecht.

Schulze-Lefert, P. (2004). Plant immunity: The origami of receptor activation. *Curr. Biol.* **14:** R22–R24.

Sela, I. (1981). Plant virus interactions related to resistance and localization of viral infections. *Adv. Virus Res.* **26:**301–337.

Sela, I., and Appelbaum, S. W. (1962). Occurrence of antiviral factors in virus-infected plants. *Virology* **17:**543–548.

Selth, L. A., Dogra, S. C., Rasheed, M. S., Healy, H., Randles, J. W., and Rezaian, M. A. (2005). A NAC domain protein interacts with *Tomato leaf curl virus* replication accessory protein and enhances viral replication. *Plant Cell* **17:**311–325.

Seo, S., Seto, H., Yamakawa, H., and Ohashi, Y. (2001). Transient accumulation of jasmonic acid during the synchronized hypersensitive cell death in tobacco mosaic virus-infected tobacco leaves. *Mol. Plant Microbe Interact.* **14:**261–264.

Seo, S., Katou, S., Seto, H., Gomi, K., and Ohashi, Y. (2007). The mitogen-activated protein kinases WIPK and SIPK regulate the levels of jasmonic and salicylic acids in wounded tobacco plants. *Plant J.* **49:**899–909.

Serrano, C., Gonzalez-Cruz, J., Jauregui, F., Medina, C., Mancilla, P., Matus, J. T., and Arce-Johnson, P. (2008). Genetic and histological studies on the delayed systemic movement of tobacco mosaic virus in *Arabidopsis thaliana. BMC Genet.* **9:**59.

Shams-Bakhsh, M., Canto, T., and Palukaitis, P. (2007). Enhanced resistance and neutralization of defense responses by suppressors of RNA silencing. *Virus Res.* **130:**103–109.

Shao, F., Golstein, C., Ade, J., Stoutemyer, M., Dixon, J. E., and Innes, R. W. (2003). Cleavage of Arabidopsis PBS1 by a bacterial type III effector. *Science* **301:**1230–1233.

Shin, R., Park, C.-J., An, J.-M., and Paek, K.-H. (2003). A novel TMV-induced hot pepper cell wall protein gene (*CaTin2*) is associated with virus-specific hypersensitive response pathway. *Plant Mol. Biol.* **51:**687–701.

Shulaev, V., Silverman, P., and Raskin, I. (1997). Airborne signalling by methyl salicylate in plant pathogen resistance. *Nature* **385:**718–721.

Siegien, I., and Bogatek, R. (2006). Cyanide action in plants—From toxicity to regulatory. *Acta Physiol. Plant.* **28:**483–497.

Simons, B. H., Millenaar, F. F., Mulder, L., van Loon, L. C., and Lambers, H. (1999). Enhanced expression and activation of the alternative oxidase during infection of *Arabidopsis* with *Pseudomonas syringae* pv. *tomato. Plant Physiol.* **120:**529–538.

Singh, D. P., Moore, C. A., Gilliland, A., and Carr, J. P. (2004). Activation of multiple antiviral defence mechanisms by salicylic acid. *Mol. Plant Pathol.* **5:**57–63.

Song, F., and Goodman, R. M. (2001). Activity of nitric oxide is dependent on, but is partially required for function of, salicylic acid in the signaling pathway in tobacco systemic acquired resistance. *Molec. Plant Microbe Interact.* **14:**1458–1462.

Spiegel, S., Gera, A., Salomon, R., Ahl, P., Harlap, S., and Loebenstein, G. (1989). Recovery of an inhibitor of virus replication from the intercellular fluid of hypersensitive tobacco infected with tobacco mosaic virus and from uninfected induced-resistant tissue. *Phytopathology* **79:**258–262.

Spoel, S. H., Mou, Z., Tada, Y., Spivey, N. W., Genschik, P., and Dong, X. (2009). Proteasome-mediated turnover of the transcription coactivator NPR1 plays dual roles in regulating plant immunity. *Cell* **137:**860–872.

Staswick, P. E., and Tiryaki, I. (2004). The oxylipin signal jasmonic acid is activated by an enzyme that conjugates it to isoleucine in Arabidopsis. *Plant Cell* **16:**2117–2127.

Strawn, M. A., Marr, S. K., Inoue, K., Inada, N., Zubieta, C., and Wildermuth, M. C. (2007). *Arabidopsis* isochorismate synthase functional in pathogen-induced salicylate biosynthesis exhibits properties consistent with a role in diverse stress responses. *J. Biol. Chem.* **282:**5919–5933.

Takabatake, R., Ando, Y., Seo, S., Katou, S., Tsuda, S., Ohashi, Y., and Mitsuhara, I. (2007). MAP kinases function downstream of HSP90 and upstream of mitochondria in TMV resistance gene *N*-mediated hypersensitive cell death. *Plant Cell Physiol.* **48:**498–510.

Takahashi, H., Miller, J., Nozaki, Y., Sukamto, Y., Takeda, M., Shah, J., Hase, S., Ikegami, M., Ehara, Y., and Dinesh-Kumar, S. P. (2002). *RCY1*, an *Arabidopsis thaliana RPP8/HRT* family resistance gene, conferring resistance to cucumber mosaic virus requires salicylic acid, ethylene and a novel signal transduction mechanism. *Plant J.* **32:**655–667.

Takahashi, A. A., Casais, C., Ichimura, K., and Shirasu, K. (2003). HSP90 interacts with RAR1 and SGT1 and is essential for RPS2-mediated disease resistance in *Arabidopsis*. *Proc. Natl. Acad. Sci. USA* **100:**11777–11782.

Takahashi, H., Kanayama, Y., Zheng, M. S., Kusano, T., Hase, S., Ikegami, M., and Shah, J. (2004). Antagonistic interactions between the SA and JA signaling pathways in Arabidopsis modulate expression of defense genes and gene-for-gene resistance to cucumber mosaic virus. *Plant Cell Physiol.* **45:**803–809.

Tameling, W. I. L., and Baulcombe, D. C. (2007). Physical association of the NB–LRR resistance protein Rx with a Ran GTPase-activating protein is required for extreme resistance to *Potato virus X*. *Plant Cell* **19:**1682–1694.

Thines, B., Katsir, L., Melotto, M., Niu, Y., Mandaokar, A., Liu, G. H., Nomura, K., He, S. Y., Howe, G. A., and Browse, J. (2007). JAZ repressor proteins are targets of the SCFCO11 complex during jasmonate signaling. *Nature* **448:**661–665.

Ton, J., van Pelt, J. A., van Loon, L. C., and Pieterse, C. M. J. (2002). Differential effectiveness of salicylate-dependent and jasmonate/ethylene-dependent induced resistance in *Arabidopsis*. *Mol. Plant Microbe Interact.* **15:**27–34.

Torres, M. A., and Dangl, J. L. (2005). Functions of the respiratory burst oxidase in biotic interactions, abiotic stress and development. *Curr. Opin. Plant Biol.* **8:**397–403.

Truman, W., Bennettt, M. H., Kubigsteltig, I., Turnbull, C., and Grant, M. (2007). *Arabidopsis* systemic immunity uses conserved defense signaling pathways and is mediated by jasmonates. *Proc. Natl. Acad. Sci. USA* **104:**1075–1080.

Truniger, V., and Aranda, M. A. (2009). Recessive resistance to plant viruses. *Adv. Virus Res.* **75:**119–159.

Ülker, B., and Somssich, I. E. (2004). WRKY transcription factors: From DNA binding towards biological function. *Curr. Opin. Plant Biol.* **7:**491–498.

van der Hoorn, R. A. L., and Kamoun, S. (2008). From Guard to Decoy: A new model for perception of plant pathogen effectors. *Plant Cell* **20:**2009–2017.

Vanlerberghe, G. C., and McIntosh, L. (1997). Alternative oxidase: From gene to function. *Annu. Rev. Plant Physiol.* **48:**703–734.

van Loon, L. C., and van Strien, E. A. (1999). The families of pathogenesis-related proteins, their activities, and comparative analysis of PR-1 type proteins. *Physiol. Mol. Plant Pathol.* **55:**85–97.

van Verk, M. C., Pappaioannou, D., Neeleman, L., Bol, J. F., and Linthorst, J. M. (2008). A novel WRKY transcription factor is required for induction of *PR-1a* gene expression by salicylic acid and bacterial elicitors. *Plant Physiol.* **146:**1983–1995.

Vaucheret, H. (2005). MicroRNA-dependent *trans*-acting siRNA production. *Sci. STKE* **2005:** pe43.

Vernooij, B., Friedrich, L., Morse, A., Reist, R., Kolditzjawhar, R., Ward, E., Uknes, S., Kessmann, H., and Ryals, J. (1994). Salicylic acid is not the translocated signal responsible for inducing systemic acquired-resistance but is required in signal transduction. *Plant Cell* **6:**959–965.

Vlot, A. C., Liu, P. P., Cameron, R. K., Park, S. W., Yang, Y., Kumar, D., Zhou, F. S., Padukkavidana, T., Gustafsson, C., Pichersky, E., and Klessig, D. F. (2008). Identification of likely orthologs of tobacco salicylic acid-binding protein 2 and their role in systemic acquired resistance in *Arabidopsis thaliana*. *Plant J.* **56:**445–456.

Vlot, A. C., Klessig, D. F., and Park, S. W. (2008). Systemic acquired resistance: The elusive signal(s). *Curr. Opin. Plant Biol.* **11:**436–442.

Waller, F., Müller, A., Chung, K.-M., Yap, Y.-K., Nakamura, K., Weiler, E., and Sano, H. (2006). Expression of a WIPK-activated transcription factor results in increase of endogenous salicylic acid and pathogen resistance in tobacco plants. *Plant Cell Physiol.* **47:**1169–1174.

Wang, D., Pajerowska-Mukhtar, K., Culler, A. H., and Dong, X. N. (2007). Salicylic acid inhibits pathogen growth in plants through repression of the auxin signaling pathway. *Curr. Biol.* **17:**1784–1790.

Wang, J., Wang, X., Liu, C., Zhang, J. D., Zhu, C. X., and Gu, X. Q. (2008). The *NgAOX1a* gene cloned from *Nicotiana glutinosa* is implicated in the response to abiotic and biotic stresses. *Biosci. Rep.* **28:**259–266.

Wang, X., Goregaoker, S. P., and Culver, J. N. (2009). Interaction of the tobacco mosaic virus replicase protein with a NAC domain transcription factor is associated with the suppression of systemic host defenses. *J. Virol.* **83:**9720–9730.

Wassenegger, M., and Krczal, G. (2006). Nomenclature and functions of RNA-directed RNA polymerases. *Trends Plant Sci.* **11:**142–151.

Watanabe, Y., Kishibayashi, N., Motoyoshi, F., and Okada, Y. (1987). Characterization of *Tm-1* gene action on replication of common isolates and a resistant-breaking isolate of TMV. *Virology* **161:**527–532.

Wei, N., and Deng, X.-W. (1999). Making sense of the COP9 signalosome, a conserved regulatory protein complex from *Arabidopsis* to human. *Trends Genet.* **15:**98–103.

Wei, N., and Ding, X.-W. (1992). *COP9*: A new genetic locus involved in light-regulated development and gene expression in Arabidopsis. *Plant Cell* **4:**1507–1518.

Weigel, R. R., Bauscher, C., Pfitzner, A. J. P., and Pfitzner, U. M. (2001). NIMIN-1, NIMIN-2 and NIMIN-3, members if a novel family of proteins from Arabidopsis that interact with NPR1/NIM1, a key regulator of systemic acquired resistance in plants. *Plant Mol. Biol.* **46:**143–160.

Wendehenne, D., Durner, J., and Klessig, D. F. (2004). Nitric oxide: A new player in plant signalling and defense responses. *Curr. Opin. Plant Biol.* **7:**449–455.

Whitham, S., Dinesh-Kumar, S. P., Choi, D., Hehl, R., Corr, C., and Baker, B. (1994). The product of the tobacco mosaic virus resistance gene *N*: Similarity to Toll and the interleukin-1 receptor. *Cell* **78:**1101–1115.

Whitham, S. A., Anderberg, R. J., Chisholm, S. T., and Carrington, J. C. (2000). Arabidopsis *RTM2* gene is necessary for specific restriction of tobacco etch virus and encodes an unusual small heat shock-like protein. *Plant Cell* **12:**569–582.

Whitham, S. A., Quan, S., Chang, H. S., Cooper, B., Estes, B., Zhu, T., Wang, X., and Hou, Y. M. (2003). Diverse RNA viruses elicit the expression of common sets of genes in susceptible *Arabidopsis thaliana* plants. *Plant J.* **33:**271–283.

Wildermuth, M. C., Dewdney, J., Wu, G., and Ausubel, F. M. (2001). Isochorismate synthase is required to synthesize salicylic acid for plant defence. *Nature* **414:**562–565.

Wong, C. E., Carson, R. A., and Carr, J. P. (2002). Chemically induced virus resistance in *Arabidopsis thaliana* is independent of pathogenesis-related protein expression and the *NPR1* gene. *Mol. Plant Microbe Interact.* **15:**75–81.

Wu, L., and Belasco, J. G. (2008). Let me count the ways: Mechanisms of gene regulation by miRNAs and siRNAs. *Mol. Cell* **29:**1–7.

Wu, S.-J., and Wu, J.-Y. (2008). Extracellular ATP-induced NO production and its dependence on membrane Ca^{2+} flux in *Salvia miltiorrhiza* hairy roots. *J. Exp. Bot.* **59:**4007–4016.

Xie, D.-X., Feys, B. F., James, S., Nieto-Rostro, M., and Turner, J. G. (1998). *COI1*: An *Arabidopsis* gene required for jasmonate-regulated defense and fertility. *Science* **280:**1091–1094.

Xie, Q., Sanz-Burgos, A. P., Guo, H., Gabcia, J. A., and Gutiérrez, C. (1999). GRAB proteins, novel members of the NAC domain family, isolated by their interaction with a geminivirus protein. *Plant Mol. Biol.* **39:**647–656.

Xie, Z., Fan, B., Chen, C., and Chen, Z. (2001). An important role of an inducible RNA-dependent RNA polymerase in plant antiviral defense. *Proc. Natl. Acad. Sci. USA* **98:**6516–6521.

Xu, L., Liu, F., Lechner, E., Genschik, P., Crosby, W. L., Ma, H., Peng, W., Huang, D., and Xie, D. (2002). The SCF^{COI1} ubiquitin-ligase complexes are required for jasmonate response in Arabidopsis. *Plant Cell* **14:**1919–1935.

Yang, Y. O., and Klessig, D. F. (1996). Isolation and characterization of a tobacco mosaic virus-inducible Myb oncogene homolog from tobacco. *Proc. Natl. Acad. Sci. USA* **93:**14972–14977.

Yang, P., Chen, C., Wang, Z., Fan, B., and Chen, Z. (1999). A pathogen- and salicylic acid-induced WRKY DNA-binding activity recognizes the elicitor response element of the tobacco class I chitinase gene promoter. *Plant J.* **18:**141–149.

Yang, K.-Y., Liu, Y., and Zhang, S. (2001). Activation of a mitogen-activated protein kinase pathway is involved in disease resistance in tobacco. *Proc. Natl. Acad. Sci. USA* **98:**741–746.

Yang, S. J., Carter, S. A., Cole, A. B., Cheng, N. H., and Nelson, R. S. (2004). A natural variant of a host RNA-dependent RNA polymerase is associated with increased susceptibility to viruses by *Nicotiana benthamiana*. *Proc. Natl. Acad. Sci. USA* **101:**6297–6302.

Yang, J. Y., Iwasaki, M., Machida, C., Machida, Y., Zhou, X., and Chua, N. H. (2008). βC1, the pathogenicity factor of TYLCCNV, interacts with AS1 to alter leaf development and suppress selective jasmonic acid responses. *Genes Dev.* **22:**2564–2577.

Yap, Y.-K., Kodama, Y., Waller, F., Chung, K. M., Ueda, H., Nakamura, K., Oldsen, M., Yoda, H., Yamaguchi, Y., and Sano, H. (2005). Activation of a novel transcription factor through phosphorylation by WIPK, a wound-induced mitogen-activated protein kinase in tobacco plants. *Plant Physiol.* **139:**127–137.

Yeam, I., Cavatorta, J. R., Ripoll, D. R., Kang, B. C., and Jahn, M. M. (2007). Functional dissection of naturally occurring amino acid substitutions in eIF4E that confers recessive potyvirus resistance in plants. *Plant Cell* **19:**2913–2928.

Yip, J. Y. H., and Vanlerberghe, G. C. (2001). Mitochondrial alternative oxidase acts to dampen the generation of active oxygen species during a period of rapid respiration induced to support a high rate of nutrient uptake. *Physiol. Plant.* **112:**327–333.

Yoda, H., Ogawa, M., Yamaguchi, Y., Koizumi, N., Kusano, T., and Sano, N. (2002). Identification of early-responsive genes associated with the hypersensitive response to tobacco mosaic virus and characterization of a WRKY-type transcription factor in tobacco plants. *Mol. Genet. Genomics* **267:**154–161.

Yoshii, M., Nishikiori, M., Tomita, K., Yoshioka, N., Kozuka, R., Naito, S., and Ishikawa, M. (2004). The Arabidopsis cucumovirus multiplication 1 and 2 loci encode translation initiation factors 4E and 4G. *J. Virol.* **78:**6102–6111.

Yoshimoto, K., Jikumaru, Y., Kamiya, Y., Kusano, M., Consonni, C., Panstruga, R., Ohsumi, Y., and Shirasu, K. (2009). Autophagy negatively regulates cell death by controlling NPR1-dependent salicylic acid signaling during senescence and the innate immune response in *Arabidopsis*. *Plant Cell* **21:**2914–2927.

Yoshinari, S., and Hemenway, C. (2003). Potato virus X: A model system for virus replication, movement and gene expression. *Mol. Plant Pathol.* **4:**125–131.

Yu, D., Fan, B., MacFarlane, S. A., and Chen, Z. (2003). Analysis of the involvement of an inducible *Arabidopsis* RNA-dependent RNA polymerase in antiviral defense. *Mol. Plant Microbe Interact.* **16:**206–216.

Zarate, S. I., Kempema, L. A., and Walling, L. L. (2007). Silverleaf whitefly induces salicylic acid defenses and suppresses effectual jasmonic acid defenses. *Plant Physiol.* **143:**866–875.

Zhang, S., and Klessig, D. F. (1998). Resistance gene *N*-mediated *de novo* synthesis and activation of a tobacco mitogen-activated protein kinase by tobacco mosaic virus infection. *Proc. Natl. Acad. Sci. USA* **95:**7433–7438.

Zhang, Y., Fan, W., Kinkema, M., Li, X., and Dong, X. (1999). Interaction of NPR1 with basic leucine zipper protein transcription factors that bind sequences required for salicylic acid induction of the *PR-1* gene. *Proc. Natl. Acad. Sci. USA* **96:**6523–6528.

Zhang, Y., Tessaro, M. J., Lassner, M., and Li, X. (2003). Knockout analysis of Arabidopsis transcription factors *TGA2*, *TGA5*, and *TGA6* reveals their redundant and essential roles in systemic acquired resistance. *Plant Cell* **15:**2647–2653.

Zhang, X., Yuan, Y. R., Pei, Y., Lin, S. S., Tuschl, T., Patel, D. J., and Chua, N. H. (2006). Cucumber mosaic virus-encoded 2b suppressor inhibits *Arabidopsis* Argonaute1 cleavage activity to counter plant defense. *Genes Dev.* **20:**3255–3268.

Zhang, Y., Cheng, Y. T., Zhao, Q., Bi, D., and Li, X. (2006). Negative regulation of defense responses in Arabidopsis by two *NPR1* paralogs. *Plant J.* **48:**647–656.

Zheng, X., Zhu, J., Kapoor, A., and Zhu, J. K. (2007). Role of *Arabidopsis* AGO6 in siRNA accumulation, DNA methylation and transcriptional gene silencing. *EMBO J.* **26:**1691–1701.

Zhou, J.-M., Trifa, Y., Silva, H., Pontier, D., Lam, E., Shah, J., and Klessig, D. F. (2000). NPR1 differentially interacts with members of the TGA/OBF family of transcription factors that bind an element of the *PR-1* gene required for induction by salicylic acid. *Mol. Plant Microbe Interact.* **13:**191–202.

CHAPTER 4

Global Genomics and Proteomics Approaches to Identify Host Factors as Targets to Induce Resistance Against *Tomato Bushy Stunt Virus*

Peter D. Nagy and **Judit Pogany**

Contents

Department of Plant Pathology, University of Kentucky, Lexington, Kentucky 40546, USA,
E-mail: pdnagy2@uky.edu

Advances in Virus Research, Volume 76
ISSN 0065-3527, DOI: 10.1016/S0065-3527(10)76004-8

Abstract

The success of RNA viruses as pathogens of plants, animals, and humans depends on their ability to reprogram the host cell metabolism to support the viral infection cycle and to suppress host defense mechanisms. Plus-strand (+)RNA viruses have limited coding potential necessitating that they co-opt an unknown number of host factors to facilitate their replication in host cells. Global genomics and proteomics approaches performed with *Tomato bushy stunt virus* (TBSV) and yeast (*Saccharomyces cerevisiae*) as a model host have led to the identification of 250 host factors affecting TBSV RNA replication and recombination or bound to the viral replicase, replication proteins, or the viral RNA. The roles of a dozen host factors involved in various steps of the replication process have been validated in yeast as well as a plant host. Altogether, the large number of host factors identified and the great variety of cellular functions performed by these factors indicate the existence of a truly complex interaction between TBSV and the host cell. This review summarizes the advantages of using a simple plant virus and yeast as a model host to advance our understanding of virus–host interactions at the molecular and cellular levels. The knowledge of host factors gained can potentially be used to inhibit virus replication via gene silencing, expression of dominant negative mutants, or design of specific chemical inhibitors leading to novel specific or broad-range resistance and antiviral tools against (+)RNA plant viruses.

I. INTRODUCTION

The success of plus-strand (+)RNA viruses as pathogens of plants, animals, and humans depends on the ability of these viruses to reprogram the host cell metabolism to support the infection process and to avoid/suppress host defense mechanisms. (+)RNA viruses have limited coding potential with usually 4–10 genes, yet they can replicate efficiently in the infected host cells. They accomplish this feat by recruiting an unknown number of host factors, such as host proteins, membranes, and ribonucleotides for their replication that can produce thousands to millions of progeny viral RNAs per cell in 24 h. The virus can trick the recruited host factors to perform novel functions that are frequently targeted against the host cells. (+)RNA viruses can also induce strong responses of the infected host cells leading to the activation of the innate immune responses. Altogether, the (+)RNA virus infected cells go through major changes during the infection process. Many of the original cellular processes/pathways are getting reprogrammed by the infecting virus and these changes make the cells dramatically different from the uninfected cells. The outcome of the virus infection often resembles to the chaotic situation and destruction caused by having two opposing armies fighting a well-planned out, but expensive war for gaining full control over the same country.

In spite of the significance of virus–host interaction for human, animal, and plant health, our current understanding of the host factors involved in (+)RNA virus infections is still incomplete. Therefore, one of the major frontiers of ongoing research is to identify all changes in the infected cells that can potentially result in better, more efficient antiviral strategies and/or reduce the damage in the host cells caused by viral infections.

A. (+)RNA virus replication is a multistep process in the infected cells

After entry of virus into the cell, translation of the viral RNA leads to production of the viral replication proteins. These proteins then facilitate the rescue of the viral RNA from translation and the viral RNA is selected/recruited for replication. This is followed by the assembly of the viral replicase on subcellular membrane surfaces. The assembled replicase complex produces complementary minus-strand (−)RNA using the original (+)RNA as a template. Then, the minus-stranded RNA intermediate is used by the viral replicase to synthesize excess amount of new (+)RNA progeny, which is released from the site of replication to the cytosol and/or become encapsidated to form new viral particles (Ahlquist *et al.*, 2003; Panavas *et al.*, 2005a). One of the amazing things about (+)RNA viruses is that they complete an infection cycle within 6–24 h in the primary infected cells. Thus, (+)RNA virus infection is often a fast race between the parasite and the host to gain control over the resources of the host cell.

B. Selection of the viral RNA template for replication and the recruitment of the replication proteins to the subcellular sites of replication

Viral (+)RNA replication takes place on the cytosolic surfaces of intracellular membranes, where the viral RNA and replication proteins, together with co-opted host factors, are sequestered and reach high local concentrations that facilitate robust replication. Many RNA viruses are known to form spherules (small membrane invaginations) with small openings towards to the cytosol (Fig. 1) (Kopek *et al.*, 2007; Schwartz *et al.*, 2002). These spherules are the sites of RNA replication and frequently connected with the sites of virus assembly (Ahlquist, 2006). But how are these spherules formed and what is the role of the co-opted host factors? Although we do not yet know the answers to these fascinating questions, it is likely that efficient recruitment of all the viral and host factors are critical for replication.

The selection of the viral (+)RNA for replication is thought to be mediated by a viral-coded protein binding selectively to a specific sequence/structure in the viral (+)RNA. For example, the *Tomato bushy stunt virus* (TBSV) p33 auxiliary replication protein recognizes a C·C mismatch within an extended hairpin in the viral RNA (Monkewich *et al.*, 2005; Pogany *et al.*, 2005). The *Brome mosaic virus* (BMV) 1a helicase-like protein binds to the subgenomic promoter region in RNA3 to facilitate the selection of the RNA for replication (Wang *et al.*, 2005). The flock house virus (FHV) protein A replication protein specifically recognizes the 5′ sequence in the RNA facilitating its recruitment to the

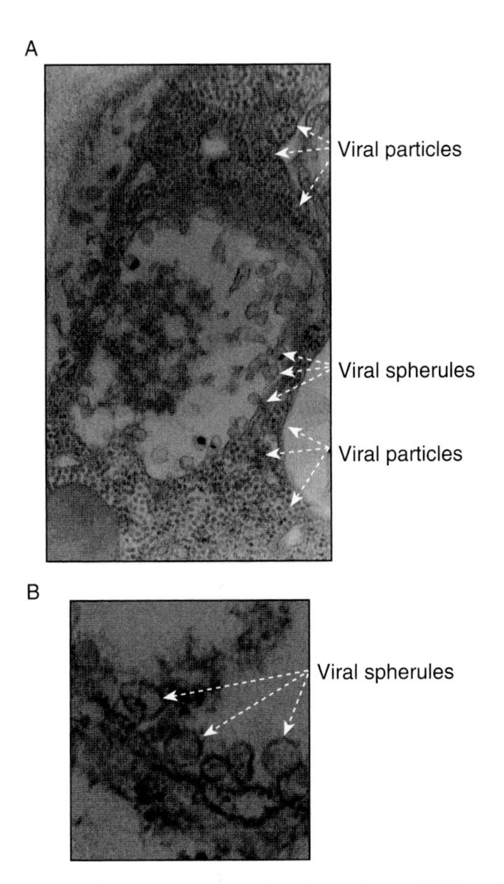

FIGURE 1 Representative electron micrographs of portions of *N. benthamiana* cells infected with the tombusvirus *Cucumber necrosis virus* (CNV). (A) The CNV-induced spherules in the center of the image and the assembled large number of virions in plant cells are depicted with arrows. Note that the entire cytosol of the portion of the cell shown is completely filled by CNV virions, demonstrating robust CNV replication. Magnification is 49,000×. (B) Several characteristic CNV-induced spherules are marked with arrowheads on the EM images. These 50–80 nm spherules are formed via membrane invagination into peroxisomal or ER-derived membranes. Narrow openings (necks) are visible likely connecting the spherules to the cytosol. Control samples lacking CNV do not show similar structures (not shown). Magnification is 98,000×. The images were taken by Dr. Barajas.

mitochondrial outer membrane (Van Wynsberghe and Ahlquist, 2009). In spite of these advances in our understanding of the template selection process, we do not yet know how the translating ribosomes are removed from the (+)RNA templates selected for replication. Models have been proposed that viral replication proteins or host factors, such as Lsm proteins, promote the switch of the RNA from translation to replication (Beckham *et al.*, 2007).

Another important step is the recruitment of the viral (+)RNA and the viral replication proteins together with co-opted host proteins to the sites of replication. The viral (+)RNA likely travels together with the viral proteins as an RNP complex. The targeting of key viral proteins in the cells is guided by signal sequences present in the replication proteins. Other viral proteins, such as the viral RdRp protein, and co-opted host factors might be "piggy-backing" on the targeted viral protein to reach the destination of the subcellular membranes.

C. The assembly of the replicase complexes of (+)RNA viruses is a complex process

A key step in viral replication is the assembly of the viral replicase on the cytosolic surfaces of intracellular membranes, which is a poorly understood process (Salonen *et al.*, 2005). The viral replicase consists of viral-coded RNA-dependent RNA polymerase (RdRp), viral auxiliary replication proteins, the subcellular membrane, and co-opted host proteins. Among these factors, the contribution of host factors to the viral replication process is the least understood. The host factors likely complement the functions of the viral replication proteins to regulate RNA replication. Moreover, the host components in the replicase complex likely provide protection from host cellular ribonucleases, including the powerful gene silencing machinery, as well as they might be involved in delaying the recognition of viral components by the host surveillance system. The host factors are also likely responsible for activation of RdRps of several (+)RNA viruses, such as p92^{pol} of TBSV, 2a^{pol} of BMV, P2 of *Alfalfa mosaic virus* (AMV), 180K of *Tomato mosaic virus* (ToMV), and the hepatitis C virus (HCV) NS5B. These viral RdRps become activated only after the assembly of the viral replicase in membranous spherules or vesicles (Panaviene *et al.*, 2004, 2005; Quadt *et al.*, 1995; Vlot *et al.*, 2001). Although the role of host factors during the assembly of the replicase and the activation of RdRp is currently unknown, they likely work together with the viral RNA as well, which also plays a key role in these processes.

D. RNA synthesis by the viral replicase is a two-step process

(+)RNA virus replication occurs within membrane-bound structures and is performed by the viral replicase complex using the recruited (+)RNA as the template (Ahlquist *et al.*, 2003). The viral RNA contains specific *cis*-acting elements, including promoters, silencers, and enhancers, which regulate *de novo* initiation of RNA synthesis and the efficiency of replication (Dreher, 1999; Kao *et al.*, 2001; Nagy and Pogany, 2006; White and

Nagy, 2004; Wu *et al.*, 2009). Interestingly, the (−)RNA produced on the (+)RNA template is the more efficient template, resulting in 20–100-fold more (+)RNA progeny than the amount of (−)RNA intermediate. It is currently thought that the (−)RNA is always present in the replicase complex and never gets released from the site of replication, whereas the majority of (+)RNA progeny is released to the cytosol or assembled into viral particles (Panavas *et al.*, 2005a). Host factors are certainly involved in these processes, although detailed mechanistic studies on their roles are not yet available.

E. Tombusviruses are simple model (+)RNA viruses of plants

TBSV and other tombusviruses are model plant RNA viruses with a single 4.8-kb genomic (g)RNA component. The gRNA codes for two replication proteins, termed p33 and $p92^{pol}$, and produces two subgenomic RNAs for the expression of three viral proteins involved in cell-to-cell movement, encapsidation, and suppression of gene silencing (Nagy and Pogany, 2008; White and Nagy, 2004). Recent advances with tombusviruses have been accelerated by the development of yeast (*Saccharomyces cerevisiae*) as a model host to study TBSV replication and recombination (Nagy, 2008; Panavas and Nagy, 2003). Yeast expressing p33 and $p92^{pol}$ replication proteins can efficiently replicate a short TBSV-derived replicon (rep)RNA (Panavas and Nagy, 2003; Panaviene *et al.*, 2004). Importantly, replication of the TBSV repRNA in yeast depends on the same *cis*-acting RNA elements and *trans*-acting p33/p92 replication proteins as in plants (Nagy, 2008; Panavas and Nagy, 2003). The DI-72 repRNA, which is derived naturally from TBSV infections of plants, does not encode proteins and it can replicate efficiently in yeast cells without maintaining an artificial selection pressure. Moreover, the tombusvirus repRNA plays several functions, including serving as a template for replication and as a platform for the assembly of the viral replicase complex (Nagy and Pogany, 2008; Panaviene *et al.*, 2005; Pogany *et al.*, 2005). The viral RNA also participates in RNA recombination (Serviene *et al.*, 2005; White and Morris, 1994; White and Nagy, 2004), which likely plays a major role in virus evolution. Altogether, the development of yeast as a model host for TBSV facilitated the application of the available genomics and proteomics tools to identify host components required or affecting TBSV replication and virus–host interactions (Jiang *et al.*, 2006; Li *et al.*, 2008, 2009; Nagy, 2008; Nagy and Pogany, 2006; Panavas *et al.*, 2005b; Serva and Nagy, 2006; Serviene *et al.*, 2005, 2006). The results generated with these genomics and proteomics tools will be described below.

II. GENOME-WIDE SCREENS FOR SYSTEMATIC IDENTIFICATION OF HOST FACTORS AFFECTING TBSV REPLICATION

An advantage to perform genome-wide screens in an organism is the availability of a collection of knockout mutants, such as the *y*east single-gene-*k*nock*o*ut (YKO) library, or a library containing gene sets with regulatable expression, such as yTHC library for the essential genes in yeast (Fig. 2). RNAi-based genome-wide screens can also be performed with large siRNA libraries, as shown for fruit fly and mammalian cells (Cherry *et al.*, 2005; Hao *et al.*, 2008; Kok *et al.*, 2009; Krishnan *et al.*, 2008). This chapter focuses on the results obtained with TBSV and the yeast libraries.

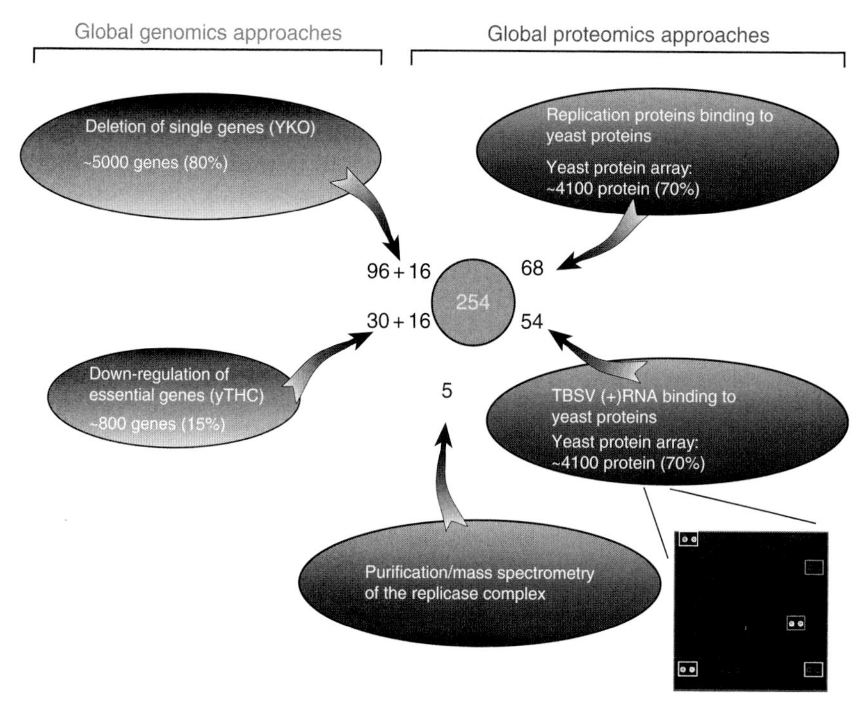

FIGURE 2 Summary of global genomics and proteomics screens performed with TBSV RNA/replication proteins. The number of host factors identified in the screens are shown, while the number behind "+" shows the number of host factors identified that affected TBSV recombination in yeast. Note that the sum of all factors from the individual screens and the total number of identified host factors (total of 254) are not the same due to identification of several common host factors in different screens. A representative subarray from the protein array is also shown. The two circled dots on the top right side of the subarray indicate a strong and a weak yeast protein interactor, while the other circled dots represent standards.

A. Single-gene-knockout YKO library

This library contains ~5000 yeast strains representing ~80% of yeast genes, and ~66% of those genes are characterized based on cellular function (Fig. 2). The advantage of the YKO library is that the strains are well defined and they lack the expression of the given gene, unlike the siRNA-based screens, where the expression of the particular gene is only knocked down. The repeatability of siRNA screens also depends on the level of knockdown, which could be influenced by several factors/parameters of the experiments.

The disadvantage of the use of the YKO library or other mutant libraries is the significant gene redundancy in biological systems. Namely, many genes have two or more homologous copies, like the heat-shock protein 70 (Hsp70), which can efficiently complement each other, leading to false negatives during the screens. Also, many proteins have pleiotropic effects in cells, affecting the functions of several other proteins and pathways. These host proteins might contribute indirectly to virus replication. Therefore, usually it is not yet known at the end of the genome-wide screens if the identified host genes affect virus replication directly or indirectly.

The systematic screen of the YKO library has revealed that TBSV repRNA replication is affected by 96 different host genes (Fig. 2) (Panavas *et al.*, 2005b). Single deletion of 90 genes reduced, while 6 increased TBSV repRNA accumulation. Grouping of the identified host genes based on their known cellular functions revealed that TBSV replication depends on a wide variety of gene functions belonging to 11 different groups. These include five genes involved in protein biosynthesis by either being part of the ribosome or acting as a translation elongation factor. The other groups are involved in protein metabolism, such as ubiquitination pathway and posttranslational modification (eight genes), RNA metabolism (five genes), and in lipid metabolism (five genes). A large group includes 20 genes implicated in vesicle-mediated transport, affecting endoplasmic reticulum (ER), Golgi, vacuole transport, or membrane fusion. Additional groups include membrane-associated proteins (seven genes) or stress responsive genes (five genes) or have variable functions in general metabolism (11 genes). The remaining groups contain genes involved in transcription and DNA remodeling or include genes with unknown function and hypothetical ORFs.

An additional screen with the YKO library and the TBSV repRNA led to the identification of 16 host genes affecting TBSV RNA recombination (Cheng *et al.*, 2006; Serviene *et al.*, 2005, 2006). The identified host genes code for proteins involved in various cellular processes, such as translation, RNA metabolism, protein modifications and intracellular transport, or membrane modifications. Since viral RNA recombination is a major mechanism for viruses in their rapid evolution, the involvement of host factors in RNA recombination suggests that the evolution of (+)RNA viruses might not depend on totally random events, but it is affected by the host.

Subsequent, more detailed studies have led to the characterization of the roles of the following host factors: Xrn1p in TBSV replication and recombination (Cheng *et al.*, 2006, 2007; Jaag and Nagy, 2009; Serviene *et al.*, 2005); Nsr1p for inhibition of TBSV replication (Jiang *et al.*, 2010); seven ESCRT (endosomal sorting complexes required for transport) proteins for affecting the quality of the assembly process for the tombusvirus replicase complex (Barajas *et al.*, 2009a); Erg4p in TBSV replication due to its affect on sterol biosynthesis (Sharma *et al.*, 2009), and Pmr1p affecting TBSV recombination and replication (Jaag *et al.*, 2010). These examples covering ~10% of the identified genes from the YKO library vindicate genome-wide screens as powerful tools for identification of host factors in TBSV replication.

Altogether, the identification of over 100 genes with rather diverse functions that affected TBSV replication/recombination via the genome-wide screens suggests that the interaction between a host and an (+)RNA virus is likely very complex and the replication of the virus is affected by many factors and pathways inside the cells. This unexpectedly high complexity in virus–host interaction validates the use of high-throughput approaches to identify all the players from the host participating in the interaction.

B. yTHC library/essential genes

The Tet-promoter-based Hughes collection (yTHC) contains ~800 out of ~1100 essential yeast genes (Mnaimneh *et al.*, 2004). In the yTHC collection, the expression of a given essential yeast gene is under the control of a Tet-titratable promoter in the genome. The expression of the essential gene can be turned off by the addition of doxycycline to the yeast growth medium (Mnaimneh *et al.*, 2004).

Using the yTHC collection, a total of 30 essential host genes have been identified that affected TBSV replication (Fig. 2) (Jiang *et al.*, 2006). The identified genes have different molecular functions in various cellular processes including RNA binding/processing (nine genes), RNA helicase/unwinding/RNA metabolism (four genes), or RNA polymerase/RNA transcription (five genes). Others are involved in protein synthesis/modification (four genes), protein transport (one gene), or lipid biosynthesis (one gene). Other genes are involved in general metabolism, in chromatin remodeling, or function as putative GTPases, while two genes have currently unknown functions.

Among the yTHC collection, 16 strains showed altered recombination frequency (Serviene *et al.*, 2006). The identified genes included five affecting RNA binding/processing/unwinding, three genes are known to code for proteins with RNA polymerase/RNA transcription function. Others are involved in protein modification/catabolism, or protein transport.

Follow-up studies characterized the roles of the following host factors: Erg25p in TBSV replication due to its affect on sterol biosynthesis (Sharma *et al.*, 2010), and the indirect role of Rpb11p transcription factor, which affected TBSV recombination via changing the ratio of p33 and p92 proteins produced in yeast (Jaag *et al.*, 2007). Overall, the host genes identified as being essential are represented almost at twice the ratio (3.75%) when compared with the nonessential genes (2%), suggesting that tombusviruses might have adapted to use and/or dependent on essential genes to higher extent than on nonessential genes. However, additional follow-up studies with the essential genes are more difficult than with the nonessential genes, due to the shared requirement of the identified essential factors in cell growth and TBSV replication/recombination.

III. PROTEOMICS-BASED SCREENS FOR SYSTEMATIC IDENTIFICATION OF HOST FACTORS AFFECTING TBSV REPLICATION

The accumulation levels and molecular functions/activities of proteins in cells are affected not only by the level of mRNA transcription, but also by many other processes, such as the efficiency of translation of a given mRNA controlled by *cis*-acting RNA sequences, stability of the protein, subcellular localization of the protein and posttranslational modifications (phosphorylation, ubiquitination, acetylation, etc.), as well as the availability of interacting protein/RNA/DNA partners and substrates. Therefore, proteome-wide screens based on proteomics approaches are needed to identify all the host factors interacting with selected viral proteins or to determine what molecular networks and cellular pathways are affected by the viral proteins.

One major advantage of proteomics approaches in general is that they are not limited by gene redundancy. Therefore, proteomics can efficiently complement the above genomics approaches to identify host factors affecting (+)RNA replication. Another advantage is that direct protein interaction networks can be established. The disadvantages of many proteomics approaches are that the abundance of a particular protein in the cell could be a critical factor, since low abundance proteins are more difficult to identify in biological samples than high abundance proteins. On the contrary, low or high expression level for a given protein is a lesser concern in the above genomics screens with gene deletions or downregulation of mRNA transcripts/protein levels. Also, weak molecular interactions important in regulatory networks are notoriously difficult to detect by many proteomics approaches. In spite of these disadvantages, the ever-improving proteomics approaches are gaining in popularity to identify host factors affecting viral replication.

Based on these considerations, we have introduced proteomics approaches to study the interaction of tombusviruses with their hosts. The two major approaches discussed in this chapter are based on protein copurification/mass spectrometry and a protein array approach.

A. Replicase purification/mass spectrometry

Identification of cellular factors recruited into the viral replicase complex for helping viral replication is important to determine the players and their functions in the replicase complex. What makes this a really challenging task is the membrane-association of the replicase complex for all eukaryotic (+)RNA viruses. Therefore, the replicase complex should be solubilized from the membranes prior to further analysis. Both the solubilization step and the following purification step likely remove proteins that are weakly/loosely associated with the replicase complex. Thus, it is highly possible that the copurified proteins in the replicase complex represent only the most abundant and strongly bound cellular proteins.

To identify the host factors present in the viral replicase complex, recent proteomics approaches revealed that 4–10 host proteins were part of the highly purified functional tombusvirus replicase (Fig. 3) (Serva and Nagy, 2006). Additional studies have determined at least seven proteins in the replicase complex, including the viral p33 and p92^{pol}, the heat-shock protein 70 chaperones (Hsp70, Ssa1/2p in yeast), glyceraldehyde-3-phosphate dehydrogenase (GAPDH, encoded by *TDH2* and *TDH3* in yeast), pyruvate decarboxylase (Pdc1p), Cdc34p ubiquitin ligase (Li *et al.*, 2008; Serva and Nagy, 2006; Wang and Nagy, 2008), and eukaryotic translation elongation factor 1A (eEF1A) (Li *et al.*, 2009). The functions of GAPDH and Hsp70 have been studied in some details (Pogany *et al.*, 2008; Wang and Nagy, 2008; Wang *et al.*, 2009a,b), but the roles of the other host proteins in the replicase complex are currently undefined. Also, the number of the identified host proteins within the tombusvirus replicase complex is likely an underestimation of the actual number of host proteins being permanent or temporally residents in the replicase complex.

B. A yeast protein microarray approach to identify host proteins interacting with the viral replication proteins

To reprogram and exploit cellular processes, tombusvirus-coded p33 and p92^{pol} replication proteins likely interact with a currently unknown number of host proteins. The recruited host proteins could be part of the

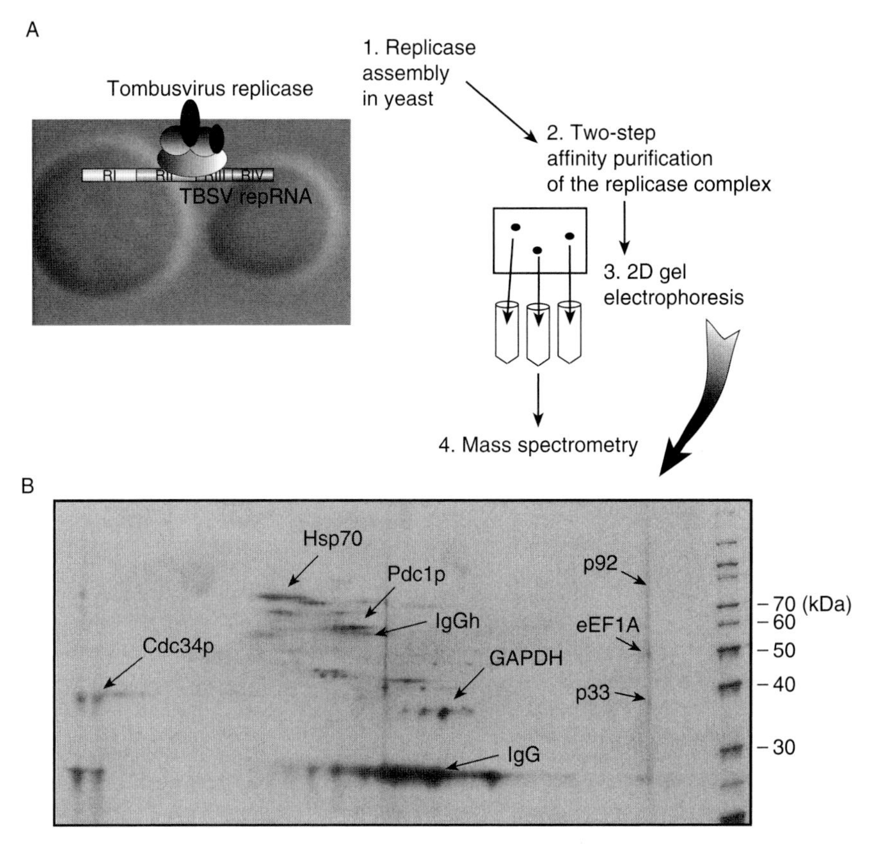

FIGURE 3 Proteomics analysis of the tombusvirus replicase complex. (A) The proteomics approach. (B) A representative silver-stained 2D gel image of the two-step affinity-purified tombusvirus replicase preparation. The identified proteins on the 2D gel are indicated. IgG and IgGh represent proteins derived from affinity purification.

replicase complex to aid viral replication. Moreover, the interacting host proteins might facilitate the transport of viral proteins in the cells or affect the assembly of the viral replicase as well as provide regulatory functions during viral replication. To catalogue the list of host proteins interacting with the viral replication proteins, we have taken a proteome-wide approach with the yeast protein array carrying ~4100 purified proteins that covers ~70% of yeast proteins (Fig. 2). This has led to the identification of 57 proteins binding to p33, whereas an additional 11 host proteins bound only to the unique portion of p92, but not to p33.

Among the identified host proteins interacting with p33, there are three protein chaperones (Gim3p, Jjj1p, and Jjj3p), five proteins involved in protein ubiquitination (Cdc34p, Rsp5p, Uba1p, Ubp10p, and Ubp15p), six translation factors involved in mRNA translation (Bfr1p, Efb1p, Hbs1p, Rpl8Ap, Tif1p, and Tif11p), and 10 proteins involved in RNA processing and metabolism (Ala1p, Bud21p, Erb1, Rib2p, Sas10p, Stm1p, Trm1p, Trz1p, Tsr2p, and Urn1p). The remaining list of identified host proteins are involved in various cellular processes and the functions of nine proteins are not yet defined. The 11 host proteins bound only to p92 includes an RNA helicase (Dpb3p), a methylase (Dot1p), an amino-peptidase (Map1p), an RNA-binding protein (Npl3p), and a translation factor (eEF1A/Tef2p).

Similar to other genome-wide approaches, the use of protein arrays might lead to false positives and false negatives as well. The false negatives could be due to many factors, including (i) the use of general binding conditions, which are not optimized for individual protein–protein interactions; (ii) the absence of cofactors or membrane surfaces under the *in vitro* conditions; and (iii) inactive or denatured proteins on the chip. Indeed, we did not detect significant binding between the purified p33 and Ssa1p, an Hsp70, which has been shown to be part of the replicase complex (Serva and Nagy, 2006). This suggests that multiple complementary approaches are needed to identify all host proteins interacting with the replication proteins.

We used additional approaches to confirm the data from the protein array, including protein pull-down experiments with purified recombinant p33 and yeast proteins as well as the split-ubiquitin yeast two-hybrid assay (Li *et al.*, 2008). The split-ubiquitin assay, unlike the original yeast two-hybrid system, allows the analysis of protein interactions on the cytosolic surfaces of membranes, which is the natural subcellular location of the membrane-bound p33 protein (McCartney *et al.*, 2005; Panavas *et al.*, 2005a).

C. A yeast protein microarray approach to identify host proteins binding to the viral RNA

Many RNA-binding host proteins likely play multiple roles during tombusvirus replication. A proteome-wide approach using the yeast protein array identified 57 host proteins bound to either TBSV or BMV RNAs. Among these host proteins, 11 proteins bound selectively to TBSV RNA, including two known helicases (*DBP2* and *YFR038W*), a translation initiation factor (*GCD2*), and two RNA modifying proteins (*DEG1* and *UTP7*). An additional 43 host proteins identified with both TBSV and BMV RNA probes are involved in a variety of cellular processes, such as translation, transcription activation, ribosomal RNA processing/binding, mRNA transport, and protein-membrane targeting with various biochemical activities—such as helicase, tRNA ligase, tRNA methyltransferase, rRNA dimethylase, ribonuclease, cochaperone, and protein kinase.

More detailed experiments with translation elongation factor eEF1A have shown that this host protein is part of the tombusvirus replicase and binds to p33/p92 replication proteins and the 3′-UTR of the TBSV (+)RNA as well. Interestingly, eEF1A has been shown to bind to the BMV RNA (Bastin and Hall, 1976). In addition, the identified pseudouridine synthase Pus4p might be involved in pseudouridinylation-based modification of TBSV RNA and BMV RNA, which has been shown to occur for the BMV RNA *in vivo* (Baumstark and Ahlquist, 2001). Interestingly, Pus4p has also been identified in a similar screen with a unique yeast protein array using a 3′-end *cis*-acting element from BMV RNA (Zhu *et al.*, 2007), although its actual function in BMV replication is currently unknown.

To validate the above proteome-wide approach for identification of host proteins binding to the viral RNA, several recombinant yeast proteins have been shown to bind to the TBSV (+)repRNA in a gel mobility shift assay and via protein/RNA copurification approach from yeast cells (Li *et al.*, 2009). Moreover, several of the identified host RNA-binding proteins affected TBSV repRNA replication, supporting the idea that a number of RNA-binding proteins play a role in the tombusvirus replication process.

IV. GROUPING OF HOST FACTORS AND IDENTIFICATION OF NETWORKS INVOLVED IN TBSV REPLICATION

The above genome-wide genomics and proteomics approaches have led to the identification of 254 host proteins that either affected TBSV repRNA accumulation or bound to the viral replicase, replication proteins, or the viral RNA (Fig. 2). The large number of factors identified and the great variety of cellular function for these factors (Fig. 4) indicate the existence of a truly complex interaction between a simple (+)RNA virus and the yeast model host. This complexity might be due to several levels of interaction taking place between the virus and the host. For example, it is possible that many recruited host factors play a direct, well-defined function for promoting virus replication; others could inhibit TBSV accumulation, while many more host proteins might have only indirect roles in TBSV replication by affecting the general metabolisms/pathways in the cells that also influence virus replication. Many of the identified host factors could be part of protein networks and pathways that are recruited in an orchestrated way for virus replication, while other factors might be recruited and function individually. Another complication is that virus replication might utilize either the known function or an unknown function of a given host factor or even multiple functions of the same protein. Moreover, some host factors might perform completely novel functions (not performed during regular cellular processes) during virus

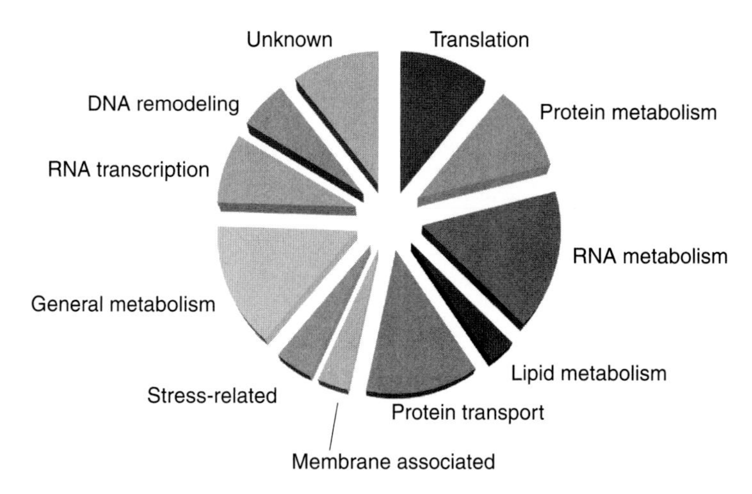

FIGURE 4 The frequency of identification of host factors representing one of the shown functional groups as described in Table I. The total number of host factors identified in the global screens is 254 (see Fig. 2).

replication. Detailed further analysis of the identified host factors based on bioinformatics in combination with biochemical, genetic, and cellular analyses will be needed to dissect the functions of the host proteins during TBSV replication. These are enormous challenges waiting for virology research to uncover the mechanism of (+)RNA virus replication in order to understand virus–host interaction.

Below, we will summarize our current grouping of host factors, view those host factors that have been characterized in more detail, and propose protein networks that could be involved in TBSV replication.

A. Translation factors and cellular proteins involved in protein biosynthesis

Translation of the viral RNA precedes viral (+)RNA replication, which uses the same RNA. Therefore, translation and replication must be coordinated to regulate temporally and spatially these processes and prevent the collision between the ribosomes and the viral replicase. Although the mechanism of the switch of the viral (+)RNA from translation to replication is currently unknown, it is likely that host factors in combination with the newly translated viral replication proteins play a role in this process. Translation factors are especially good candidates to be involved in the switch, since they are already present during translation and several of them have been identified to bind to either the replication proteins or the viral RNA (Li *et al.*, 2008, 2009). The large number of the identified host factors in this group, however, makes it difficult to narrow down the actual candidates. Also, these factors could play direct roles in subsequent steps in replication.

Additional experiment was performed on eEF1A to dissect its role in TBSV replication. eEF1A is a highly abundant cellular protein and its best-known function is to deliver aminoacyl-tRNA to the elongating ribosome in a GTP-dependent manner. eEF1A has many additional functions, such as quality control of newly produced proteins, ubiquitin-dependent protein degradation, and organization of the actin cytoskeleton (Chuang *et al.*, 2005; Gross and Kinzy, 2005).

The experiments with eEF1A revealed that it is a permanent resident of the tombusvirus replicase complex (Li *et al.*, 2009). Mutational analysis of eEF1A suggests that it might be involved in promoting (−)RNA synthesis by the replicase complex (Z. Li and P.D. Nagy, unpublished data). Another function of eEF1A is to stabilize the p33 replication protein, while it did not affect the half-life of the less abundant $p92^{Pol}$ replication protein in yeast (Li *et al.*, 2009). Interestingly, a prokaryotic homolog of eEF1A, called Tu translation elongation factor, plays a role in replication of bacteriophage Qβ (Blumenthal *et al.*, 1976). In addition, eEF1A was found to bind to many viral (+)RNAs, including the 3′-UTR of *Turnip yellow mosaic virus* (TYMV) (Dreher, 1999), of West Nile virus (WNV), Dengue virus, *Tobacco mosaic virus* (TMV) and *Turnip mosaic virus* (De Nova-Ocampo *et al.*, 2002; Nishikiori *et al.*, 2006; Thivierge *et al.*, 2008; Zeenko *et al.*, 2002). In addition, eEF1A has also been shown to interact with the NS5A replication protein of Bovine viral diarrhea virus (BVDV) (Johnson *et al.*, 2001), NS4A of HCV (Kou *et al.*, 2006), the TMV replicase (Yamaji *et al.*, 2006), and the Gag polyprotein of HIV-1 (Cimarelli and Luban, 1999). The biochemical functions provided by eEF1A for (+) RNA virus replication are currently poorly understood. It has been proposed to affect minus-strand synthesis for WNV (Davis *et al.*, 2007), albeit it repressed minus-strand synthesis of TYMV *in vitro* (Dreher, 1999; Dreher *et al.*, 1999; Matsuda *et al.*, 2004). Overall, the interaction of eEF1A with viral RNAs and viral replication proteins and its high abundance in cells might facilitate recruitment of eEF1A into virus replication.

The identification of 28 host factors (11% of all host factors) (Fig. 4) suggests that translation factors and/or the ribosome itself might be involved in correct folding of the viral RNA or the newly made viral replication proteins. Several proteins might also be involved in modification of the viral (+)RNA, similar to tRNAs and rRNAs. These processes could affect the stability and/or subsequent localization of the viral proteins/RNA. Albeit additional translation factors/ribosomal proteins, similar to eEF1A, might be recruited for replication, it is unlikely that all 28 proteins in this group would be directly targeted for assisting replication.

B. Protein modification enzymes

Posttranslational modification serves as an important means switching protein molecules between active and inactive forms, in regulating their stability, their subcellular localization, and their interactions with other proteins, nucleic acids, or membranes. The TBSV p33 replication protein can be phosphorylated (Shapka *et al.*, 2005; Stork *et al.*, 2005) and ubiquitinated (Barajas *et al.*, 2009; Li *et al.*, 2008) with both modifications likely playing roles in TBSV replication. Accordingly, Cka1p, Mob1p, Mps1p, Sln1p kinases and Siw14p phosphatase have been identified in the genome-wide screens (Table I). Albeit it is yet unknown what host kinases phosphorylate p33 in cells, protein kinase C (PKC) was shown to phosphorylate purified p33 *in vitro*. Interestingly, the sites of p33 phosphorylation are located proximal to an essential RNA-binding domain (Shapka *et al.*, 2005; Stork *et al.*, 2005). Since the phosphorylated form of p33 lost its ability to bind to (+)rep RNA and *in vitro* phosphorylation of the p33: repRNA complex led to the release of the RNA from the complex, it has been proposed that phosphorylation of p33 and possibly p92, which carries the same p33 sequence at its N-terminus due to its overlapping ORF, might lead to the release of the viral RNA from the replicase complex (Stork *et al.*, 2005).

Ubiquitination also plays a role in TBSV replication. Ubiquitination of host proteins by the highly conserved 76 aa ubiquitin (Ub) regulates many cellular processes, such as protein degradation, protein trafficking, transcription, immune response, signal transduction, and autophagy. Protein ubiquitination/deubiquitination requires four types of enzymes. E1 proteins activate ubiquitin, E2s (Ub-conjugating enzymes) function in transferring Ub to the client proteins, whereas E3s are involved in substrate selection, while DUBs remove the Ub from the proteins.

The genome-wide screens with TBSV identified 11 host proteins involved in the ubiquitin-dependent pathway of protein modification/degradation. These proteins include E2 ubiquitin-conjugating enzymes (*CDC34* and *RAD6*), ubiquitin-protein ligases (*RSP5* and *BRE1*), an ubiquitin-activating enzyme (*UBA1*), and four ubiquitin-specific proteases (*DOA4*, *UBP3*, *UBP10*, and *UBP15*), while *LGE1* is involved in protein monoubiquitination, and *BRE5* is a ubiquitin protease cofactor (Table I) (Jiang *et al.*, 2006; Li *et al.*, 2008; Panavas *et al.*, 2005b; Serviene *et al.*, 2005, 2006). Binding of p33 with ubiquitin-specific proteins suggests that p33 could be modified posttranslationally by ubiquitination as shown *in vivo* and *in vitro* (Barajas *et al.*, 2009b; Li *et al.*, 2008).

Among the identified proteins in the ubiquitin pathway, Cdc34p, Rsp5p, and Doa4p (described in Section E) have been characterized in more details (Table I). Cdc34p (also called Ubc3p) was found to bind to p33 and is a permanent resident of the tombusvirus replicase complex

TABLE I Functional grouping and roles of the identified host genes affecting TBSV RNA replication and recombination[a]

Gene	Cellular function	Viral replication[b]	Viral recombination	Interaction
1. Translation/protein biosynthesis				
BFR1	mRNP complexes/polyribosomes			p33/RNA
DED1	DEAD-box RNA helicase, translation	Required		
EFB1	Translation elongation factor 1β			p33
ERB1	Maturation of ribosomal RNAs	Stimulatory[c]		p33, p92
GCD2	δ-subunit of eIF2B	Stimulatory[c]		RNA
HBS1	GTP binding, similarity to EF-1α	Stimulatory[c]		p33
IPI3	Rix1 complex, pre-rRNA processing			p33
MRPL32	Protein biosynthesis	Inhibitory		
NOG1	Putative GTPase, ribosome biogenesis	Inhibitory		
NOG2	Putative GTPase, ribosome biogenesis	Inhibitory		
NOP53	Processing of 27S pre-rRNA	Inhibitory[c]		RNA
RPL1B	Protein biosynthesis	Inhibitory		
RPL4A	Component of the large ribosomal subunit			RNA
RPL7A	Protein biosynthesis	Inhibitory		
RPL8A	Ribosomal protein L4			p33/p92/RNA
RPL17A	Structural constituent of ribosome	Inhibitory		
RPL26B	Component of the large ribosomal subunit			RNA
RPS21B	Protein biosynthesis	Required		

(*continued*)

TABLE I (*continued*)

Gene	Cellular function	Viral replication[b]	Viral recombination	Interaction
SAS10	Ribosomal processome			p33/p92/RNA
SSF2	rRNA binding			RNA
STM1	Required for optimal translation	Stimulatory[c]		p33/p92/RNA
TEF1[d]	Translational elongation factor eEF1A	Required		p33/p92/RNA/ replicase
TEF2[d]	Translational elongation factor eEF1A	Required		p33/p92/RNA/ replicase
TEF4	Translation elongation factor	Required		
TIF1	Translation initiation factor eIF4A			p33
TIF11	Translation initiation factor eIF1A			p33/p92
TSR2	pre-rRNA processing			p33
YCR016W	Ribosome biogenesis (predicted)			p33/p92/RNA
2. Protein metabolism, posttranslation modification				
ARO1	Aromatic amino acid synthesis	Required		
BRE1	Ubiquitin-protein ligase	Required		
CDC34[d]	Ubiquitin-conjugating enzyme or E2	Required, stimulatory[c]		p33/replicase
CKA1	α-subunit of protein kinase CK2			RNA
DOA4[d]	Protein deubiquitination	Required		
EPL1	Histone acetyltransferase activity	Required		
LGE1	Protein monoubiquitination	Required		
MAK3	Protein amino acid acetylation	Required		

MAP1	Methionine aminopeptidase			p92/RNA
MET1	Uroporphyrin methyltransferase	Required		
MOB1	Protein amino acid phosphorylation	Required		
MPS1	Protein threonine/tyrosine kinase		Accelerator	
NOB1	Protein involved in proteasome maturation			p92/RNA
OTU2	Predicted cysteine proteases	Inhibitory[c]		p33/p92/RNA
RAD6	Ubiquitin-conjugating enzyme	Required		
RPT4	Endopeptidase		Accelerator	
RSP5[d]	Ubiquitin-protein ligase	Inhibitory		p33/p92
SLN1	Protein histidine kinase activity	Required		
SIW14	Protein tyrosine phosphatase	Required		
UBA1	Ubiquitin-activating enzyme			p33
UBP3	Ubiquitin-specific protease		Suppressor	
UBP10	Ubiquitin-specific protease			p33/p92/RNA
UBP15	Ubiquitin-specific protease			p33
YDR161W	ER-associated protein degradation			p33
3. RNA-binding proteins/RNA metabolism				
BUD21	snoRNA binding	Inhibitory, inhibitory[c]		p33/RNA
CCR4	3′–5′ exoribonuclease	Required		
CTL1	Polynucleotide 5′-phosphatase		Suppressor	
CWC25	pre-mRNA splicing			RNA
DBP2	RNA helicase of the DEAD-box protein family	Stimulatory[c]		RNA

(continued)

TABLE I *(continued)*

Gene	Cellular function	Viral replication[b]	Viral recombination	Interaction
DBP3	Putative RNA helicase/DEAD-box family			p92
DIM1	Essential 18S rRNA dimethylase			RNA
DEG1	Nonessential tRNA: pseudouridine synthase	Inhibitory[c]		RNA
GLO3	GTPase activation, ER-Golgi transport			RNA
GRC3	Possibly involved in rRNA processing	Inhibitory		
HAS1	Putative ATP-dependent RNA helicase	Inhibitory[c]		RNA
IRC5	DEAD-box helicase			RNA
LHP1	RNA-binding protein/maturation of tRNA			RNA
LRP1	Nuclear cofactor for exosome activity			RNA
MET22/HAL2	3′(2′),5′-bisphosphate nucleotidase		Suppressor	
MEX67	Poly(A)RNA-binding protein	Required		
MSE1	Glutamate-tRNA ligase activity			RNA
NAB2	Polyadenylated RNA binding; hnRNPs	Required		
NOP4	RNA binding, ribosomal RNA processing	Inhibitory		
NOP10	RNA binding, pseudouridylation, 18S rRNA		Modifier	
NPL3	mRNA binding	Required, inhibitory[c]		p92/RNA

NSR1[d]	RNA binding/rRNA processing	Required		
PRP5	RNA helicase in the DEAD-box family	Inhibitory		
PRP39	RNA binding, nuclear mRNA splicing	Required		
PUS4	Pseudouridine synthase	Inhibitory[c]		RNA
RIB2	Cytoplasmic tRNA pseudouridine synthase			p33
RNA14	RNA binding/mRNA cleavage	Required		
RNY1	RNAse; endoribonucleases	Inhibitory[c]		RNA
RPL15A	Binds to 5.8S rRNA	Inhibitory		
RPM2	Ribonuclease P activity		Modifier	
RRP9	RNA binding, pre-rRNA processing	Required	Accelerator	
RRP42	3′–5′ exoribonuclease activity	Required		
SEN1	RNA helicase, processing of tRNA, rRNA	Required	Accelerator	
TRM1	tRNA methyltransferase			p33/p92/RNA
TRZ1	tRNase Z, involved in RNA processing	Stimulatory[c]		p33
URN1	Pre-mRNA splicing factor			p33/p92
UTP7	Small subunit (SSU) processome	Inhibitory[c]		p92/RNA
UTP9	snoRNA binding, interacts with UTP15	Required		
UTP15	snoRNA binding, interacts with UTP9	Required		
XRN1/KEM1[d]	5′–3′ exoribonuclease	Required	Suppressor	
YBL055C	3′–5′ exoribonuclease, endoribonuclease			RNA
YKL023W	mRNA degradation			p92/RNA

(*continued*)

TABLE I *(continued)*

Gene	Cellular function	Viral replication[b]	Viral recombination	Interaction
4. Lipid metabolism				
ERG4[d]	δ24(24-1)-sterol reductase	Required/sterol level		
ERG25[d]	Ergosterol biosynthesis	Required/sterol level		
FAS2	α-subunit of fatty acid synthetase		Modifier	
FOX2	Peroxisomal fatty acid β-oxidation pathway			p33/RNA
INO2	Phospholipid biosynthesis	Required		
MCT1	*S*-malonyltransferase/fatty acid metabolism	Required		
POX1	Acyl-CoA oxidase/fatty acid β-oxidation	Required		
TGL2	Triacylglycerol lipase/lipid metabolism	Required		
5. Protein and vesicle-mediated transport				
APM2	Vesicle-mediated transport	Inhibitory[c]		RNA
ARL3	Small monomeric GTPase	Required		
BRE5	Vesicle-mediated transport	Required		
COP1	Protein transporter, COPI vesicle	Required	Suppressor	
DID2[d]	ESCRT/protein–vacuolar targeting	Required		
GOS1	v-SNARE activity/intra-Golgi transport	Required		

MCH5	Transporter/membrane associated	Required		
MON1	Protein–vacuolar targeting	Required		
NUP53	Subunit of the nuclear pore complex			RNA
PEP3	Transporter/vacuolar membrane	Required		
PEP7/ VPS19	Unknown/Golgi to vacuole transport		Accelerator	
PEX19[d]	Chaperone/import to peroxisome			p33
PTH1/ VAM3	Golgi to vacuole transport		Accelerator	
RIC1	Guanyl-nucleotide exchange factor	Required		
SEC62	SRP-dependent/protein-membrane targeting			RNA
SNF7[d]	ESCRT/late endosome	Required		
SNL1	Nuclear pore organization and biogenesis			RNA
SRP40	Nucleocytoplasmic transport/ chaperone			RNA
TLG2	t-SNARE, v-SNARE/vesicle fusion	Required		
TOM71	Component of the TOM translocase			p33/p92/RNA
VPS4[d]	ESCRT/ATPase/late endosome	Required		
VPS23/ STP22[d]	ESCRT/protein–vacuolar targeting	Required		
VPS24[d]	ESCRT/late endosome	Required		
VPS28[d]	Protein–vacuolar targeting	Required		

(*continued*)

TABLE I *(continued)*

Gene	Cellular function	Viral replication[b]	Viral recombination	Interaction
VPS29	Retrograde/endosome to Golgi/ transport	Required	Accelerator	
VPS35	Endosome to Golgi transport		Accelerator	
VPS41	Rab guanyl-nucleotide exchange factor	Required		
VPS43/ VAM7	Golgi to vacuole transport		Accelerator	
VPS51	Protein–vacuolar targeting	Required		
VPS61	Protein–vacuolar targeting	Required		
VPS66	Cytoplasmic protein/vacuolar protein sorting			p92/RNA
VPS69	Protein–vacuolar targeting	Required		
YOS9	Protein transporter/ER to Golgi transport	Required		
6. Membrane associated				
KEG1	Integral membrane protein of the ER	Inhibitory[c]		RNA
MSP1	ATPase/mitochondrial translocation	Required		
OPT1	Oligopeptide transporter	Required		
PMR1 (HUR1)[d]	Ca^{2+}/Mn^{2+} ion pump		Suppressor	
SAC1	Inositol/phosphatidylinositol phosphatase	Required		
SNF4	Protein kinase activator	Required		

STE14	Isoprenylcysteine methyltransferase	Required		
STV1	Hydrogen-transporting ATPase	Required		
TOK1	Potassium channel	Required		
7. Stress-related/chaperone				
DDR48	DNA damage-response, heat-shock stress	Inhibitory[c]		p33
JJJ1	Cochaperone of Ssa1p	Stimulatory[c]		p33/p92/RNA
JJJ3	Contains J-domain			p33
GIM3	Heterohexameric cochaperone prefoldin complex			p33
GRE3	Aldehyde reductase	Required		
GTT1	Glutathione transferase	Required		
IRA2	Ras GTPase activator	Required		
SSA1[d]	HSP70 chaperone	Required		p33/p92/replicase
SSA2[d]	HSP70 chaperone	Required		p33/p92/replicase
UGA2	Glutamate catabolism	Required		
WHI2	Phosphatase activator	Required		
8. General metabolism				
ALA1	Cytoplasmic alanyl-tRNA synthetase			p33
BEM4	Rho protein signal transduction	Required		
COX12	Cytochrome *c* oxidase	Required		
CHO2/PEM1	Phosphatidylethanolamine *N*-methyltransferase		Accelerator	
DCI1	Dodecenoyl-CoA δ-isomerase		Accelerator	
DSE1	Cell wall organization and biogenesis	Required		

(continued)

TABLE I *(continued)*

Gene	Cellular function	Viral replication[b]	Viral recombination	Interaction
ERR2	Phosphopyruvate hydratase			p33
GLO2	Hydroxyacylglutathione hydrolase	Required		
GPH1	Glycogen phosphorylase	Required		p33
GSY2	Glycogen synthase			p33/p92
HAP3	Regulation of carbohydrate metabolism	Required		
HOR2	DL-glycerol-3-phosphatase			p33
IPK1	Inositol/phosphatidylinositol kinase		Accelerator	
ISN1	Inosine 5′-monophosphate 5′-nucleotidase			p33
LPD1	Pyruvate dehydrogenase	Required		
MAM33	Mitochondrial matrix/oxidative phosphorylation			p33
MDH3	Cytoplasmic malate dehydrogenase			RNA
MDM38	Mitochondrial inner membrane protein	Stimulatory[c]		RNA
MSB1	Establishment of cell polarity	Required		
NAP1	Regulation of microtubule dynamics			p33
PCS60	Peroxisomal AMP-binding protein			RNA
PDC1	Pyruvate decarboxylase			Replicase
PDI1	Protein disulfide isomerase, ER lumen			p33
PHD1	Pseudohyphal growth	Required		
PLP2	Actin binding/similarity to phosducins			p33
PYC1	Pyruvate carboxylase isoform			p33

QCR6	Ubiquinol–cytochrome *c* reductase complex			p33/p92
RIB7	Deaminase, riboflavin biosynthesis		Modifier	
RMD7	Cell wall organization and biogenesis	Required		
SHO1	Transmembrane osmosensor			p33
SPE3	Spermidine synthase		Modifier	
TDH2[d]	Glyceraldehyde-3-phosphate dehydrogenase	Required		RNA/replicase
TDH3[d]	Glyceraldehyde-3-phosphate dehydrogenase	Required		RNA/replicase
THI3	Carboxy-lyase/thiamin biosynthesis	Required		
TUM1	Mitochondrial, similar to rhodanase			p33
YJL218W	Acetyltransferase activity			RNA
YIL064W	*S*-adenosylmethionine methyltransferase	Required		
9. RNA transcription				
ARP9	RNA polymerase, actin-related protein	Inhibitory	Accelerator	
CDC50	Transcription regulator	Required		
HAA1	Transcriptional activator	Inhibitory[c]		RNA
MED6	RNA polymerase II transcription mediator	Required		
ELF1	A zinc finger transcription elongation factor			p33
IWR1	Affects transcription by pol II			p33/p92
NGG1	Transcription cofactor		Accelerator	

(*continued*)

TABLE I *(continued)*

Gene	Cellular function	Viral replication[b]	Viral recombination	Interaction
POL1	α-DNA polymerase, synthesis of RNA primer		Suppressor	
RDS2	Zinc cluster transcription activator			p92
RGR1	Transcription mediator		Suppressor	
RPB11[d]	RNA polymerase II subunit B12.5	Required	Accelerator	
RPO21	RNA polymerase	Required		
ROX3	RNA polymerase II transcription mediator	Required		
SUB1	Transcriptional coactivator	Inhibitory[c]		RNA
SPT3	Transcription cofactor		Modifier	
SPT16	Pol II transcription elongation factor			p33/p92
SRB8	RNA polymerase II transcription mediator	Required		
SWI3	General RNA polymerase II transcription factor	Required		
TEA1	Transcription regulator	Required		
TFA2	General RNA polymerase II transcription factor	Required		
UME6	Transcription regulator	Required		
10. DNA remodeling, metabolism				
ADA2[d]	Chromatin modification, histone acetylation	Required		

ARP8	Nuclear actin-related, chromatin remodeling	Stimulatory[c]		p33
DOT1	Nucleosomal histone methylase			p92
DPB4	ϵ-DNA polymerase	Required		
HEX3	DNA recombination	Required		
NGG1	Chromatin modification, histone acetylation	Required		
ORC6	DNA replication		Modifier	
POL30	Proliferating cell nuclear antigen (PCNA)	Stimulatory[c]		p33
RSC8	Chromatin remodeling	Required		
RTT106	Histone chaperone/Ty transposition			p33
SAS3	Acetyltransferase/chromatin silencing	Required		
SIN3	Histone deacetylase	Required		
SLX8	DNA metabolism	Required		
SLX9	DNA metabolism	Inhibitory		
SNF6	Chromatin modeling/SWI/SNF complex	Required		
11. Function unknown				
BSC2	Unknown	Required		
EMI2	Protein of unknown function			p33
FMP40	Protein of unknown function			p33
LDB7	Unknown	Required		
YBR007C	Unknown	Required		
YBR032W	Unknown	Required		
YCR099C	Unknown	Required		

(continued)

TABLE I *(continued)*

Gene	Cellular function	Viral replication[b]	Viral recombination	Interaction
YDR327W	Unknown	Inhibitory	Modifier	
YFL043C	Unknown	Required		
YGL140C	Unknown	Required		
YGL242C	Unknown			p33/p92
YGR017W	Unknown			p33/p92
YGR026W	Unknown	Inhibitory[c]		RNA
YGR027W	Retrotransposon TYA gag gene			p33
YGR064W	Unknown	Required		
YHR009C	Unknown			p33
YHR029C	Unknown	Required		
YIL090W	Unknown	Required		
YJL175W	Unknown	Required		
YKL033W	Cytoplasmic protein with unknown function		Modifier	
YLR125W	Unknown/Ty3 transposition			p33
YLR358C	Unknown	Required		
YNL196C	Unknown/leucine zipper protein			RNA
YNL321W	Unknown	Required		
YOR309C	Hypothetical protein			p92/RNA
YPR050C	Unknown	Required		
YPR174C	Unknown			RNA

[a] The shown data are from Jiang *et al.* (2006), Li *et al.* (2008, 2009), Panavas *et al.* (2005b), Serva and Nagy (2006), and Serviene *et al.* (2005, 2006).
[b] "Required" is based on more than twofold drop in TBSV replication when the host gene is deleted or its expression is downregulated.
[c] Based on protein overexpression in yeast.
[d] Host genes, whose roles/functions have been characterized in details in TBSV replication.

(Li *et al.*, 2008). A purified preparation of Cdc34p ubiquitinated p33 *in vitro*, indicating that Cdc34p is active on the p33 substrate in the absence of an E3 enzyme. Downregulation of Cdc34p level decreased TBSV repRNA accumulation and the activity of the tombusvirus replicase by three- to fivefold. Interestingly, a Cdc34p mutant inactive in ubiquitin conjugation could not complement the reduced amount of wt Cdc34p based on the activity of the isolated tombusvirus replicase (Li *et al.*, 2008), suggesting that the ubiquitination activity of Cdc34p is critical for TBSV replication. However, the actual function of Cdc34p within the replicase complex is not yet known.

Rsp5p E3 ubiquitin ligase has been shown to bind to p33 and $p92^{pol}$ replication proteins and it can ubiquitinate p33 in the presence of E1 and E2 proteins *in vitro* (Barajas *et al.*, 2009b). However, unlike Cdc34p, Rsp5p inhibits TBSV replication by binding via its three WW repeats to $p92^{pol}$ and destabilizing $p92^{pol}$ and reducing the replicase activity (Fig. 5). Surprisingly, the HECT domain involved in protein ubiquitination is not required for the inhibitory activity of Rsp5p (Barajas *et al.*, 2009b). Future experiments should address what is the role of Rsp5p in regulation of TBSV replication.

Other viruses are also known to take advantage of the ubiquitination pathway (Barry and Fruh, 2006; Shackelford and Pagano, 2004, 2005; Taylor and Barry, 2006), by using it to regulate protein stability via ubiquitination and deubiquitination of viral proteins (Geoffroy *et al.*, 2006; Mechali *et al.*, 2004; Miller *et al.*, 2004; Nerenberg *et al.*, 2005; Ott *et al.*, 2000; Poon *et al.*, 2006; Wang *et al.*, 2006; Woo and Berk, 2007). For example, ubiquitination has been documented for replication proteins of *Turnip mosaic virus*, HCV, coxsackievirus, and coronaviruses (Barretto *et al.*, 2005; Hericourt *et al.*, 2000; Ratia *et al.*, 2006; Sulea *et al.*, 2005; Wong *et al.*, 2007). Moreover, a host ubiquitin gene sequence was found inserted in the bovine viral diarrhea virus genomic RNA via RNA recombination (Baroth *et al.*, 2000; Tautz and Thiel, 2003). In spite of intensive efforts, the current knowledge on the roles of ubiquitination in (+)RNA replication and infections is incomplete.

Altogether, the identification of many host proteins involved in the ubiquitin pathway that affected TBSV replication/recombination suggests that ubiquitination plays a critical role in TBSV replication. This was indeed demonstrated in studies with p33 and the so-called ESCRT proteins described in Section E.

C. RNA-binding proteins, RNA modification enzymes, and proteins involved in RNA metabolism

This group of proteins constitutes the largest group among the identified host proteins from our genome-wide studies. The 42 host proteins in this group represent ~17% of all the identified proteins (Fig. 4), suggesting

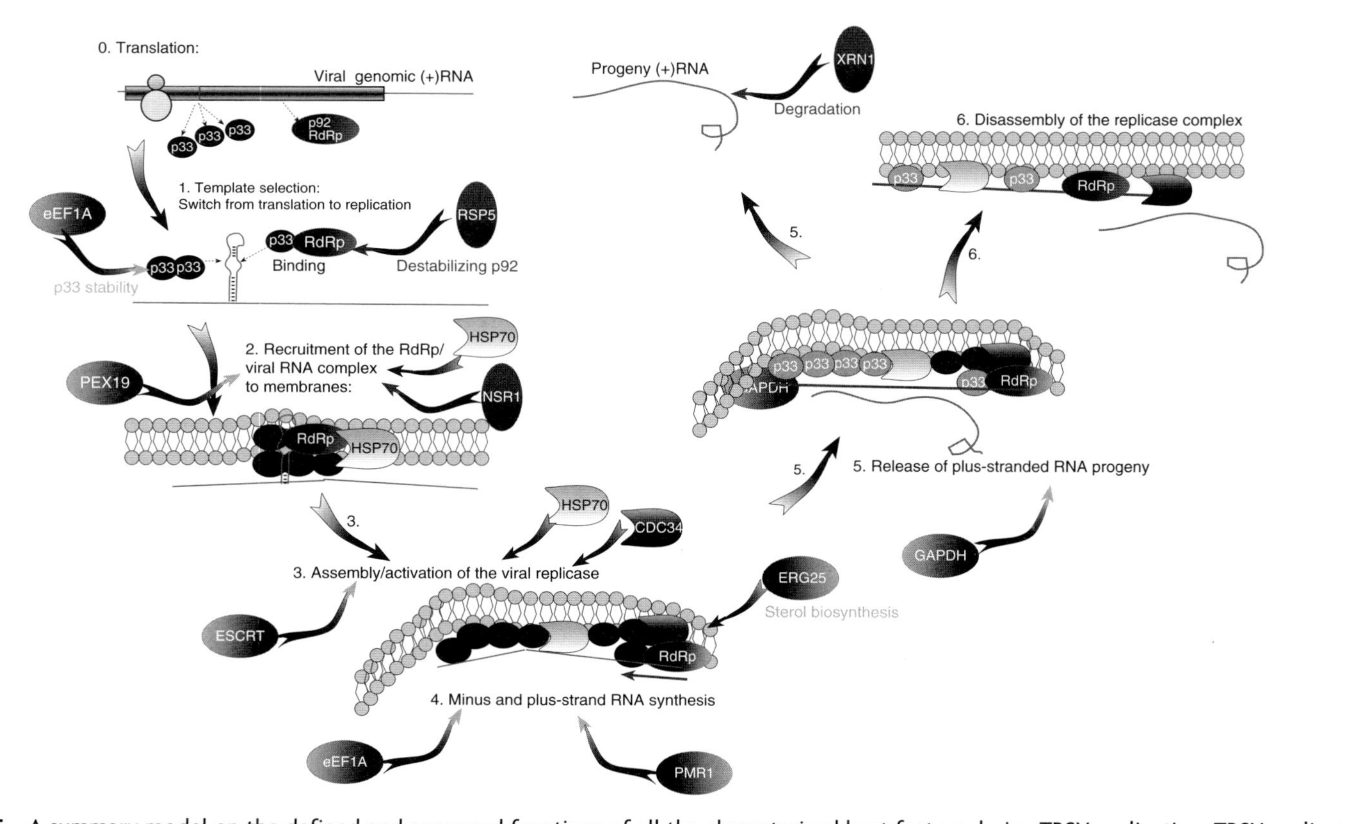

FIGURE 5 A summary model on the defined and proposed functions of all the characterized host factors during TBSV replication. TBSV replication is divided into six sequential steps and preceded by translation of the TBSV replication proteins. Host factors shown in green circles and green arrows are required, while factors shown in blue circles and blue arrows are inhibitory for TBSV replication. Note that the ESCRT protein circle represents seven of the identified ESCRT proteins. Hsp70 is colored uniquely due to the essential role of Hsp70 in several steps in TBSV replication. The active p33 proteins are represented by black circles, while the inactive (phosphorylated form) is shown with pink circles. Also, the inactive p92 protein is shown as a red circle, while the activated p92 is represented by a green circle. (See Page 1 in Color Section at the back of the book.)

that RNA-binding host proteins likely play important as well as diverse roles during TBSV replication. For example, the RNA-binding host proteins might affect (i) translation of the viral (+)RNA, (ii) selection and recruitment of the viral (+)RNA template for replication, (iii) the quality or efficiency of RNA synthesis, and/or (iv) stability of the viral RNA (Ahlquist *et al.*, 2003; Brinton, 2001; Cristea *et al.*, 2006; Nagy, 2008; Nagy and Pogany, 2006; Shi and Lai, 2005).

Since the known cellular functions of this group of proteins are amazingly diverse as well as some functions are possibly redundant, it seems that dissecting of the functions of these proteins during TBSV replication will be of a great challenge. Nevertheless, follow-up experiments with this group of proteins provided some insight into the possible functions of two proteins in TBSV replication, namely Nsr1p and Xrn1p (Fig. 5) as summarized below.

Nsr1p, also known as nucleolin, binds to the TBSV RNA and exerts its effect on viral replication directly (Jiang *et al.*, 2010). Nucleolin/Nsr1p is found in various cell compartments and it is especially abundant in the nucleolus. Nucleolin is a ubiquitous and abundant protein involved in multiple processes, such as ribosome biogenesis, transcription of rDNA, processing and modification of rRNA and nuclear to cytosolic transport of ribosomal protein, and ribosomal subunits by shuttling between the nucleus and the cytoplasm (Tuteja and Tuteja, 1998).

Nsr1p was discovered during screening of the yeast YKO library for TBSV replication (Panavas *et al.*, 2005b). Nsr1p seems to be an inhibitor of TBSV replication since virus replication was boosted threefolds in the absence of *NSR1*. Nsr1p binds to the upstream portion of the 3′-UTR, known as RIII in DI-72 (+)RNA (Jiang *et al.*, 2010). The binding of Nsr1p to RIII(+) is relevant since the inhibitory effect of Nsr1p on DI-72 repRNA accumulation *in vivo* was lost when DI-72 repRNA lacked RIII sequence. Regulated overexpression of Nsr1p revealed that Nsr1p must be present at the beginning of viral replication for efficient inhibition. Moreover, the purified recombinant Nsr1p inhibited the *in vitro* replication of the viral RNA in a yeast cell-free assay only when preincubated with the viral RNA before the *in vitro* replication assay (Jiang *et al.*, 2010). These data suggest that Nsr1p likely inhibits an early step, such as RNA recruitment, in the replication process. We propose that Nsr1p may inhibit TBSV replication via specific binding to the viral RNA and, thus, resulting in inefficient viral RNA recruitment for replication.

Nucleolin is also involved in replication/pathogenesis of various RNA and DNA viruses. Similar to its inhibitory role in tombusvirus replication, nucleolin also inhibits replication of simian virus 40 (SV40) DNA virus by interfering with the unwinding of SV40 origin (Daniely and Borowiec, 2000). In several other cases, nucleolin stimulates viral infections by, for example, interacting with the 3′-UTR of poliovirus and stimulating an early step of virus replication *in vitro* (Waggoner and Sarnow, 1998). The NS1 protein of influenza A virus binds to nucleolin and colocalizes with nucleolin in the nucleolus, possibly affecting cellular events, such as shut down of host protein synthesis (Murayama *et al.*, 2007).

Another RNA-binding host protein identified during the YKO screens is Xrn1p/Kem1p 5′–3′ exoribonuclease (Xrn4p in plants/mammals) (Panavas *et al.*, 2005b; Serviene *et al.*, 2005). Xrn1p inhibits tombusvirus replication and might be a component of the host innate immunity. Xrn1p is a major enzyme in the RNA degradation pathway in yeast (Johnson, 1997; Sheth and Parker, 2003). Xrn1p is involved in degradation of tombusvirus RNA, including partially degraded viral RNAs generated by endoribonucleases (Cheng *et al.*, 2006, 2007; Jaag and Nagy, 2009). In the absence of Xrn1p/Xrn4p, accumulation of tombusvirus RNA increased several fold as well as novel viral recombinant RNAs emerged rapidly in yeast and in plants. Moreover, based on a yeast cell-free TBSV replication assay, which supports authentic replication and recombination of TBSV, it has been shown that the purified recombinant Xrn1p efficiently inhibited the accumulation of recombinants and partly degraded viral RNAs. Altogether, the data from yeast and plant hosts and a cell-free assay confirmed a central role for the cytosolic 5′–3′ exoribonuclease in TBSV replication, recombination and viral RNA degradation (Cheng *et al.*, 2006, 2007; Jaag and Nagy, 2009).

D. Proteins involved in lipid/membrane biosynthesis and metabolism

Many tombusviruses replicate on the cytosolic surface of peroxisomes, where the replicase complexes form (McCartney *et al.*, 2005; Pathak *et al.*, 2008). Electron microscopic images of cells replicating tombusviruses have revealed extensive remodeling of membranes and indicated active lipid biosynthesis (Fig. 1) (Barajas *et al.*, 2009; McCartney *et al.*, 2005; Navarro *et al.*, 2006). Therefore, it was expected that the genome-wide screens would identify lipid biosynthesis genes affecting TBSV replication. Indeed, the systematic genome-wide screens in yeast identified a list of 14 host genes involved in lipid biosynthesis/metabolism, which affected tombusvirus replication and recombination (Fig. 4) (Jiang *et al.*, 2006; Panavas *et al.*, 2005b; Serviene *et al.*, 2005, 2006). The 14 identified host genes involved in lipid biosynthesis/metabolism included eight genes affecting phospholipid biosynthesis, four genes affecting fatty acid biosynthesis/metabolism, and two genes affecting ergosterol synthesis (Table I).

Further studies have been performed with Erg25p, a critical enzyme in the sterol biosynthesis pathway. Sterols are ubiquitous and essential membrane components in all eukaryotes, affecting membrane rigidity, fluidity and permeability by interacting with other lipids and proteins within the membranes (Bloch, 1983, 1992). Sterols are important for the organization of detergent-resistant lipid rafts (Roche *et al.*, 2008). Erg25p in yeast and the orthologous *SMO1* (sterol-4α-methyl-oxidase) and *SMO2* in plants perform the removal of two methyl groups at C4 position, which is critical and rate limiting during sterol synthesis (Darnet and Rahier, 2004). Indeed, sterol molecules become functional structural components of membranes only after the removal of the two methyl groups at C_4. Down-regulation or pharmacological inhibition of *ERG25* in yeast led to four- to fivefold decreased TBSV RNA accumulation (Sharma *et al.*, 2010). Among the functions provided by sterols during tombusvirus replication, two roles for sterols have been identified. The first function is to facilitate the assembly of the viral replicase complex based on the reduced *in vitro* activity of the tombusvirus replicase when isolated from yeast cells with reduced level of sterols. The second function is related to the stability of $p92^{pol}$ viral replication protein, which showed approximately threefold reduced half-life when expressed in yeast treated with a chemical inhibitor of *ERG25*. The bulky $p92^{pol}$ replicase protein might be exposed to cytosolic proteases in sterol-poor microenvironment. Alternatively, the structure of $p92^{pol}$ is different under sterol-depleted condition, leading to premature degradation of $p92^{pol}$. Moreover, the subcellular localization of $p92^{pol}$ could be different if less than normal level of sterols was available in cells.

Replication of other viruses, such as Dengue virus, Norwalk virus, and HCV, also depends on sterols (Chang, 2009; Kapadia and Chisari, 2005; Rothwell *et al.*, 2009; Sagan *et al.*, 2006). For example, infection with WNV has been shown to result in redistribution of cholesterol to the sites of virus replication, possibly from the plasma membrane, and reduce anti-viral responses (Mackenzie *et al.*, 2007). The HCV replicase complex has been shown to be associated with cholesterol-rich lipid rafts (Aizaki *et al.*, 2004). These findings are expected to promote further studies on dissecting the functional/structural roles of sterols during virus replication.

E. Cellular proteins involved in vesicle-mediated transport/ intracellular protein targeting

After translation of the replication proteins from the TBSV RNA by the host ribosome, the replication proteins together with the viral RNA must be localized to the peroxisomal membranes. The replication proteins might have additional functions in the infected cells that could require targeting and transportation. Therefore, it is interesting that 33 host genes have been identified, which are involved in intracellular protein targeting and vesicle-mediated transport (Fig. 4). However, the functions of the identified host proteins during TBSV replication are currently not understood.

Follow-up experiments have been conducted with the so-called ESCRT proteins (Table I), since the YKO screen revealed the involvement of seven ESCRT proteins in TBSV replication (Panavas *et al.*, 2005b). The large number of ESCRT proteins identified during the screens suggests that TBSV might hijack the ESCRT proteins to assist replication. For example, recruitment of ESCRT proteins for TBSV replication could facilitate the assembly of the replicase complex, including the formation of TBSV-induced spherules and vesicles in infected cells (Fig. 1) (McCartney *et al.*, 2005). Indeed, induction of membranous spherule-like replication structures in infected cells might be common for many plus-stranded RNA viruses (Kopek *et al.*, 2007).

ESCRT proteins are known to be involved in the endosome pathway, which is a major protein-sorting pathway in eukaryotic cells. The endosome pathway is used to downregulate plasma membrane proteins; and sort newly synthesized membrane proteins from *trans*-Golgi vesicles to the lysosome or the plasma membrane (Hurley and Emr, 2006; Katzmann *et al.*, 2002; Slagsvold *et al.*, 2006). The ESCRT proteins have a major role in sorting of cargo proteins from the endosomal limiting membrane to multivesicular bodies (MVBs) via membrane invagination and vesicle formation. Defects in the endosome/MVB pathway can cause serious diseases, including early embryonic lethality, defect in growth control, and cancer (Bowers and Stevens, 2005; Hurley and Emr, 2006; Katzmann *et al.*, 2002; Slagsvold *et al.*, 2006).

TBSV replication was inhibited in the absence of the following ESCRT proteins: Vps23p and Vps28p (ESCRT-I complex), Snf7p and Vps24p (ESCRT-III complex); Doa4p ubiquitin isopeptidase, Did2p having Doa4p-related function; and Vps4p AAA-type ATPase (Table I) (Panavas *et al.*, 2005b). Intriguingly, the ubiquitinated TBSV p33 replication protein was found to interact with Vps23p ESCRT-I and Bro1p accessory ESCRT factors (Barajas *et al.*, 2009a). The interaction has been shown to lead to the recruitment of Vps23p and possibly Bro1p to the peroxisomes, the sites of TBSV replication. This is followed by the recruitment of ESCRT-III proteins, Snf7p and Vps24p, which could help the optimal assembly of the replicase complex, facilitate the grouping of p33/p92 molecules together in the membrane and/or promote the formation of viral spherules by deforming the membrane (membrane invagination). Then, Doa4p deubiquitination enzyme is predicted to remove ubiquitin from the ubiquitinated p33, while Vps4p ATPase likely recycles the ESCRT proteins from the replicase complex at the end of the assembly (Barajas *et al.*, 2009a).

The above model is strongly supported by experimental data, such as the reduced activity of the tombusviral replicase when derived from *vps23Δ* or *vps24Δ* yeast (Barajas *et al.*, 2009a). Moreover, the minus-stranded viral RNA in the replicase from *vps23Δ* or *vps24Δ* yeast became more accessible to ribonuclease, suggesting that the protection of the viral RNA is compromised within the replicase complex assembled in the absence of ESCRT proteins. Thus, the role of ESCRT proteins seems to control the quality of the replicase complex assembly, making the viral RNAs within replicase complex well protected from ribonucleases. Based on these observations, we propose that ESCRT proteins help tombus-viruses hide from host defense recognition and avoid the attack by the host defense machinery during viral replication. Similar role for ESCRT proteins or other host factors might help other (+)RNA viruses, which are also known to deform membranes and form spherules during replication.

The ESCRT proteins are also recruited by various viruses, such as enveloped retro-, filo-, arena-, rhabdo-, and paramyxoviruses to the plasma membrane, leading to budding and fission of the viral particles from infected cells (Morita and Sundquist, 2004; Perlman and Resh, 2006).

F. Membrane-associated cellular proteins

Since tombusvirus replication takes place on the peroxisomal and alternatively on the ER membranes, it is possible that some membrane-bound host proteins could affect TBSV replication directly or indirectly. Among the identified nine host proteins in this group that affected TBSV replication, only the role of Pmr1p has been characterized in detail. Inactivation of *PMR1*, which codes for the highly conserved Ca^{2+}/Mn^{2+} pump in yeast, led to greatly increased level of TBSV RNA recombination as well as higher viral RNA accumulation (Jaag *et al.*, 2010).

Pmr1p (for "*p*lasma *m*embrane ATPase *r*elated") is an ATPase-driven Ca^{2+}/Mn^{2+} exporter/pump in yeast (Ton and Rao, 2004). Pmr1p controls Ca^{2+} and Mn^{2+} influx to the Golgi from the cytosol, which is important for signal transduction and protein sorting in yeast.

Inactivation of *PMR1* has been shown to lead to an ~160-fold increase in TBSV RNA recombination (Jaag *et al.*, 2010). Expression of separation-of-function mutants of Pmr1p revealed that the ability of Pmr1p to control the Mn^{2+} concentration in the cytosol is a key factor in viral RNA recombination. Based on the known cellular function of Pmr1p and *in vitro* and *in vivo* TBSV recombination assays, it has been proposed that the Pmr1p Ca^{2+}/Mn^{2+} ion pump regulates TBSV RNA recombination by keeping the Mn^{2+} concentration low in the cytosol (Jaag *et al.*, 2010). When Mn^{2+} concentration is low in the cell, then the RdRp within the viral replicase utilizes the far more abundant Mg^{2+} over Mn^{2+}, leading to low-frequency RNA recombination. On the other hand, deletion/inhibition of the Pmr1p Ca^{2+}/Mn^{2+} pump leads to an increased level of cytosolic Mn^{2+} (Mandal *et al.*, 2000), promoting the more efficient use of Mn^{2+} by the viral RdRp, which leads to high-frequency RNA recombination. Thus, Pmr1p activity in the cell affects TBSV RNA replication and recombination through regulating the cytosolic Mn^{2+} level (Jaag *et al.*, 2010). Overall, the emerging picture from the genome-wide studies and the more detailed studies on *PMR1* is that complex interactions between TBSV and its host affect not only TBSV replication, but viral adaptation and evolution as well.

Interestingly, high Mn^{2+} also affects the activity of the reverse transcriptase (Bolton *et al.*, 2002; Vartanian *et al.*, 1999) and the template activity of several RNA virus RdRps, making the polymerase action less specific for templates, and stimulating nucleotide misincorporation (Alaoui-Lsmaili *et al.*, 2000; Arnold *et al.*, 2004; Hardy *et al.*, 2003; Poranen *et al.*, 2008; Yi *et al.*, 2003). Thus, the roles of Ca^{2+}/Mn^{2+} ion pumps could be general and widespread among viruses.

G. Proteins with stress-related functions

Viruses are known to induce cellular stress during the infection process. The stress stimuli then lead to the activation and high-level expression of stress-related proteins, such as chaperones, including heat-shock proteins (Brodsky and Chiosis, 2006; Mayer, 2005). The genome-wide screens identified 11 proteins in this group that affected TBSV replication (Fig. 4). Among the stress-related proteins, further studies were conducted with Ssa1/2p Hsp70.

The tombusvirus replicase complex contains Hsp70, an abundant cytosolic chaperone, which is required for TBSV replication (Pogany *et al.*, 2008; Serva and Nagy, 2006). The interaction between Hsp70 and the tombusvirus replication proteins occurs in the functional replicase (Fig. 3), since affinity purification of Ssa1p from the solubilized membrane fraction of yeast resulted in copurification of tombusvirus replicase activity (Wang *et al.*, 2009b). Hsp70 chaperone seems to play multiple and essential roles during TBSV replication (Fig. 5). For example, using a temperature-sensitive mutant of Hsp70 at nonpermissive temperature has led to cytosolic localization of p33 replication protein (Wang *et al.*, 2009a). Shifting down from nonpermissive to permissive temperature resulted in relocalization of p33 to the peroxisome membrane surface in yeast. Subcellular fractionation experiments have shown that the viral replication proteins are mostly cytosolic in *ssa1ssa2* yeast at the early time point (Wang *et al.*, 2009b). Interestingly, the viral proteins become partly membrane-bound at a latter time point likely due to partial complementation by the cytosolic, stress-inducible Ssa3p and Ssa4p, which operate at higher levels in *ssa1ssa2* cells (Becker *et al.*, 1996; Werner-Washburne *et al.*, 1987). This suggests that Hsp70 is involved in localization/transportation of the viral replication proteins. It has been proposed that the binding of p33 and $p92^{pol}$ replication proteins to Hsp70 results in shielding the hydrophobic transmembrane domains in the replication protein that could prevent their aggregation and promote binding to

Pex19p transport protein (see below). The latter interaction is needed for peroxisomal targeting of the replication proteins (Pathak *et al.*, 2008).

The second demonstrated function of Hsp70 is the insertion of the replication proteins into intracellular membranes (Wang *et al.*, 2009b). Integration of p33 replication protein into subcellular membranes, such as peroxisomal and ER (Jonczyk *et al.*, 2007), is thought to be critical for tombusvirus replication. This is because p33 mutants localized in the cytosol do not support TBSV replication in yeast or in plant cells (McCartney *et al.*, 2005; Panavas *et al.*, 2005a). The insertion of the replication proteins into the membrane might require additional cellular factors as shown for cellular membrane-associated proteins (Brodsky and Chiosis, 2006; Young *et al.*, 2004).

The third function of Hsp70 is to assist the assembly of the TBSV replicase. This was demonstrated by Pogany *et al.* (2008) using a yeast extract depleted in Hsp70. The addition of purified recombinant Hsp70 to the cell-free assay was necessary for the assembly of the viral replicase complex and active replication of the repRNA *in vitro* (Pogany *et al.*, 2008).

Taken together, the available data support a functional role for the cytosolic Hsp70 in several steps of tombusvirus replication, including subcellular localization, membrane insertion, and assembly of the viral replicase complex (Fig. 5) (Pogany *et al.*, 2008).

The above findings with TBSV and Hsp70 might be common among other viruses as well. Accordingly, host-coded heat-shock proteins, such as Hsp70 chaperone family, the J-domain chaperones, and Hsp90, are implicated in replication of HCV, FHV, influenza, vesicular stomatitis virus, retroviruses (HIV), hepatitis B virus, and other RNA viruses (Brown *et al.*, 2005; Castorena *et al.*, 2007; Connor *et al.*, 2007; Kumar and Mitra, 2005; Momose *et al.*, 2002; Naito *et al.*, 2007; Nakagawa *et al.*, 2007; Okamoto *et al.*, 2006; Qanungo *et al.*, 2004; Sohn *et al.*, 2006; Tomita *et al.*, 2003; Weeks and Miller, 2008). The host chaperones have been proposed to stimulate polymerase (RdRp) activity (Momose *et al.*, 2002), and activate the reverse transcriptase for hepadnaviruses (Hu *et al.*, 2004; Tavis *et al.*, 1998). The cytosolic Hsp70 proteins might also affect stability and function of viral proteins during infections since a subset of *HSP70* genes are expressed at enhanced levels in plants infected by various plant viruses (Aparicio *et al.*, 2005; Aranda *et al.*, 1996; Whitham *et al.*, 2003, 2006).

H. Proteins involved in general metabolism of the cell

Virus replication absolutely depends on the resources provided by the host cells. Thus, many proteins involved in general cellular metabolism could indirectly affect virus replication. However, most host proteins have multiple functions and viruses might exploit alternative, less characterized functions of cellular proteins as explained below for a host metabolic enzyme co-opted for TBSV replication. Overall, surprisingly large number of host proteins has been identified (15% of all identified proteins; Fig. 4) in this group that affected TBSV replication. Future studies will be needed to dissect the direct or indirect roles of these proteins for tombusvirus replication.

The metabolic enzyme studied in more detail is called glyceraldehyde-3-phosphate dehydrogenase (GAPDH, coded by *TDH1* and *TDH2* genes in yeast), which was discovered as a component of the tombusvirus replicase complex via a proteomics analysis of a purified viral replicase preparation (Fig. 3) (Serva and Nagy, 2006). The cellular distribution of GAPDH changed dramatically due to relocalization from the cytosol to the site of TBSV repRNA replication (peroxisome) in yeast cells (Wang and Nagy, 2008). Downregulation of GAPDH levels in yeast correlated with reduced level of (+)-strand repRNA, suggesting an important role for this protein in TBSV replication.

GAPDH is a ubiquitous, highly conserved, very abundant protein (Sirover, 1999) with glyceraldehyde-3-phosphate dehydrogenase function, which is a key component of cytosolic energy production. However, GAPDH displays many additional activities that are unrelated to its glycolytic function. These cellular activities include roles in modulation of the cytoskeleton, vesicular secretary transport, endocytosis, nuclear membrane fusion, nuclear tRNA transport, apoptosis, DNA replication and repair, maintenance of telomere structure, and transcriptional control of histone gene expression (Sirover, 1999, 2005). An emerging new function of GAPDH is to bind to various RNAs, such as AU-rich sequences at the 3′-terminus of mRNAs, which can lead to stabilization of the RNA in the cell (Bonafe *et al.*, 2005).

In addition to relocalization from the cytosol to the peroxisomal membrane surface during TBSV replication in yeast, and being part of the replicase complex, GAPDH has been shown to affect viral RNA synthesis. For example, downregulation of GAPDH inhibited TBSV replication, resulting in an ~1:1 ratio of (+) and (−)RNAs, instead of the hallmark asymmetric RNA synthesis leading to excess (+)RNA. Since GAPDH binds to an AU pentamer sequence in the TBSV (−)RNA, GAPDH has been proposed to play a role in asymmetric viral RNA synthesis by selectively retaining the TBSV (−)RNA template in the replicase complex (Fig. 5) (Wang and Nagy, 2008).

The novel functional role of GAPDH in TBSV RNA replication expands the battery of activities for this multifunctional enzyme. GAPDH is likely co-opted by RNA viruses due to its ability to bind to AU-rich sequences (Nagy and Rigby, 1995; Nagy *et al.*, 2000). Accordingly, GAPDH has been shown to bind to AU-rich sequences present in various RNA viruses, including hepatitis A virus (HAV), HCV, and human parainfluenza virus type 3 (De *et al.*, 1996; Dollenmaier and

Weitz, 2003; Randall *et al.*, 2007). The functional role of binding of GAPDH to the above viruses is not yet clear. It has been proposed that binding of GAPDH to the internal entry site (IRES) element in HAV could suppress cap-independent translation of HAV RNA (Yi *et al.*, 2000). GAPDH may also be involved in the posttranscriptional regulation of hepatitis B virus gene expression (Zang *et al.*, 1998).

I. Cellular transcription factors

It is currently not yet known if the identified 21 host factors in this group would have direct or indirect roles in TBSV replication (Fig. 4). It is possible that a given transcription factor affects mRNA levels for a set of host proteins that are involved in TBSV replication. It is also feasible that the TBSV replication proteins and/or RNA could interact with cellular transcription factors to reprogram host transcription, for example, in order to increase lipid biosynthesis that could be beneficial for TBSV. The interaction with a cellular transcription factor could also inhibit the production of the components of the innate antiviral pathways, thus reducing host responses to TBSV infections. Future experiments will be needed to answer these questions.

Experiments with Rpb11p, which is part of the pol II complex, revealed that this transcription factor affected the levels of p33 and $p92^{pol}$ in yeast (Jaag *et al.*, 2007). As predicted, downregulation of Rpb11p inhibited TBSV repRNA replication and altered RNA recombination. An *in vitro* tombusvirus replicase assay supported that Rpb11p affects TBSV replication and recombination only indirectly, via regulating p33 and $p92^{pol}$ levels. A model has been proposed that the local concentration of replication proteins, described as molecular crowdedness, within the viral replicase is a factor affecting viral replication and recombination (Jaag *et al.*, 2007).

J. Cellular proteins involved in DNA remodeling/metabolism

The functions of any of these 15 host proteins (Fig. 4) in TBSV replication are not yet known. It is possible that they might affect TBSV replication by regulating transcription of cellular genes important for TBSV replication, similar to the above transcription factors.

K. Cellular and hypothetical proteins with unknown functions

The functions of any of these 27 host proteins in this group (10% of all proteins identified during the screens; Fig. 4) in TBSV replication are not yet known. However, the number of genes in this group keeps decreasing due to advances in yeast research in general and TBSV host factors in particular. For example, the original YKO screens on TBSV replication and recombination led to the identification of *HUR1* with unknown function (Panavas *et al.*, 2005b; Serviene *et al.*, 2005). Subsequent research on *HUR1*, however, revealed that its effect on TBSV replication and recombination was due to the *PMR1* gene (see Section F), which overlaps with *HUR1* on the yeast chromosome (Jaag *et al.*, 2010). Overall, additional research on the cellular functions of this group of genes and bioinformatics analysis to predict their putative functions/activities will likely help understanding if these factors affect TBSV RNA replication directly or indirectly.

L. Host factors missed during the global genomics and proteomics screens

In spite of the systematic and multiple genome-wide screens to identify factors affecting TBSV replication, it is possible that there are still additional host factors not yet identified. Indeed, it has recently been shown that the host shuttle protein Pex19p, which is involved in peroxisomal membrane protein transport, plays a role in TBSV protein transportation to the site of replication (Fig. 5) (Pathak *et al.*, 2008). Pex19p not only binds to the peroxisomal targeting signals in p33 *in vitro*, but also it is temporarily associated with the replicase complex. When Pex19p was mistargeted to mitochondrial membranes, the wt p33 was also colocalized to the same mitochondrial membranes (Pathak *et al.*, 2008). However, the role of Pex19p is not essential for TBSV replication since TBSV repRNA replicated as efficiently in *pex19Δ* yeast defective in peroxisome biogenesis as in the wt yeast (Jonczyk *et al.*, 2007; Panavas *et al.*, 2005b). Indeed, confocal microscopy-based approach revealed that the wt tombusvirus p33 replication protein accumulated in the ER in *pex3Δ* or *pex19Δ* yeast lacking peroxisomes, suggesting that tombusvirus replication could occur on the surface of the ER membrane. Moreover, the activity of the isolated tombusvirus replicase from wt, *pex3Δ*, or *pex19Δ* yeasts was comparable, indicating that the assembly of the replicase was as efficient in the ER as in the authentic subcellular environment (Jonczyk *et al.*, 2007). Overall, these data demonstrated that TBSV, relying on the wt replication proteins, could efficiently replicate on an alternative intracellular membrane. Thus, RNA viruses might have remarkable flexibility for using various host membranes for their replication.

It is currently not known how many other host proteins, similar to Pex19p, have been "missed" in the previous screens. Performing additional genome-wide screens will tell us if we are getting closer to identification of all host factors affecting TBSV replication and recombination at the cellular level.

V. VALIDATION OF HOST FACTORS IN A PLANT HOST AND INDUCTION OF RESISTANCE AGAINST TBSV

Viruses are cellular pathogens that utilize abundant cellular resources for their replication. A small and simple eukaryotic cell, such as yeast, likely can provide most host factors needed for virus replication as demonstrated in case of TBSV (Nagy, 2008). Yet, it is important to demonstrate the relevance of the host factors identified in yeast in a natural plant host as well. The validation of the identified yeast factors has now been done for a growing number of host genes as discussed below. Moreover, the gained knowledge on these host factors can be used to interfere with their functions during TBSV replication, resulting in induction of resistance or development of antiviral treatments.

Knocking down the expression level of the plant *NbGAPDH*, the ortholog of yeast *TDH1* and *TDH2* genes, in *Nicotiana benthamiana* host led to 85–90% reduction in TBSV and *Cucumber necrosis virus* (CNV, another tombusvirus) genomic RNA accumulation (Wang and Nagy, 2008). Importantly, the GAPDH-silenced plants showed resistance against tombusviruses, indicating that this approach could lead to a new antiviral approach. The same strategy did not interfere with the accumulation of an unrelated plant RNA virus (i.e., TMV), which could be due to the lack of function for GAPDH in TMV replication. This is not surprising since TMV is more similar to BMV than to TBSV and it has been shown previously that TBSV is affected by a vastly different set of host factors as BMV (Kushner *et al.*, 2003; Panavas *et al.*, 2005b).

The second example that TBSV replication requires similar host factors in yeast and in plants is based on the cytosolic Hsp70. Knockdown and chemical inhibition experiments showed that Hsp70 is required for TBSV genomic RNA replication in plant cells and whole *N. benthamiana* host (Wang *et al.*, 2009b). Interestingly, the pharmacologic inhibition of Hsp70 also inhibited the accumulation of other plant RNA viruses, suggesting that several plant viruses also depend on Hsp70 during their infection cycles (Wang *et al.*, 2009b). Accordingly, Hsp70 was shown to be part of the CNV and TMV replicase complexes (Nishikiori *et al.*, 2006; Serva and Nagy, 2006). Altogether, the chemical inhibition of Hsp70 functions in plant hosts could be a broad antiviral approach.

The third example is that inhibition of sterol biosynthesis in plant protoplasts or in plant leaves with chemical inhibitors or silencing of *SMO1/SMO2* genes (involved in phytosterol synthesis) in *N. benthamiana* also resulted in reduced TBSV RNA accumulation. These data strongly support the role of sterols and host membranes in tombusvirus replication in plants as well (Sharma *et al.*, 2010). Moreover, using chlorpromazine (CPZ) to alter membrane properties of the host cells led to strong inhibition of TBSV accumulation in plants (Sasvari *et al.*, 2009). Interestingly, CPZ was also an effective inhibitor of other plant viruses, including TMV and *Turnip crinkle virus*, suggesting that CPZ has a broad range of antiviral activity.

Another example in plants is the successful inhibition of tombusvirus replication in *N. benthamiana* via overexpression of dominant negative mutants of ESCRT-III and Vps4p (Barajas *et al.*, 2010). This inhibition by the dominant negative ESCRT mutants seems to be specific for tombusviruses, since the distantly related *Tobacco rattle virus* (TRV) RNA accumulation was not inhibited in these plants. The inhibitory effect on tombusvirus replication by the overexpressed dominant negative ESCRT mutants seems to be direct, since the activity of the tombusvirus replicase was also reduced when isolated from these plants (Barajas *et al.*, 2010).

It seems that similar to replication factors, plant recombination factors, such as the 5′–3′ exoribonuclease and the Ca^{2+}/Mn^{2+} pump, affect TBSV RNA recombination. For example, knockdown of *XRN4* (Jaag and Nagy, 2009), the ortholog of the yeast *XRN1*, or cosilencing of *LCA1* and *ECA3* Ca^{2+}/Mn^{2+} pumps (Jaag *et al.*, 2010), orthologs of the yeast *PMR1*, in *N. benthamiana* plants resulted in enhanced TBSV recombination and replication, similar to the picture obtained in yeast with knockout mutants.

In conclusion, the above examples provide strong evidence that host factors identified and characterized in the yeast model host are also relevant for TBSV replication in a native plant host infected with the full-length wt TBSV RNA. The discussed examples also showed convincingly that inhibition of host factors could lead to the development of specific or broad-range resistance or other antiviral strategies against TBSV and possibly other plant viruses as well.

VI. SUMMARY AND OUTLOOK

Multiple genome-wide screens with TBSV and intensive research on individual host genes using yeast as a model host led to the identification of over 250 host genes. The roles of several of these host genes in TBSV replication and recombination have been validated. However, one of the surprising observations is that the genome-wide screens led to the identification of mostly unique host factors that were not identified in other genome-wide screens. It is possible that many genes involved in TBSV repRNA replication and recombination are functionally redundant. Indeed, the host proteins that were identified in the highly purified functional tombusvirus replicase complex (such as Hsp70, GAPDH, and eEF1A) are coded by two or more genes in the yeast genome. This functional redundancy in host genes justifies the need for additional genome- or proteome-wide approaches in the identification of host genes affecting TBSV replication and recombination. Characterization of the functions of all host factors in the infected plant cells will likely result in better, more efficient antiviral strategies and/or reduce the damage to the plants caused by viral infections.

ACKNOWLEDGMENTS

We are grateful to Dr. Daniel Barajas for the unpublished EM images in Fig. 1. This work was supported by NSF (MCB0078152), NIH, and by the University of Kentucky to PDN.

REFERENCES

Ahlquist, P. (2006). Parallels among positive-strand RNA viruses, reverse-transcribing viruses and double-stranded RNA viruses. *Nat. Rev. Microbiol.* **4**:371–382.

Ahlquist, P., Noueiry, A. O., Lee, W. M., Kushner, D. B., and Dye, B. T. (2003). Host factors in positive-strand RNA virus genome replication. *J. Virol.* **77**:8181–8186.

Aizaki, H., Lee, K. J., Sung, V. M., Ishiko, H., and Lai, M. M. (2004). Characterization of the hepatitis C virus RNA replication complex associated with lipid rafts. *Virology* **324**:450–461.

Alaoui-Lsmaili, M. H., Hamel, M., L'Heureux, L., Nicolas, O., Bilimoria, D., Labonte, P., Mounir, S., and Rando, R. F. (2000). The hepatitis C virus NS5B RNA-dependent RNA polymerase activity and susceptibility to inhibitors is modulated by metal cations. *J. Hum. Virol.* **3**:306–316.

Aparicio, F., Thomas, C. L., Lederer, C., Niu, Y., Wang, D., and Maule, A. J. (2005). Virus induction of heat shock protein 70 reflects a general response to protein accumulation in the plant cytosol. *Plant Physiol.* **138**:529–536.

Aranda, M. A., Escaler, M., Wang, D., and Maule, A. J. (1996). Induction of HSP70 and polyubiquitin expression associated with plant virus replication. *Proc. Natl. Acad. Sci. USA* **93**:15289–15293.

Arnold, J. J., Gohara, D. W., and Cameron, C. E. (2004). Poliovirus RNA-dependent RNA polymerase (3Dpol): Pre-steady-state kinetic analysis of ribonucleotide incorporation in the presence of Mn^{2+}. *Biochemistry* **43**:5138–5148.

Barajas, D., Jiang, Y., and Nagy, P. D. (2009a). A unique role for the host ESCRT proteins in replication of Tomato bushy stunt virus. *PLoS Pathog* **5**:e1000705.

Barajas, D., Li, Z., and Nagy, P. D. (2009b). The Nedd4-type Rsp5p ubiquitin ligase inhibits tombusvirus replication by regulating degradation of the p92 replication protein and decreasing the activity of the tombusvirus replicase. *J. Virol.* **83**:11751–11764.

Baroth, M., Orlich, M., Thiel, H. J., and Becher, P. (2000). Insertion of cellular NEDD8 coding sequences in a pestivirus. *Virology* **278**:456–466.

Barretto, N., Jukneliene, D., Ratia, K., Chen, Z., Mesecar, A. D., and Baker, S. C. (2005). The papain-like protease of severe acute respiratory syndrome coronavirus has deubiquitinating activity. *J. Virol.* **79**:15189–15198.

Barry, M., and Fruh, K. (2006). Viral modulators of cullin RING ubiquitin ligases: Culling the host defense. *Sci. STKE* **2006**:pe21.

Bastin, M., and Hall, T. C. (1976). Interaction of elongation factor 1 with aminoacylated *Brome mosaic virus* and tRNA's. *J. Virol.* **20**:117–122.

Baumstark, T., and Ahlquist, P. (2001). The *Brome mosaic virus* RNA3 intergenic replication enhancer folds to mimic a tRNA TpsiC-stem loop and is modified in vivo. *RNA* **7**:1652–1670.

Becker, J., Walter, W., Yan, W., and Craig, E. A. (1996). Functional interaction of cytosolic hsp70 and a DnaJ-related protein, Ydj1p, in protein translocation in vivo. *Mol. Cell. Biol.* **16:**4378–4386.

Beckham, C. J., Light, H. R., Nissan, T. A., Ahlquist, P., Parker, R., and Noueiry, A. (2007). Interactions between *Brome mosaic virus* RNAs and cytoplasmic processing bodies. *J. Virol.* **81:**9759–9768.

Bloch, K. E. (1983). Sterol structure and membrane function. *CRC Crit. Rev. Biochem.* **14:**47–92.

Bloch, K. (1992). Sterol molecule: Structure, biosynthesis, and function. *Steroids* **57:**378–383.

Blumenthal, T., Young, R. A., and Brown, S. (1976). Function and structure in phage Qbeta RNA replicase. Association of EF-Tu-Ts with the other enzyme subunits. *J. Biol. Chem.* **251:**2740–2743.

Bolton, E. C., Mildvan, A. S., and Boeke, J. D. (2002). Inhibition of reverse transcription in vivo by elevated manganese ion concentration. *Mol. Cell* **9:**879–889.

Bonafe, N., Gilmore-Hebert, M., Folk, N. L., Azodi, M., Zhou, Y., and Chambers, S. K. (2005). Glyceraldehyde-3-phosphate dehydrogenase binds to the AU-rich 3′ untranslated region of colony-stimulating factor-1 (CSF-1) messenger RNA in human ovarian cancer cells: Possible role in CSF-1 posttranscriptional regulation and tumor phenotype. *Cancer Res.* **65:**3762–3771.

Bowers, K., and Stevens, T. H. (2005). Protein transport from the late Golgi to the vacuole in the yeast *Saccharomyces cerevisiae. Biochim. Biophys. Acta* **1744:**438–454.

Brinton, M. A. (2001). Host factors involved in West Nile virus replication. *Ann. NY Acad. Sci.* **951:**207–219.

Brodsky, J. L., and Chiosis, G. (2006). Hsp70 molecular chaperones: Emerging roles in human disease and identification of small molecule modulators. *Curr. Top. Med. Chem.* **6:**1215–1225.

Brown, G., Rixon, H. W., Steel, J., McDonald, T. P., Pitt, A. R., Graham, S., and Sugrue, R. J. (2005). Evidence for an association between heat shock protein 70 and the respiratory syncytial virus polymerase complex within lipid-raft membranes during virus infection. *Virology* **338:**69–80.

Castorena, K. M., Weeks, S. A., Stapleford, K. A., Cadwallader, A. M., and Miller, D. J. (2007). A functional heat shock protein 90 chaperone is essential for efficient flock house virus RNA polymerase synthesis in Drosophila cells. *J. Virol.* **81:**8412–8420.

Chang, K. O. (2009). Role of cholesterol pathways in norovirus replication. *J. Virol.* **83:**8587–8595.

Cheng, C. P., Serviene, E., and Nagy, P. D. (2006). Suppression of viral RNA recombination by a host exoribonuclease. *J. Virol.* **80:**2631–2640.

Cheng, C. P., Jaag, H. M., Jonczyk, M., Serviene, E., and Nagy, P. D. (2007). Expression of the Arabidopsis Xrn4p 5′–3′ exoribonuclease facilitates degradation of tombusvirus RNA and promotes rapid emergence of viral variants in plants. *Virology* **368:**238–248.

Cherry, S., Doukas, T., Armknecht, S., Whelan, S., Wang, H., Sarnow, P., and Perrimon, N. (2005). Genome-wide RNAi screen reveals a specific sensitivity of IRES-containing RNA viruses to host translation inhibition. *Genes Dev.* **19:**445–452.

Chuang, S. M., Chen, L., Lambertson, D., Anand, M., Kinzy, T. G., and Madura, K. (2005). Proteasome-mediated degradation of cotranslationally damaged proteins involves translation elongation factor 1A. *Mol. Cell. Biol.* **25:**403–413.

Cimarelli, A., and Luban, J. (1999). Translation elongation factor 1-alpha interacts specifically with the human immunodeficiency virus type 1 Gag polyprotein. *J. Virol.* **73:**5388–5401.

Connor, J. H., McKenzie, M. O., Parks, G. D., and Lyles, D. S. (2007). Antiviral activity and RNA polymerase degradation following Hsp90 inhibition in a range of negative strand viruses. *Virology* **362:**109–119.

Cristea, I. M., Carroll, J. W., Rout, M. P., Rice, C. M., Chait, B. T., and MacDonald, M. R. (2006). Tracking and elucidating alphavirus–host protein interactions. *J. Biol. Chem.* **281:**30269–30278.

Daniely, Y., and Borowiec, J. A. (2000). Formation of a complex between nucleolin and replication protein A after cell stress prevents initiation of DNA replication. *J. Cell Biol.* **149:**799–810.

Darnet, S., and Rahier, A. (2004). Plant sterol biosynthesis: Identification of two distinct families of sterol 4alpha-methyl oxidases. *Biochem. J.* **378**(Pt. 3)**:**889–898.

Davis, W. G., Blackwell, J. L., Shi, P. Y., and Brinton, M. A. (2007). Interaction between the cellular protein eEF1A and the 3′-terminal stem-loop of West Nile virus genomic RNA facilitates viral minus-strand RNA synthesis. *J. Virol.* **81:**10172–10187.

De, B. P., Gupta, S., Zhao, H., Drazba, J. A., and Banerjee, A. K. (1996). Specific interaction in vitro and in vivo of glyceraldehyde-3-phosphate dehydrogenase and LA protein with *cis*-acting RNAs of human parainfluenza virus type 3. *J. Biol. Chem.* **271:**24728–24735.

De Nova-Ocampo, M., Villegas-Sepulveda, N., and del Angel, R. M. (2002). Translation elongation factor-1alpha, La, and PTB interact with the 3′ untranslated region of dengue 4 virus RNA. *Virology* **295:**337–347.

Dollenmaier, G., and Weitz, M. (2003). Interaction of glyceraldehyde-3-phosphate dehydrogenase with secondary and tertiary RNA structural elements of the hepatitis A virus 3′ translated and non-translated regions. *J. Gen. Virol.* **84**(Pt. 2)**:**403–414.

Dreher, T. W. (1999). Functions of the 3′-untranslated regions of positive strand RNA viral genomes. *Annu. Rev. Phytopathol.* **37:**151–174.

Dreher, T. W., Uhlenbeck, O. C., and Browning, K. S. (1999). Quantitative assessment of EF-1alpha.GTP binding to aminoacyl-tRNAs, aminoacyl-viral RNA, and tRNA shows close correspondence to the RNA binding properties of EF-Tu. *J. Biol. Chem.* **274:**666–672.

Geoffroy, M. C., Chadeuf, G., Orr, A., Salvetti, A., and Everett, R. D. (2006). Impact of the interaction between herpes simplex virus type 1 regulatory protein ICP0 and ubiquitin-specific protease USP7 on activation of adeno-associated virus type 2 rep gene expression. *J. Virol.* **80:**3650–3654.

Gross, S. R., and Kinzy, T. G. (2005). Translation elongation factor 1A is essential for regulation of the actin cytoskeleton and cell morphology. *Nat. Struct. Mol. Biol.* **12:** 772–778.

Hao, L., Sakurai, A., Watanabe, T., Sorensen, E., Nidom, C. A., Newton, M. A., Ahlquist, P., and Kawaoka, Y. (2008). Drosophila RNAi screen identifies host genes important for influenza virus replication. *Nature* **454:**890–893.

Hardy, R. W., Marcotrigiano, J., Blight, K. J., Majors, J. E., and Rice, C. M. (2003). Hepatitis C virus RNA synthesis in a cell-free system isolated from replicon-containing hepatoma cells. *J. Virol.* **77:**2029–2037.

Hericourt, F., Blanc, S., Redeker, V., and Jupin, I. (2000). Evidence for phosphorylation and ubiquitinylation of the *Turnip yellow mosaic virus* RNA-dependent RNA polymerase domain expressed in a baculovirus-insect cell system. *Biochem. J.* **349**(Pt. 2)**:**417–425.

Hu, J., Flores, D., Toft, D., Wang, X., and Nguyen, D. (2004). Requirement of heat shock protein 90 for human hepatitis B virus reverse transcriptase function. *J. Virol.* **78:**13122–13131.

Hurley, J. H., and Emr, S. D. (2006). The ESCRT complexes: Structure and mechanism of a membrane-trafficking network. *Annu. Rev. Biophys. Biomol. Struct.* **35:**277–298.

Jaag, H. M., and Nagy, P. D. (2009). Silencing of *Nicotiana benthamiana* Xrn4p exoribonuclease promotes tombusvirus RNA accumulation and recombination. *Virology* **386:**344–352.

Jaag, H. H. M., Pogany, J., and Nahy, P. D. (2010). A host Ca2+/Mn2+ ion pump is a factor in the emergence of viral RNA recombinants. *Cell Host Microbe.* **7:**74–81.

Jaag, H. M., Stork, J., and Nagy, P. D. (2007). Host transcription factor Rpb11p affects tombusvirus replication and recombination via regulating the accumulation of viral replication proteins. *Virology* **368:**388–404.

Jiang, Y., Serviene, E., Gal, J., Panavas, T., and Nagy, P. D. (2006). Identification of essential host factors affecting tombusvirus RNA replication based on the yeast Tet promoters Hughes Collection. *J. Virol.* **80:**7394–7404.

Jiang, Y., Li, Z., and Nagy, P. D. (2010). Nucleolin/Nsr1p binds to the 3′ noncoding region of the tombusvirus RNA and inhibits replication. *Virology* **396:**10–20.

Johnson, A. W. (1997). Rat1p and Xrn1p are functionally interchangeable exoribonucleases that are restricted to and required in the nucleus and cytoplasm, respectively. *Mol. Cell. Biol.* **17:**6122–6130.

Johnson, C. M., Perez, D. R., French, R., Merrick, W. C., and Donis, R. O. (2001). The NS5A protein of bovine viral diarrhoea virus interacts with the alpha subunit of translation elongation factor-1. *J. Gen. Virol.* **82**(Pt. 12):2935–2943.

Jonczyk, M., Pathak, K. B., Sharma, M., and Nagy, P. D. (2007). Exploiting alternative subcellular location for replication: Tombusvirus replication switches to the endoplasmic reticulum in the absence of peroxisomes. *Virology* **362:**320–330.

Kao, C. C., Singh, P., and Ecker, D. J. (2001). De novo initiation of viral RNA-dependent RNA synthesis. *Virology* **287:**251–260.

Kapadia, S. B., and Chisari, F. V. (2005). Hepatitis C virus RNA replication is regulated by host geranylgeranylation and fatty acids. *Proc. Natl. Acad. Sci. USA* **102:**2561–2566.

Katzmann, D. J., Odorizzi, G., and Emr, S. D. (2002). Receptor downregulation and multivesicular-body sorting. *Nat. Rev. Mol. Cell Biol.* **3:**893–905.

Kok, K. H., Lei, T., and Jin, D. Y. (2009). siRNA and shRNA screens advance key understanding of host factors required for HIV-1 replication. *Retrovirology* **6:**78.

Kopek, B. G., Perkins, G., Miller, D. J., Ellisman, M. H., and Ahlquist, P. (2007). Three-dimensional analysis of a viral RNA replication complex reveals a virus-induced mini-organelle. *PLoS Biol.* **5:**e220.

Kou, Y. H., Chou, S. M., Wang, Y. M., Chang, Y. T., Huang, S. Y., Jung, M. Y., Huang, Y. H., Chen, M. R., Chang, M. F., and Chang, S. C. (2006). Hepatitis C virus NS4A inhibits cap-dependent and the viral IRES-mediated translation through interacting with eukaryotic elongation factor 1A. *J. Biomed. Sci.* **13:**861–874.

Krishnan, M. N., Ng, A., Sukumaran, B., Gilfoy, F. D., Uchil, P. D., Sultana, H., Brass, A. L., Adametz, R., Tsui, M., Qian, F., Montgomery, R. R., Lev, S., *et al.* (2008). RNA interference screen for human genes associated with West Nile virus infection. *Nature* **455:**242–245.

Kumar, M., and Mitra, D. (2005). Heat shock protein 40 is necessary for human immunodeficiency virus-1 Nef-mediated enhancement of viral gene expression and replication. *J. Biol. Chem.* **280:**40041–40050.

Kushner, D. B., Lindenbach, B. D., Grdzelishvili, V. Z., Noueiry, A. O., Paul, S. M., and Ahlquist, P. (2003). Systematic, genome-wide identification of host genes affecting replication of a positive-strand RNA virus. *Proc. Natl. Acad. Sci. USA* **100:**15764–15769.

Li, Z., Barajas, D., Panavas, T., Herbst, D. A., and Nagy, P. D. (2008). Cdc34p ubiquitin-conjugating enzyme is a component of the tombusvirus replicase complex and ubiquitinates p33 replication protein. *J. Virol.* **82:**6911–6926.

Li, Z., Pogany, J., Panavas, T., Xu, K., Esposito, A. M., Kinzy, T. G., and Nagy, P. D. (2009). Translation elongation factor 1A is a component of the tombusvirus replicase complex and affects the stability of the p33 replication co-factor. *Virology* **385:**245–260.

Mackenzie, J. M., Khromykh, A. A., and Parton, R. G. (2007). Cholesterol manipulation by West Nile virus perturbs the cellular immune response. *Cell Host Microbe* **2:**229–239.

Mandal, D., Woolf, T. B., and Rao, R. (2000). Manganese selectivity of pmr1, the yeast secretory pathway ion pump, is defined by residue gln783 in transmembrane segment 6. Residue Asp778 is essential for cation transport. *J. Biol. Chem.* **275:**23933–23938.

Matsuda, D., Yoshinari, S., and Dreher, T. W. (2004). eEF1A binding to aminoacylated viral RNA represses minus strand synthesis by TYMV RNA-dependent RNA polymerase. *Virology* **321:**47–56.

Mayer, M. P. (2005). Recruitment of Hsp70 chaperones: A crucial part of viral survival strategies. *Rev. Physiol. Biochem. Pharmacol.* **153:**1–46.

McCartney, A. W., Greenwood, J. S., Fabian, M. R., White, K. A., and Mullen, R. T. (2005). Localization of the *Tomato bushy stunt virus* replication protein p33 reveals a peroxisome-to-endoplasmic reticulum sorting pathway. *Plant Cell* **17:**3513–3531.

Mechali, F., Hsu, C. Y., Castro, A., Lorca, T., and Bonne-Andrea, C. (2004). Bovine papillomavirus replicative helicase E1 is a target of the ubiquitin ligase APC. *J. Virol.* **78:**2615–2619.

Miller, C. L., Parker, J. S., Dinoso, J. B., Piggott, C. D., Perron, M. J., and Nibert, M. L. (2004). Increased ubiquitination and other covariant phenotypes attributed to a strain- and temperature-dependent defect of reovirus core protein mu2. *J. Virol.* **78:**10291–10302.

Mnaimneh, S., Davierwala, A. P., Haynes, J., Moffat, J., Peng, W. T., Zhang, W., Yang, X., Pootoolal, J., Chua, G., Lopez, A., Trochesset, M., Morse, D., *et al.* (2004). Exploration of essential gene functions via titratable promoter alleles. *Cell* **118:**31–44.

Momose, F., Naito, T., Yano, K., Sugimoto, S., Morikawa, Y., and Nagata, K. (2002). Identification of Hsp90 as a stimulatory host factor involved in influenza virus RNA synthesis. *J. Biol. Chem.* **277:**45306–45314.

Monkewich, S., Lin, H. X., Fabian, M. R., Xu, W., Na, H., Ray, D., Chernysheva, O. A., Nagy, P. D., and White, K. A. (2005). The p92 polymerase coding region contains an internal RNA element required at an early step in tombusvirus genome replication. *J. Virol.* **79:**4848–4858.

Morita, E., and Sundquist, W. I. (2004). Retrovirus budding. *Annu. Rev. Cell Dev. Biol.* **20:**395–425.

Murayama, R., Harada, Y., Shibata, T., Kuroda, K., Hayakawa, S., Shimizu, K., and Tanaka, T. (2007). Influenza A virus non-structural protein 1 (NS1) interacts with cellular multifunctional protein nucleolin during infection. *Biochem. Biophys. Res. Commun.* **362:**880–885.

Nagy, P. D. (2008). Yeast as a model host to explore plant virus–host interactions. *Annu. Rev. Phytopathol.* **46:**217–242.

Nagy, P. D., and Pogany, J. (2006). Yeast as a model host to dissect functions of viral and host factors in tombusvirus replication. *Virology* **344:**211–220.

Nagy, P. D., and Pogany, J. (2008). Multiple roles of viral replication proteins in plant RNA virus replication. *Methods Mol. Biol.* **451:**55–68.

Nagy, E., and Rigby, W. F. (1995). Glyceraldehyde-3-phosphate dehydrogenase selectively binds AU-rich RNA in the NAD(+)-binding region (Rossmann fold). *J. Biol. Chem.* **270:**2755–2763.

Nagy, E., Henics, T., Eckert, M., Miseta, A., Lightowlers, R. N., and Kellermayer, M. (2000). Identification of the NAD(+)-binding fold of glyceraldehyde-3-phosphate dehydrogenase as a novel RNA-binding domain. *Biochem. Biophys. Res. Commun.* **275:**253–260.

Naito, T., Momose, F., Kawaguchi, A., and Nagata, K. (2007). Involvement of Hsp90 in assembly and nuclear import of influenza virus RNA polymerase subunits. *J. Virol.* **81:**1339–1349.

Nakagawa, S., Umehara, T., Matsuda, C., Kuge, S., Sudoh, M., and Kohara, M. (2007). Hsp90 inhibitors suppress HCV replication in replicon cells and humanized liver mice. *Biochem. Biophys. Res. Commun.* **353:**882–888.

Navarro, B., Russo, M., Pantaleo, V., and Rubino, L. (2006). Cytological analysis of *Saccharomyces cerevisiae* cells supporting cymbidium ringspot virus defective interfering RNA replication. *J. Gen. Virol.* **87**(Pt. 3)**:**705–714.

Nerenberg, B. T., Taylor, J., Bartee, E., Gouveia, K., Barry, M., and Fruh, K. (2005). The poxviral RING protein p28 is a ubiquitin ligase that targets ubiquitin to viral replication factories. *J. Virol.* **79:**597–601.

Nishikiori, M., Dohi, K., Mori, M., Meshi, T., Naito, S., and Ishikawa, M. (2006). Membrane-bound *Tomato mosaic virus* replication proteins participate in RNA synthesis and are associated with host proteins in a pattern distinct from those that are not membrane bound. *J. Virol.* **80:**8459–8468.

Okamoto, T., Nishimura, Y., Ichimura, T., Suzuki, K., Miyamura, T., Suzuki, T., Moriishi, K., and Matsuura, Y. (2006). Hepatitis C virus RNA replication is regulated by FKBP8 and Hsp90. *EMBO J.* **25:**5015–5025.

Ott, D. E., Coren, L. V., Chertova, E. N., Gagliardi, T. D., and Schubert, U. (2000). Ubiquitination of HIV-1 and MuLV Gag. *Virology* **278:**111–121.

Panavas, T., and Nagy, P. D. (2003). Yeast as a model host to study replication and recombination of defective interfering RNA of *Tomato bushy stunt virus*. *Virology* **314:**315–325.

Panavas, T., Hawkins, C. M., Panaviene, Z., and Nagy, P. D. (2005a). The role of the p33:p33/p92 interaction domain in RNA replication and intracellular localization of p33 and p92 proteins of *Cucumber necrosis* tombusvirus. *Virology* **338:**81–95.

Panavas, T., Serviene, E., Brasher, J., and Nagy, P. D. (2005b). Yeast genome-wide screen reveals dissimilar sets of host genes affecting replication of RNA viruses. *Proc. Natl. Acad. Sci. USA* **102:**7326–7331.

Panaviene, Z., Panavas, T., Serva, S., and Nagy, P. D. (2004). Purification of the *Cucumber necrosis virus* replicase from yeast cells: Role of coexpressed viral RNA in stimulation of replicase activity. *J. Virol.* **78:**8254–8263.

Panaviene, Z., Panavas, T., and Nagy, P. D. (2005). Role of an internal and two 3′-terminal RNA elements in assembly of tombusvirus replicase. *J. Virol.* **79:**10608–10618.

Pathak, K. B., Sasvari, Z., and Nagy, P. D. (2008). The host Pex19p plays a role in peroxisomal localization of tombusvirus replication proteins. *Virology* **379:**294–305.

Perlman, M., and Resh, M. D. (2006). Identification of an intracellular trafficking and assembly pathway for HIV-1 gag. *Traffic* **7:**731–745.

Pogany, J., White, K. A., and Nagy, P. D. (2005). Specific binding of tombusvirus replication protein p33 to an internal replication element in the viral RNA is essential for replication. *J. Virol.* **79:**4859–4869.

Pogany, J., Stork, J., Li, Z., and Nagy, P. D. (2008). In vitro assembly of the *Tomato bushy stunt virus* replicase requires the host heat shock protein 70. *Proc. Natl. Acad. Sci. USA* **105:**19956–19961.

Poon, A. P., Gu, H., and Roizman, B. (2006). ICP0 and the US3 protein kinase of herpes simplex virus 1 independently block histone deacetylation to enable gene expression. *Proc. Natl. Acad. Sci. USA* **103:**9993–9998.

Poranen, M. M., Salgado, P. S., Koivunen, M. R., Wright, S., Bamford, D. H., Stuart, D. I., and Grimes, J. M. (2008). Structural explanation for the role of Mn^{2+} in the activity of phi6 RNA-dependent RNA polymerase. *Nucleic Acids Res.* **36:**6633–6644.

Qanungo, K. R., Shaji, D., Mathur, M., and Banerjee, A. K. (2004). Two RNA polymerase complexes from vesicular stomatitis virus-infected cells that carry out transcription and replication of genome RNA. *Proc. Natl. Acad. Sci. USA* **101:**5952–5957.

Quadt, R., Ishikawa, M., Janda, M., and Ahlquist, P. (1995). Formation of *Brome mosaic virus* RNA-dependent RNA polymerase in yeast requires coexpression of viral proteins and viral RNA. *Proc. Natl. Acad. Sci. USA* **92:**4892–4896.

Randall, G., Panis, M., Cooper, J. D., Tellinghuisen, T. L., Sukhodolets, K. E., Pfeffer, S., Landthaler, M., Landgraf, P., Kan, S., Lindenbach, B. D., Chien, M., Weir, D. B., *et al.* (2007). Cellular cofactors affecting hepatitis C virus infection and replication. *Proc. Natl. Acad. Sci. USA* **104:**12884–12889.

Ratia, K., Saikatendu, K. S., Santarsiero, B. D., Barretto, N., Baker, S. C., Stevens, R. C., and Mesecar, A. D. (2006). Severe acute respiratory syndrome coronavirus papain-like protease: Structure of a viral deubiquitinating enzyme. *Proc. Natl. Acad. Sci. USA* **103:**5717–5722.

Roche, Y., Gerbeau-Pissot, P., Buhot, B., Thomas, D., Bonneau, L., Gresti, J., Mongrand, S., Perrier-Cornet, J. M., and Simon-Plas, F. (2008). Depletion of phytosterols from the plant plasma membrane provides evidence for disruption of lipid rafts. *FASEB J.* **22:**3980–3991.

Rothwell, C., Lebreton, A., Young Ng, C., Lim, J. Y., Liu, W., Vasudevan, S., Labow, M., Gu, F., and Gaither, L. A. (2009). Cholesterol biosynthesis modulation regulates dengue viral replication. *Virology* **389**(1–2):8–19.

Sagan, S. M., Rouleau, Y., Leggiadro, C., Supekova, L., Schultz, P. G., Su, A. I., and Pezacki, J. P. (2006). The influence of cholesterol and lipid metabolism on host cell structure and hepatitis C virus replication. *Biochem. Cell Biol.* **84:**67–79.

Salonen, A., Ahola, T., and Kaariainen, L. (2005). Viral RNA replication in association with cellular membranes. *Curr. Top. Microbiol. Immunol.* **285:**139–173.

Sasvari, Z., Bach, S., Blondel, M., and Nagy, P. D. (2009). Inhibition of RNA recruitment and replication of an RNA virus by acridine derivatives with known anti-prion activities. *PLoS One* **4:**e7376.

Schwartz, M., Chen, J., Janda, M., Sullivan, M., den Boon, J., and Ahlquist, P. (2002). A positive-strand RNA virus replication complex parallels form and function of retrovirus capsids. *Mol. Cell* **9:**505–514.

Serva, S., and Nagy, P. D. (2006). Proteomics analysis of the tombusvirus replicase: Hsp70 molecular chaperone is associated with the replicase and enhances viral RNA replication. *J. Virol.* **80:**2162–2169.

Serviene, E., Shapka, N., Cheng, C. P., Panavas, T., Phuangrat, B., Baker, J., and Nagy, P. D. (2005). Genome-wide screen identifies host genes affecting viral RNA recombination. *Proc. Natl. Acad. Sci. USA* **102:**10545–10550.

Serviene, E., Jiang, Y., Cheng, C. P., Baker, J., and Nagy, P. D. (2006). Screening of the yeast yTHC collection identifies essential host factors affecting tombusvirus RNA recombination. *J. Virol.* **80:**1231–1241.

Shackelford, J., and Pagano, J. S. (2004). Tumor viruses and cell signaling pathways: Deubiquitination versus ubiquitination. *Mol. Cell. Biol.* **24:**5089–5093.

Shackelford, J., and Pagano, J. S. (2005). Targeting of host-cell ubiquitin pathways by viruses. *Essays Biochem.* **41:**139–156.

Shapka, N., Stork, J., and Nagy, P. D. (2005). Phosphorylation of the p33 replication protein of *Cucumber necrosis* tombusvirus adjacent to the RNA binding site affects viral RNA replication. *Virology* **343:**65–78.

Sharma, M., Sasvari, Z., and Nagy, P. D. (2009). Inhibition of sterol biosynthesis reduces tombusvirus replication in yeast and plants. *J Virol.* (published on line).

Sharma, M., Sasvari, Z., and Nagy, P. D. (2010). Inhibition of sterol biosynthesis reduces tombusvirus replication in yeast and plants. *J Virol.* **84:**2270–2281.

Sheth, U., and Parker, R. (2003). Decapping and decay of messenger RNA occur in cytoplasmic processing bodies. *Science* **300:**805–808.

Shi, S. T., and Lai, M. M. (2005). Viral and cellular proteins involved in coronavirus replication. *Curr. Top. Microbiol. Immunol.* **287:**95–131.

Sirover, M. A. (1999). New insights into an old protein: The functional diversity of mammalian glyceraldehyde-3-phosphate dehydrogenase. *Biochim. Biophys. Acta* **1432:**159–184.

Sirover, M. A. (2005). New nuclear functions of the glycolytic protein, glyceraldehyde-3-phosphate dehydrogenase, in mammalian cells. *J. Cell. Biochem.* **95:**45–52.

Slagsvold, T., Pattni, K., Malerod, L., and Stenmark, H. (2006). Endosomal and non-endosomal functions of ESCRT proteins. *Trends Cell Biol.* **16:**317–326.

Sohn, S. Y., Kim, S. B., Kim, J., and Ahn, B. Y. (2006). Negative regulation of hepatitis B virus replication by cellular Hsp40/DnaJ proteins through destabilization of viral core and X proteins. *J. Gen. Virol.* **87**(Pt. 7):1883–1891.

Stork, J., Panaviene, Z., and Nagy, P. D. (2005). Inhibition of in vitro RNA binding and replicase activity by phosphorylation of the p33 replication protein of *Cucumber necrosis* tombusvirus. *Virology* **343:**79–92.

Sulea, T., Lindner, H. A., Purisima, E. O., and Menard, R. (2005). Deubiquitination, a new function of the severe acute respiratory syndrome coronavirus papain-like protease? *J. Virol.* **79:**4550–4551.

Tautz, N., and Thiel, H. J. (2003). Cytopathogenicity of pestiviruses: Cleavage of bovine viral diarrhea virus NS2–3 has to occur at a defined position to allow viral replication. *Arch. Virol.* **148:**1405–1412.

Tavis, J. E., Massey, B., and Gong, Y. (1998). The duck hepatitis B virus polymerase is activated by its RNA packaging signal, epsilon. *J. Virol.* **72:**5789–5796.

Taylor, J. M., and Barry, M. (2006). Near death experiences: Poxvirus regulation of apoptotic death. *Virology* **344:**139–150.

Thivierge, K., Cotton, S., Dufresne, P. J., Mathieu, I., Beauchemin, C., Ide, C., Fortin, M. G., and Laliberte, J. F. (2008). Eukaryotic elongation factor 1A interacts with *Turnip mosaic virus* RNA-dependent RNA polymerase and VPg-Pro in virus-induced vesicles. *Virology* **377:**216–225.

Tomita, Y., Mizuno, T., Diez, J., Naito, S., Ahlquist, P., and Ishikawa, M. (2003). Mutation of host DnaJ homolog inhibits *Brome mosaic virus* negative-strand RNA synthesis. *J. Virol.* **77:**2990–2997.

Ton, V. K., and Rao, R. (2004). Functional expression of heterologous proteins in yeast: Insights into Ca^{2+} signaling and Ca^{2+}-transporting ATPases. *Am. J. Physiol. Cell Physiol.* **287:**C580–C589.

Tuteja, R., and Tuteja, N. (1998). Nucleolin: A multifunctional major nucleolar phosphoprotein. *Crit. Rev. Biochem. Mol. Biol.* **33:**407–436.

Van Wynsberghe, P. M., and Ahlquist, P. (2009). 5′ *cis* elements direct nodavirus RNA1 recruitment to mitochondrial sites of replication complex formation. *J. Virol.* **83:** 2976–2988.

Vartanian, J. P., Sala, M., Henry, M., Wain-Hobson, S., and Meyerhans, A. (1999). Manganese cations increase the mutation rate of human immunodeficiency virus type 1 ex vivo. *J. Gen. Virol.* **80**(Pt. 8)**:**1983–1986.

Vlot, A. C., Neeleman, L., Linthorst, H. J., and Bol, J. F. (2001). Role of the 3′-untranslated regions of *Alfalfa mosaic virus* RNAs in the formation of a transiently expressed replicase in plants and in the assembly of virions. *J. Virol.* **75:**6440–6449.

Waggoner, S., and Sarnow, P. (1998). Viral ribonucleoprotein complex formation and nucleolar–cytoplasmic relocalization of nucleolin in poliovirus-infected cells. *J. Virol.* **72:**6699–6709.

Wang, R. Y., and Nagy, P. D. (2008). *Tomato bushy stunt virus* co-opts the RNA-binding function of a host metabolic enzyme for viral genomic RNA synthesis. *Cell Host Microbe* **3:**178–187.

Wang, X., Lee, W. M., Watanabe, T., Schwartz, M., Janda, M., and Ahlquist, P. (2005). *Brome mosaic virus* 1a nucleoside triphosphatase/helicase domain plays crucial roles in recruiting RNA replication templates. *J. Virol.* **79:**13747–13758.

Wang, J., Loveland, A. N., Kattenhorn, L. M., Ploegh, H. L., and Gibson, W. (2006). High-molecular-weight protein (pUL48) of human cytomegalovirus is a competent deubiquitinating protease: Mutant viruses altered in its active-site cysteine or histidine are viable. *J. Virol.* **80:**6003–6012.

Wang, R. Y., Stork, J., Pogany, J., and Nagy, P. D. (2009a). A temperature sensitive mutant of heat shock protein 70 reveals an essential role during the early steps of tombusvirus replication. *Virology* **394:**28–38.

Wang, R. Y., Stork, J., and Nagy, P. D. (2009b). A key role for heat shock protein 70 in the localization and insertion of tombusvirus replication proteins to intracellular membranes. *J. Virol.* **83:**3276–3287.

Weeks, S. A., and Miller, D. J. (2008). The heat shock protein 70 cochaperone YDJ1 is required for efficient membrane-specific flock house virus RNA replication complex assembly and function in *Saccharomyces cerevisiae*. *J. Virol.* **82:**2004–2012.

Werner-Washburne, M., Stone, D. E., and Craig, E. A. (1987). Complex interactions among members of an essential subfamily of hsp70 genes in *Saccharomyces cerevisiae*. *Mol. Cell. Biol.* **7**:2568–2577.

White, K. A., and Morris, T. J. (1994). Recombination between defective tombusvirus RNAs generates functional hybrid genomes. *Proc. Natl. Acad. Sci. USA* **91**:3642–3646.

White, K. A., and Nagy, P. D. (2004). Advances in the molecular biology of tombusviruses: Gene expression, genome replication, and recombination. *Prog. Nucleic Acid Res. Mol. Biol.* **78**:187–226.

Whitham, S. A., Quan, S., Chang, H. S., Cooper, B., Estes, B., Zhu, T., Wang, X., and Hou, Y. M. (2003). Diverse RNA viruses elicit the expression of common sets of genes in susceptible *Arabidopsis thaliana* plants. *Plant J.* **33**:271–283.

Whitham, S. A., Yang, C., and Goodin, M. M. (2006). Global impact: Elucidating plant responses to viral infection. *Mol. Plant Microbe Interact.* **19**:1207–1215.

Wong, J., Zhang, J., Si, X., Gao, G., and Luo, H. (2007). Inhibition of the extracellular signal-regulated kinase signaling pathway is correlated with proteasome inhibitor suppression of coxsackievirus replication. *Biochem. Biophys. Res. Commun.* **358**:903–907.

Woo, J. L., and Berk, A. J. (2007). Adenovirus ubiquitin-protein ligase stimulates viral late mRNA nuclear export. *J. Virol.* **81**:575–587.

Wu, B., Pogany, J., Na, H., Nicholson, B. L., Nagy, P. D., and White, K. A. (2009). A discontinuous RNA platform mediates RNA virus replication: Building an integrated model for RNA-based regulation of viral processes. *PLoS Pathog.* **5**:e1000323.

Yamaji, Y., Kobayashi, T., Hamada, K., Sakurai, K., Yoshii, A., Suzuki, M., Namba, S., and Hibi, T. (2006). In vivo interaction between *Tobacco mosaic virus* RNA-dependent RNA polymerase and host translation elongation factor 1A. *Virology* **347**:100–108.

Yi, M., Schultz, D. E., and Lemon, S. M. (2000). Functional significance of the interaction of hepatitis A virus RNA with glyceraldehyde 3-phosphate dehydrogenase (GAPDH): Opposing effects of GAPDH and polypyrimidine tract binding protein on internal ribosome entry site function. *J. Virol.* **74**:6459–6468.

Yi, G. H., Zhang, C. Y., Cao, S., Wu, H. X., and Wang, Y. (2003). De novo RNA synthesis by a recombinant classical swine fever virus RNA-dependent RNA polymerase. *Eur. J. Biochem.* **270**:4952–4961.

Young, J. C., Agashe, V. R., Siegers, K., and Hartl, F. U. (2004). Pathways of chaperone-mediated protein folding in the cytosol. *Nat. Rev. Mol. Cell Biol.* **5**:781–791.

Zang, W. Q., Fieno, A. M., Grant, R. A., and Yen, T. S. (1998). Identification of glyceraldehyde-3-phosphate dehydrogenase as a cellular protein that binds to the hepatitis B virus post-transcriptional regulatory element. *Virology* **248**:46–52.

Zeenko, V. V., Ryabova, L. A., Spirin, A. S., Rothnie, H. M., Hess, D., Browning, K. S., and Hohn, T. (2002). Eukaryotic elongation factor 1A interacts with the upstream pseudoknot domain in the 3′ untranslated region of *Tobacco mosaic virus* RNA. *J. Virol.* **76**:5678–5691.

Zhu, J., Gopinath, K., Murali, A., Yi, G., Hayward, S. D., Zhu, H., and Kao, C. (2007). RNA-binding proteins that inhibit RNA virus infection. *Proc. Natl. Acad. Sci. USA* **104**: 3129–3134.

CHAPTER 5

Resistance to Aphid Vectors of Virus Disease

Jack H. Westwood[*,†] and **Mark Stevens**[†]

Contents

Abstract

The majority of plant viruses rely on vectors for their transmission and completion of their life cycle. These vectors comprise a diverse range of life forms including insects, nematodes, and fungi with the most common of these being insects. The geographic range of many of these vectors is continually expanding due to climate change. The viruses that they carry are therefore also expanding their range to exploit novel and naïve plant hosts. There are many forms of naturally occurring vector resistance ranging from broad nonhost resistance to more specific types of inducible resistance. Understanding and exploiting the many and varied forms of natural resistance to virus vectors is therefore extremely important for

* Department of Plant Sciences, University of Cambridge CB2 3EA, Cambridge, United Kingdom
† Broom's Barn Research Centre, Higham, Bury St Edmunds, Suffolk IP28 6NP, United Kingdom,
E-mail: jhw37@cam.ac.uk

Advances in Virus Research, Volume 76
ISSN 0065-3527, DOI: 10.1016/S0065-3527(10)76005-X

current and future agricultural production systems. To demonstrate the range and extent of these resistance mechanisms, this chapter will primarily focus on aphids to highlight key developments appropriate to plant–insect–virus interactions.

I. RESISTANCE TO APHIDS

Aphids are the most prevalent insect vectors of virus disease with over 200 species known to transmit plant pathogenic viruses. They are responsible for the transmission of 50% of the insect vectored viruses (Ng and Perry, 2004). Compared to other insects, aphids cause relatively little mechanical damage to their host plant due to their specialized mode of feeding. They use stylets to target the vascular tissue of their host and probe plant tissue intercellularly, thus preventing extensive tissue damage. The impact of this "piercing-sucking" mode of feeding on their plant host is therefore mainly through the transmission of one of the approximately 300 viruses which are aphid transmissible, although the direct effects of assimilate withdrawal from the vasculature cannot be discounted completely.

Since the introduction of pesticides in the mid-1900s, control of aphid pests has traditionally relied on insecticides to prevent colonization. However, over the past 30 years the occurrence of insecticide-resistant aphids, particularly the peach-potato aphid (*Myzus persicae*) has increased, so control strategies need to include, wherever possible, integrated pest management strategies (Smith and Furk, 1989). The cause of resistance has been found to be based on at least three coexisting resistance mechanisms including an overproduction of carboxylesterase conferring resistance to organophosphates, carbamates, and partial resistance to pyrethroids, an altered acetylcholinesterase (AChE) producing resistance to carbamates and target site (kdr) resistance to pyrethroids (Foster *et al.*, 2007). In certain instances, for example, sugar beet production in the United Kingdom, this has left the neonicotinoid class of insecticides as the only effective alternative for the control of virus-carrying aphids. Up to 20-fold levels of aphid resistance to neonicotinoids have been detected in *M. persicae* clones collected from the United Kingdom, mainland Europe, United States, Zimbabwe, and Japan (Devine *et al.*, 1996; Foster *et al.*, 2003; Nauen and Denholm, 2005). However, there is considerable variation in the levels of resistance exhibited, although when neonicotinoids have been applied at recommended field rates there have been no difficulties in the control of *M. persicae* or the viruses they transmit (Foster *et al.*, 2007). Nevertheless, it is clear that the intense selection pressure placed on aphids through insecticide use is enough to result in the continued development of aphid resistance. Hence, it is vital that alternative methods of control are developed for contingency before there is a significant build up of neonicotinoid resistance in natural aphid populations.

To achieve this, a greater understanding of natural plant resistance mechanisms is required with the goal of implementing this knowledge in protecting crop species from aphids and subsequent virus infection. Aphid host-plant-resistance mechanisms are classified as antixenosis if they prevent aphids from colonizing or as antibiosis if they affect aphids which have already colonized the plant. Antibiosis includes mechanisms which reduce life history traits such as longevity, reproduction, and fecundity or increasing mortality (Van Emden, 2007). This chapter will look at aspects of natural plant–aphid defense ranging from basal and nonhost resistance to induced genetic resistance and then consider current and future approaches to implementing novel methods of aphid resistance into agricultural practice.

II. NATURAL RESISTANCE

Natural aphid resistance mechanisms range from constitutive basal defenses to inducible species-specific defenses under high levels of regulation.

A. Basal resistance

1. Physical defenses

Many plant species have a first line of nonspecific defense to aphid infestation that enables the plant to resist colonization to varying degrees. At the leaf surface, trichomes play an important role in basal defense and may confer both antixenotic and antibiotic properties to the plant. Nonglandular trichomes provide a barrier to aphid infestation in many plant species. It was first reported by Johnson (1953) that the hooked nonglandular trichomes of the French bean (*Phaseolus vulgaris*) had a detrimental effect on populations of the cowpea aphid (*Aphis craccivora*; Johnson, 1953). High densities of nonglandular trichomes in wheat and crosses of tomato with wild potato have also been demonstrated to deter feeding by the yellow sugarcane aphid (*Siphus flava*) and *M. persicae*, respectively (Simmons *et al.*, 2005; Webster *et al.*, 1994). However, not all plants with high trichome densities deter aphid colonization, as glabrous varieties of cotton are more resistant to colonization by *Aphis gossypii* than their pubescent counterparts (Weathersbee and Hardee, 1994; Weathersbee *et al.*, 1994, 1995).

Glandular trichomes are also an important basal defense system against aphid infestation. Polyphenolic fluids either contained in the tips of the hairs or encased in a spherical head have significant antibiotic effects against aphid infestation (Van Emden, 2007). Polyphenolic compounds can repel aphids and when secreted, they harden on the leaf surface often trapping and disabling aphid mouthparts and tarsi (Gibson, 1976; Simmons *et al.*, 2003, 2005; Van Emden, 2007). Alkaloids such as nicotine, nornicotine, and anabasine secreted by glandular trichomes in tobacco have been shown to cause leg paralysis in *M. persicae* (Levin, 1973). These physical defense structures are therefore a crucial aspect of anti-aphid basal defense.

Upon herbivore attack or wounding, plants regenerate new leaves that possess increased trichome density (Agrawal, 1999; Traw and Bergelson, 2003; Traw and Dawson, 2002), indicating a role for trichomes as an inducible basal defense. Jasmonic acid (JA) defense signaling is an important plant defense pathway that is induced in response to wounding and herbivory, but only recently was it demonstrated to have a role in the herbivore- and wound-induced trichome development (Boughton *et al.*, 2005; Traw and Bergelson, 2003; Yoshida *et al.*, 2009). Experiments with the *Arabidopsis* mutants *coi1* and *aos,* which are compromised in their ability to perceive and initiate JA signaling, showed that they were unable to increase trichome density on new leaves following wounding (Yoshida *et al.*, 2009). Conversely, in experiments where methyl-jasmonic acid (MeJA) was applied exogenously to plants trichome density on developing leaves was significantly increased (Boughton *et al.*, 2005; Traw and Bergelson, 2003). However, despite the confirmed role in basal aphid resistance for trichomes, it has yet to be demonstrated that aphid feeding induces their formation in increased numbers in new leaves.

Waxiness of leaf surface may also be a physical factor in basal aphid resistance, with less waxy (glossy) varieties often exhibiting increased resistance (Van Emden, 2007). Several studies in diverse species such as wheat, pea, and brassicas have shown that less waxy varieties are more aphid resistant, with reductions in aphid populations as high as 95% in the glossy varieties being reported (Ellis *et al.*, 1996; Lowe *et al.*, 1985; Stoner, 1992; White and Eigenbrode, 2000). The mechanism of resistance is chemical, with glossy varieties containing higher levels of aphid-deterrent compounds such as dihydroketones, despite having a reduced wax layer (Stoner, 1992).

2. Chemical defense

Many recent studies have examined host plant transcriptomes before and after aphid feeding and identified defense responses common to both compatible and incompatible interactions (Delp *et al.*, 2009; Gao *et al.*, 2007; Moran and Thompson, 2001; Thompson and Goggin, 2006). Basal defenses that are induced by aphid feeding include cell wall modification and upregulation of anti-insect compounds such as proteinase inhibitors, secondary metabolites, and volatiles. Transcriptome changes induced by aphid feeding are often mediated by one or several of well-characterized plant hormones including JA, salicylic acid (SA), and ethylene. There is significant cross-talk, both synergistic and antagonistic between these defense pathways, which allows the plant to optimize its defense strategy to particular threats (Koornneef and Pieterse, 2008). However, the intricate interplay between plant defense responses provides the opportunity for insects to manipulate the pathways in order to enhance their own success and there is a growing evidence to show that this is the case (Walling, 2008).

JA-mediated plant defense responses have been traditionally associated with the plant response to wounding and necrotizing pathogens. However, there has recently been a great deal of interest in characterizing the impact of JA and JA-induced plant responses to aphid infestation. Several studies have shown that exogenous application of JA or a derivative of JA (e.g. MeJA, a JA-isoleucine conjugate [JA-Ile] or *cis*-jasmone) significantly reduces aphid colonization (Bruce *et al.*, 2003; Ellis *et al.*, 2002; Gao *et al.*, 2007; Zhu-Salzman *et al.*, 2004). Furthermore, aphid colony development was significantly inhibited on the *Arabidopsis* constitutive JA biosynthesis mutant *cev1*, but increased on the jasmonate-insensitive *coi1* mutant (Ellis *et al.*, 2002; Mewis *et al.*, 2005). These results clearly implicate JA-induced plant defenses in aphid resistance and there is a growing body of evidence, in diverse species, that aphid feeding induces JA signaling. Lipoxygenases (LOX) are enzymes involved in the biosynthesis of JA and its derivatives and their induction is a marker of the activation of JA-mediated signaling pathways. *LOX* genes were found to be induced by the feeding of *Myzus nicotianae* on *Nicotiana attenuata*, *Macrosiphum euphorbiae* on tomato, and *M. persicae* on *Arabidopsis* (Fidantsef *et al.*, 1999; Moran and Thompson, 2001; Voelckel *et al.*, 2004), thus indicating that aphid feeding has the capacity to modestly induce JA-mediated defenses. However, the induction of *LOX* genes and other JA biosynthetic enzymes following aphid infestation was much weaker than seen when plants are subject to chewing insects or other forms of mechanical wounding and indeed weaker than the induction of SA-inducible transcripts (Fidantsef *et al.*, 1999; Heidel and Baldwin, 2004). A schematic diagram of the JA biosynthesis and signaling is shown in Fig. 1.

In addition to the modest induction of JA-induced defense responses in some species, aphid feeding is known to induce SA-dependent defense responses across a number of plant species including *Arabidopsis*, tomato, *Medicago*, and wheat (Fidantsef *et al.*, 1999; Gao *et al.*, 2007; Moran and Thompson, 2001; van der Westhuizen *et al.*, 1998a,b). Gene expression analysis of SA-inducible pathogenesis-related (PR) proteins indicates that aphid feeding induces SA-related defenses and are upregulated more strongly than JA-induced defenses (Moran and Thompson, 2001). The increase in SA-induced defense pathways is more indicative of a pathogen defense response than an insect defense response and there is evidence to suggest that aphids may manipulate the plant's perception of the threat in order that it responds in this manner. The subtle manner of aphid feeding means that it causes relatively little damage to the plant

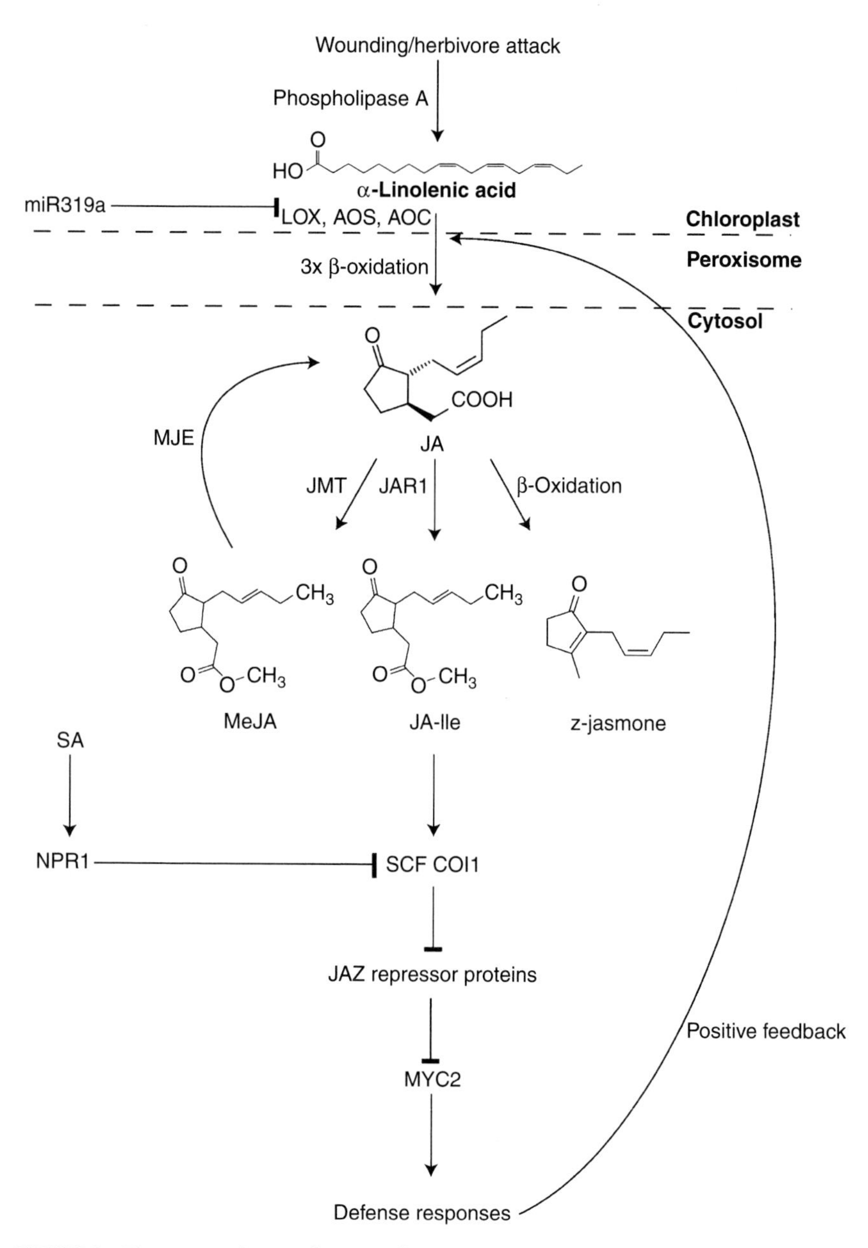

FIGURE 1 The JA signaling pathway. Following wounding or herbivore attack, fatty acid JA precursors, including α-linolenic acid are released from cell membranes through the action of phospholipase enzymes (Wasternack, 2007). These precursors are converted to JA via the action of lipoxygenase (LOX), allene oxidase synthase (AOS), allene oxidase cyclase (AOC), and three rounds of β-oxidation. JA is converted to bioactive forms including MeJA, jasmonyl isoleucine (JA-Ile), and *cis*-jasmone by JMT,

when compared to feeding by other herbivorous insects, which may cause significant wounding. This is likely to be part of the reason that JA-dependent defense responses are induced less than SA-dependent responses. Additionally, due to the antagonistic cross-talk between parts of the SA and JA signaling pathways, the upregulation of SA-dependent responses is itself likely to limit the extent of induction of JA-dependent responses (Thompson and Goggin, 2006; Walling, 2008; Zarate *et al.*, 2007). This is significant as exogenous application of JA as well as aphid performance experiments on JA signaling mutants as outlined above have indicated an important role for JA in aphid defense. It has therefore been suggested that aphids (and indeed other phloem-feeding insects such as whitefly) are able to manipulate plant defense pathways in order that they avoid JA-regulated defenses which inhibit aphid performance (Thompson and Goggin, 2006; Walling, 2008; Zarate *et al.*, 2007).

Foliar application of JA and its derivatives causes large-scale transcriptional reprogramming in diverse plant species indicating that it is a highly conserved signaling pathway (Pauwels *et al.*, 2008; Salzman *et al.*, 2005). The downstream effects of JA-induced defense signaling include upregulation of the phenylpropanoid and octadecanoid pathways as well as biosynthesis of glucosinolates, a class of anti-insect defence compounds specific to the Brassicaceae (Fahey *et al.*, 2001). The phenylpropanoid pathway synthesizes an array of compounds with diverse properties ranging from the structural, for example, lignin, to the antimicrobial and anti-insecticidal (Dixon *et al.*, 2002; Hahlbrock and Scheel, 1989) whereas the octadecanoid pathway leads to the accumulation of JA and has been shown to mediate insect resistance (Gao *et al.*, 2007).

Glucosinolates are a class of secondary metabolites found mainly in the order Brassicales where they function in defence against pathogens

---

JAR1, and β-oxidation, respectively. MeJA is volatile and is thought to be involved in inter- and intraplant signaling, however, in order for it to elicit gene expression changes it must first be demethylated by methyl-jasmonate esterase (MJE; Stuhlfelder *et al.*, 2004; Wu *et al.*, 2008). *cis*-jasmone has been shown to serve as an attractant for aphid predators (Bruce *et al.*, 2008). JA-Ile signals through its receptor, COI1. This allows transcription factors such as MYC2 to function in the upregulation of target genes, including those involved in defense and JA biosynthesis. JA-Ile promotes physical interaction between COI1 and JAZ (JASMONATE-ZIM-domain) proteins (Chini *et al.*, 2007; Thines *et al.*, 2007). JAZ proteins repress jasmonate responses by binding transcription factors such as MYC2. COI1 is part of an ubiquitin E3 ligase complex that ubiquinates JAZ proteins, targeting them for proteolysis. Recently, it has been shown that JA biosynthesis is negatively regulated by a small RNA (Schommer *et al.*, 2008). MicroRNA319 (miR319a) targets transcripts of several transcription factors that activate promoters of JA biosynthesis genes such as *LOX2*. Cross-talk between JA and SA signaling pathways is mediated by NPR1 (Spoel *et al.*, 2003) and although its precise node of action is unknown, it is thought that it may interfere with COI1 activity (Spoel and Dong, 2008). Figure based on Farmer *et al.* (2009).

and herbivores (De Vos *et al.*, 2007). These compounds are sulphonated thioglycosides comprising a common glycone moiety with a variable aglycone side chain (Mewis *et al.*, 2005). Their biosynthesis is complex and regulated by a combination of SA, JA and ET levels (Mewis *et al.*, 2005). They are a highly potent anti-insect defense and plant mutants with reduced glucosinolate levels showed enhanced susceptibility to herbivores, including aphids (Mewis *et al.*, 2005). Glucosinolates and their hydrolysing enzymes, myrosinases, are compartmentalised into adjacent cells (Halkier and Gershenzon, 2006). Upon tissue damage, glucosinolates are brought into contact with myrosinases and their hydrolysis results in highly toxic breakdown products such as isothiocyanates and nitriles (De Vos *et al.*, 2007). However, aphids are able to subvert this defense mechanism due to the lack of cellular tissue damage during feeding. This means that glucosinolates are not brought into contact with myrosinases and no toxic breakdown products are released. In fact, in many cases total glucosinolate levels have been shown to decrease following aphid feeding (Kim and Jander, 2007; Mewis *et al.*, 2006), thus creating a more hospitable environment for the aphid. However, while aphids are able to avoid inducing most foliar glucosinolates, one indole glucosinolate, 4M13M, an effective aphid deterrent even in the absence of myrosinase activation, is found at elevated levels following aphid feeding (Kim and Jander, 2007), indicating that not all glucosinolate-based plant defenses can be subverted by aphids.

The putative lipase PAD4 has recently been shown to be important in basal *Arabidopsis* resistance against *M. persicae* (Pegadaraju *et al.*, 2007). The *PAD4* transcript was induced following aphid infestation and using electrical monitoring techniques, it was demonstrated that *PAD4* mediates a phloem-based defense mechanism as aphids spent significantly more time feeding from the sieve elements of *pad4* mutants compared to wild type (Pegadaraju *et al.*, 2007). This was also correlated with an increase in aphid colony size (Pegadaraju *et al.*, 2005). *PAD4* controls the synthesis of SA and the indolic compound camelexin in response to pathogen infection and as such plays a central role in the establishment of systemic acquired resistance (SAR) in incompatible interactions (Wiermer *et al.*, 2005). *pad4* mutants exhibit reduced SA accumulation (Glazebrook *et al.*, 1997); however, *PAD4*-mediated aphid resistance was shown to be *EDS1* and SA-independent (Pegadaraju *et al.*, 2005, 2007) and instead attributed to *PAD4*-modulated leaf senescence (Pegadaraju *et al.*, 2005).

Cell wall structure plays an important role in influencing plant–aphid interactions. Several transcript profiling studies have shown that following aphid infestation on several plant species, there is a significant upregulation of cell wall remodeling enzymes. Genes commonly affected include those encoding pectin esterase, cellulose synthase, and xyloglucan endotransglycosylase (XTH) (Divol *et al.*, 2005; Heidel and Baldwin, 2004; Thompson and Goggin, 2006; Voelckel *et al.*, 2004). An *Arabidopsis* line mutated in an XTH-encoding gene (*xth33*) was shown to be more susceptible to infestation by *M. persicae* (Divol *et al.*, 2007), although ectopic expression of this gene was not sufficient to confer aphid resistance. Nevertheless, this finding indicates that cell wall modification contributes to aphid resistance in plants. The nature of the enhanced resistance to aphids conferred through cell wall modification is as yet unknown. It has been suggested that modifying the cell wall may either create an enhanced mechanical barrier to stylet penetration or contribute to defensive signaling pathways through the release of oligosaccharides (Goggin, 2007).

A crucial plant defense response to wounding is the ability to seal sieve elements quickly and efficiently when they become punctured. Experiments that used microcapillary tubes to pierce sieve elements demonstrated that plants respond rapidly to these events by plugging the sieve plates by mobilizing P-proteins and synthesizing callose in order to limit the loss of organic nutrients from the phloem (Knoblauch and Van Bel, 1998; Will and Van Bel, 2006). These rapid wound responses to sieve element injury pose a large problem for phloem-feeding insects such as aphids as the plugs impede the flow of sap and therefore nutrition through the phloem. Forisomes are spindle-like protein bodies which occur in the phloem of the Fabacae and undergo rapid dispersal to plug sieve elements in response to an influx and release of calcium ions following wounding (Knoblauch and Peters, 2004; Knoblauch *et al.*, 2001, 2003).

Aphids are able to prevent the plugging of sieve elements in response to stylet penetration of the phloem through physical and chemical mechanisms. These mechanisms are thoroughly reviewed by Will and van Bel (2006). Upon stylet penetration, the secretion of sheath saliva by the feeding aphid forms a gel that surrounds the stylet, sealing the sieve elements that were wounded upon stylet insertion (Will *et al.*, 2007). This physically prevents influx of Ca^{2+} and also helps to maintain the turgor pressure of the phloem which further prevents calcium influx. The small volume of an aphid stylet is an important physical factor in limiting Ca^{2+} influx and means that the loss of phloem pressure following stylet insertion is negligible (Will and Van Bel, 2006). This means that mechanosensitive Ca^{2+} channels are not stimulated, thus further preventing calcium influx and limiting the extent of sieve element sealing.

In addition to the physical measures preventing calcium influx into penetrated sieve elements, several Ca^{2+}-binding proteins have been identified in aphid watery saliva (Will *et al.*, 2007). These are likely to help prevent the dispersal of forisomes and other P-proteins as well as preventing sieve element proteins from coagulating and blocking the food canal of the aphid's stylet. Therefore, it is clear that successful aphid colonization is dependent on the insects being able to prevent the dispersal of P-proteins and subsequent clogging of sieve elements. In plant–aphid interactions where the plant is able to seal its punctured sieve elements more quickly than the aphid can prevent Ca^{2+} influx, the aphid will be unable to feed and thus the plant will be resistant to infestation. For example, the *Vat* gene from melon appears to enhance sieve element wound healing and this is likely to be a crucial characteristic in conferring resistance to aphids (Martin *et al.*, 2003).

B. *R* gene-mediated aphid resistance

Plant-encoded resistance (*R*) genes are involved in gene-for-gene resistance and recognize (directly or indirectly) avirulence (*Avr*) gene products in attacking parasites (Flor, 1971). Traditionally, this type of resistance has been associated with defense against plant pathogens; however, it is now widely accepted that *R* genes play a role in the recognition and defense against insect pests, including aphids. Gene-for-gene resistance was first identified in the flax/flax rust fungus system (Flor, 1955) but it has since been shown to be a highly conserved mechanism in conferring resistance to many pathogenic organisms. The majority of *R* genes encode R-proteins that are highly conserved across species, and share major structural similarity. They are and characterized by the presence of both nucleotide-binding (NB) and leucine-rich repeat (LRR) motifs (Dangl and Jones, 2001; Jones and Dangl, 2006; Takken *et al.*, 2006). *R* genes form part of a molecular surveillance system allowing plants to specifically recognize pathogens that are able to overcome the basal resistance mechanisms outlined above (Jones and Dangl, 2006).

R genes associated with aphid resistance have been identified and mapped in several species including lettuce, soybean, and *Medicago* (Klingler *et al.*, 2007; Li *et al.*, 2007; Wroblewski *et al.*, 2007). However, as yet the only aphid *R* genes to be cloned and characterized are two from tomato and melon (De Ilarduya *et al.*, 2003; Villada *et al.*, 2009). The NB-LRR *Mi-1* (resistance to *Meloidogyne incognita*) gene, cloned from tomato confers resistance to a number of insects including the phloem-feeding potato aphid (*Macrosiphum euphorbiae*), tomato psyllids (*Bactericera cockerelli*), two biotypes of whitefly (*Bemisia tabaci*) and root-knot nematodes (*M. incognita;* Casteel *et al.*, 2006; Milligan *et al.*, 1998; Nombela *et al.*, 2003; Rossi *et al.*, 1998). Electronic monitoring of aphid feeding behavior in *Mi-1*-mediated incompatible interactions has shown that this resistance has strong antixenotic effects. Despite aphids being able to freely access the phloem in these interactions, they do not ingest significant amounts of phloem sap thus leading to their starvation and/or desiccation (Kaloshian *et al.*, 2000). This has large and rapid effects on aphid colonization of resistant tomato plants with aphids dying at just 24 h postinfestation (Kaloshian *et al.*, 1997).

The mechanism of resistance that the *Mi-1* locus confers differs for nematodes and aphids. The *Mi-1* transcript is constitutively expressed in both roots and shoots, however, only resistance to nematodes is constitutive. Resistance to aphids and whiteflies is not expressed until plants are 5 weeks old indicating that there is developmental aspect to the regulation of resistance and that regulation may occur at the posttranscriptional level (De Ilarduya *et al.*, 2003; Kaloshian *et al.*, 1995). This is further exemplified by the fact that in adult plants, fully developed leaves exhibit the resistance phenotype, whereas expanding leaves remain susceptible to aphids throughout the plant's lifetime. Additionally, resistant interactions with nematodes are associated with localized cell death and a hypersensitive response (HR), whereas in incompatible interactions with aphids no cell death has been observed (De Ilarduya *et al.*, 2003). These differences indicate that *Mi-1*-mediated aphid and nematode resistance may utilize different factors downstream of recognition which would provide an explanation for the fact that when eggplant was transformed with *Mi-1.2*, no resistance to aphids was observed despite the transgenic line showing nematode resistance to similar levels as seen during incompatible interactions with wild-type tomato (Goggin *et al.*, 2006). The downstream signaling events leading to the aphid-resistance phenotype are therefore of great current and future interest.

SA often plays a key role in *R* gene-mediated resistance against plant pathogens where its accumulation leads to SAR in a host plant. *Mi-1*-mediated aphid resistance also appears to be SA-dependent as SA-dependent transcripts of pathogenesis-related proteins such as *PR-1* accumulated faster and to higher levels in incompatible than in compatible potato aphid–tomato interactions (De Ilarduya *et al.*, 2003). This finding of SA-dependence in *Mi-1*-mediated aphid resistance was illustrated again when Li *et al.* (2006) showed that in transgenic tomato plants expressing *NahG*, a bacterial enzyme that metabolizes SA, this resistance was attenuated. Moreover, the *Mi-1* resistance phenotype can be rescued in *NahG* tomato by the exogenous application of an SA analog, thus further implicating SA as a key component in *Mi-1*-mediated resistance. The role of JA and, more recently, ethylene as intermediaries in *Mi-1*-mediated aphid resistance have also been investigated. Despite the well-documented role of JA in inducing aphid resistance in compatible interactions (Bruce *et al.*, 2003; Ellis *et al.*, 2002; Gao *et al.*, 2007; Zhu-Salzman *et al.*, 2004), it was found to have little effect in enhancing resistance phenotypes observed in *Mi-1*-mediated aphid–tomato incompatible interactions (Cooper and Goggin, 2005; Cooper *et al.*, 2004). Similarly, an investigation into the effects of ethylene on incompatible plant–aphid interactions by Mantelin *et al.* (2009) demonstrated that silencing genes involved in ethylene biosynthesis had no effect on *Mi-1*-mediated aphid resistance. One explanation for this is that *Mi-1*-mediated resistance and JA/ethylene-mediated resistance act at different ends of the antixenotic/antibiotic spectrum. As previously mentioned, *Mi-1*-mediated resistance has antixenotic properties, preventing aphids from ingesting sap from the phloem. Conversely, it is thought that JA/ethylene-mediated resistance pathways induce the expression and synthesis of toxic or antifeedant compounds and thus the primary effects are antibiotic. By decreasing ingestion from the phloem, *Mi-1*-mediated resistance therefore masks the resistance-inducing effects of JA/ethylene (Cooper and Goggin, 2005).

The *Vat* (*v*irus *a*phid *t*ransmission) gene cloned from melon (*Cucumis melo* L.) is a NB-LRR gene conferring resistance to *A. gossypii* and virus transmission (Villada *et al.*, 2009). Villada *et al.* (2009) observed that on resistant melon varieties carrying the *Vat* gene, cell collapse, loss of plasmalemma integrity and an increase in peroxidise activity occurred soon after infestation with *A. gossypii*. These are characteristics of the HR and thus this type of *R* gene-mediated resistance differs from aphid resistance mediated by the *Mi-1* gene where no HR is seen. This response was specific to *A. gossypii* and not seen in *Vat*-carrying lines in response to whitely or *M. persicae* feeding (Villada *et al.*, 2009). The deposition of callose and lignin in cell walls is a classic and well-characterized response to wounding and cell damage in many plant species (Garcia-Brugger *et al.*, 2006; Wright *et al.*, 2000) and may occur in compatible interactions. Callose deposition also has a well-characterized role in limiting movement of plant viruses as extracellular deposition around plasmodesmata acts to prevent movement of virions between cells (Iglesias and Meins, 2000; Pennazio *et al.*, 1981). However, the work of Villada and colleagues (2009) demonstrated that the deposition of callose, lignin, and other phenolic compounds occurs much more rapidly in response to aphid feeding on resistant lines compared to in compatible interactions and thus forms part of *Vat* gene-mediated resistance. This indicates that *Vat*-mediated resistance occurs in epidermal or mesophyll cell layers after recognition of the attacking herbivore following individual cell punctures. The rapid induction of an HR and other resistance responses following infestation is probably also crucial in mediating the resistance to virus transmission phenotype also conferred by this gene. By initiating a HR soon after a cell puncture, this response is likely to impede the release, replication, and spread of nonpersistently transmitted viruses which are not limited to the phloem (Villada *et al.*, 2009).

Several *R* gene loci have been identified in the model legume species *Medicago truncatula* which is a natural host for many aphid species (Nair *et al.*, 2003) including the bluegreen aphid (*Acyrthosiphon kondoi*) and the pea aphid (*Acyrthosiphon pisum*). As mentioned above, the pea aphid is also a model species and this makes the *M. truncatula*–pea aphid interaction of particular interest and high potential to further the understanding of plant–aphid interactions, especially since *Arabidopsis* is not a host for the pea aphid. So far, four aphid-resistance genes have been reported in various lines of *M. truncatula*. Recently, *AIN* was identified in the Jemalong-A17 line as conferring resistance to the bluegreen aphid (Klingler *et al.*, 2009). An *AIN*-dependent HR accompanies this resistance which is also seen following infestation by the pea aphid, although the plant remains susceptible to the pea aphid (Klingler *et al.*, 2009).

Further work with the resistant Jemalong genoptype by Stewart *et al.* (2009) identified a locus, *AIL*, which also mediates an HR in response to feeding by the pea aphid but does not confer resistance. It is therefore possible that the *AIN* and *AIL* loci are equivalent, although this has not yet been proven. This work also identified a new gene, *RAP1*, which conferred effective resistance to the pea aphid independent of the HR mediated by *AIL*. *RAP1* mapped to chromosome 3, but was clearly genetically distinct from an already known pea aphid-resistance gene, *AKR* (Gao *et al.*, 2008; Klingler *et al.*, 2005; Stewart *et al.*, 2009).

Despite having a well-characterized role in resistance to pathogens, little is known about the role of the HR in insect resistance. Hypersensitivity has been observed in other plant–aphid resistant interactions, the best examples being the resistance against the Russian wheat aphid (RWA; *Diuraphis noxia*) in barley and wheat (Belefantmiller *et al.*, 1994; Moloi and Van Der Westhuizen, 2006); however, the role of the HR in these instances remains unclear particularly since on inducing an HR, aphids are able to walk across the leaf to an area not affected by the HR.

The other *M. truncatula* aphid resistance genes which have been reported are TTR, which confers resistance to spotted alfalfa aphid (*Therioaphis trifolii* f. *maculate*) (Klingler *et al.*, 2007) and APR, which provides protection against the pea aphid (Guo *et al.*, 2009). Interestingly, of the five aphid resistance genes reported in *M. truncatula* at least four have been mapped to close proximity of each other on the north arm of chromosome 3 (Klingler *et al.*, 2009) where over 80 NB-LRR-like sequences are found (Ameline-Torregrosa *et al.*, 2008). Combined with that fact that the two aphid-specific *R* genes cloned thus far, *Mi-1* and *Vat* are both NB-LRR genes, the tight genetic linkage between these resistance loci and NB-LRR gene sequences makes it likely that the *M. truncatula* aphid resistance loci encode NB-LRR-type gene products.

In contrast to what is known about *R* gene-mediated pathogen resistance, little is currently known about the mechanisms of *R* gene-mediated insect resistance. To understand the links between *R* gene-mediated aphid resistance and inducible plant defense pathways, Gao *et al.* (2007) profiled a selection of defense-related transcripts in compatible and incompatible interactions of *M. truncatula* with bluegreen aphid in order to determine if the SA, JA, and ethylene defense pathways had an impact on *AKR*-mediated aphid resistance. They found that in both compatible and incompatible interactions, SA- and ethylene-responsive transcripts were induced. Most strikingly, however, JA-responsive transcripts associated with the octadecanoid pathway were induced only in incompatible interactions with the resistant line. JA-mediated defenses are well-characterized as anti-insect and have large effects on aphid colonization (Section I.A.1.b), it therefore makes sense that resistant interactions would employ this pathway in order to combat aphid infestation. The lack of induction of JA-responsive transcripts in susceptible plants is intriguing and consistent with studies previously described which found that JA-inducible transcripts were markedly less induced than SA-inducible transcripts in compatible interactions (Fidantsef *et al.*, 1999; Heidel and Baldwin, 2004). This indicates that *M. truncatula* plants with no *AKR* gene and therefore no *R* gene capable of recognizing bluegreen aphid infestation are unable to induce the octadecanoid pathway and JA biosynthesis upon bluegreen aphid infestation. This may be further evidence to support the "decoy" hypothesis where activation of the SA pathway leads to suppression of the vigorously anti-aphid JA pathway through signal cross-talk (Thompson and Goggin, 2006; Zhu-Salzman *et al.*, 2004).

From these examples, it is clear that *R* gene-mediated anti-aphid defense may act through several different mechanisms. These mechanisms are as yet not fully understood; however, it is likely that the downstream effects of aphid recognition rely on complex defense signaling pathways with significant levels of cross-talk. Further downstream, the specific outputs of these pathways which cause resistance is also unsure but are likely to involve a diverse arsenal of secondary metabolites and volatile compounds such as phenylpropanoids and terpenoids. The recognition process which induces resistance is also unknown, mainly due to our lack of knowledge concerning components of aphid saliva. It is therefore not known whether these recognition events are direct or indirect (as in the guard hypothesis (Kaloshian, 2004)), but it is likely that these details will vary greatly depending on the interaction.

C. Indirect resistance and extrinsic factors

1. Volatile-mediated plant defense

Through modification and induction of plant defensive signaling pathways as outlined above, aphid feeding often results in the release and synthesis of volatile compounds by the host plant (comprehensively reviewed by Arimura *et al.*, 2005). Compounds emitted include methylsalicylate, ethylene, C_6 volatiles, *cis*-jasmone, MeJA, terpenoids and of particular interest, (*E*)-*β*-farnesene (EBF). Such volatiles may either act in direct or indirect defense against aphids. Acting directly, plant volatiles may work as insect repellents, decrease fecundity, deter feeding, or provide information to colonizing aphids about the population density on a particular host. EBF is a common aphid alarm pheromone which is released by aphids under stress and causes other aphids in close proximity to cease feeding and find a new host (Beale *et al.*, 2006; De Vos *et al.*, 2007; Unsicker *et al.*, 2009). EBF is present in some aphid-induced plant volatile blends, most notably in wild potato where it has a large repellent effect on feeding aphids (Gibson and Pickett, 1983). *Arabidopsis* that was transformed to express EBF at high levels repelled *M. persicae* (Beale *et al.*,

2006) and in another study, peach-potato aphids were repelled from *Arabidopsis* plants transformed with a strawberry terpene synthase (Aharoni *et al.*, 2003). These studies demonstrate the efficacy of plant-derived volatile compounds in direct defense and revealing the potential for engineering volatile-based aphid resistance into plant species.

Aphid-induced plant volatiles also play an important role in indirect defense against aphids. Among the volatile blend released upon aphid infestation are chemicals including terpenoids that serve as attractants and host-location factors for natural enemies of herbivores such as parasitoids and predators (De Vos *et al.*, 2007; Kessler and Baldwin, 2002). Plant interactions with parasitic wasps such as *Diaeretiella rapae* and *Aphidus ervi* are well studied and much research has focused on the behavioral responses of these organisms to aphid-induced plant volatiles. *D. rapae* primarily attacks aphids which feed on cruciferous plants and several studies have demonstrated that this parasitic wasp is able to distinguish between aphid-infested and uninfested plants on *Arabidopsis* as well as on other agriculturally important crucifers (Blande *et al.*, 2007; Girling *et al.*, 2006; Read *et al.*, 1970, Reed *et al.*, 1995). In one study, *Arabidopsis* plants transformed with a terpene synthase gene from maize emitted high quantities of sesquiterpene products, including EBF, which could be used as a signal to attract the parasitic wasp *Cotesia marginiventris* (Schnee *et al.*, 2006). The release of volatile isothiocyanates during glucosinolate hydrolysis following insect feeding not only has direct toxic effects on the feeding insect, but they have also been shown to be crucial in acting as volatile cues for *D. rapae* in locating aphid-infested plants (Blande *et al.*, 2007; Pope *et al.*, 2008).

In addition to increasing aphid mortality, parasitism may also modify the feeding behavior of the host aphids. For example, electrical monitoring of pea aphids parasitized by *A. ervi* revealed a significant increase in ingestion of xylem fluids and decreased ingestion of phloem sap compared to unparasitized aphids (Ramirez and Niemeyer, 2000). Aphid-induced volatiles provide specific information to predators and parasitoids and the volatile blend may differ both qualitatively and quantitatively depending on the specific plant–aphid interaction. The wasp *A. ervi* is able to discriminate between the volatile blends induced in the same plant species by host and nonhost species of aphid (Guerrieri *et al.*, 1999).

The quantity of volatiles produced by plants infested by aphids is usually much less than when attacked by more aggressive chewing insects. This is likely to be due to the restricted amount of cellular and tissue damage caused by the stealthy nature of feeding by aphids. This provides an obvious advantage for the aphid as reducing the plant's response lessens the negative effects on population of aphids. However, this also provides a problem for parasitoids and predators of aphids as it makes detection of subtle changes in the volatile blend much more difficult. Most experiments investigating the olfactory responses of parasitoids to infested plants and particular chemicals have been lab-based; however, there have recently been experiments demonstrating that the principle also holds true at the field scale (Cai *et al.*, 2009; Kessler and Baldwin, 2001).

Indirect resistance provides a significant level of inducible defense against aphid colonization. As demonstrated in a number of studies outlined previously, there is large potential for engineering crop species to express particular genes for the synthesis of parasitoid/predator-attractant compounds in order to limit aphid infestation.

2. The influence of pathogens on aphid resistance

It is well known that plant pathogens are able to manipulate a plant's defensive signal transduction networks involving SA and JA. These networks play pivotal roles in mediating a plant's resistance response to aphids and thus perturbation of these pathways in infected plants is likely to have effects on plant–aphid interactions. It is estimated that aphids are responsible for the transmission of approximately half of the insect transmissible plant viruses (Ng and Perry, 2004) and thus virus-induced changes to the host plant can subsequently have significant consequences for the success and transmission of the pathogen itself. Therefore, it is likely that viruses and other pathogens are under selection to induce changes to their hosts which provide a more favorable environment for the insect vector.

Aphid performance on hosts infected with viruses of various transmission modes has been assessed in several studies. Infection of potato plants with *Potato leaf roll virus* (PLRV), which is transmitted in a persistent manner, has been shown to improve its quality as an aphid host (Alvarez *et al.*, 2007; Castle and Berger, 1993; Mowry and Ophus, 2006). Aphids are preferentially attracted to settle on potato leaves infected with PLRV (Eigenbrode *et al.*, 2002). This has been shown to be due to an alteration in the volatile blend emitted by infected leaves as aphids aggregated preferentially on screening placed over PLRV-infected leaves compared to healthy plants or plants infected with a nonaphid transmissible virus or a nonpersistently transmitted virus (Alvarez *et al.*, 2007; Eigenbrode *et al.*, 2002). This preference has therefore been attributed to the persistent transmission mode of this virus. Acquisition and inoculation of persistently transmitted viruses requires a relatively long period of plant access (Katis *et al.*, 2007) and by increasing the host suitability of plants infected with PLRV, they facilitate their vector and hence the likelihood of transmission. This phenomenon has also been demonstrated for some other plant viruses including the persistently transmitted barley yellow dwarf virus (BYDV)-infected wheat (Jimenez-Martinez *et al.*, 2004). Other examples are provided in the review of Colvin *et al.* (2006).

Virus-induced alteration of a plant's attractiveness to aphid vectors is not, however, conserved across all virus–plant combinations. *Vicia faba* plants infected with *Pea enation mosaic virus* (PEMV), an association between a circulatively transmitted luteovirus (PEMV-1) and a non-persistently transmitted umbravirus (PEMV-2) showed no increased attractiveness or host suitability to the pea aphid (Hodge and Powell, 2008a,b). There are many other examples of plant virus infections having neutral or negative effects on the performance of aphids and it is therefore clear that although some plant viruses have the ability to alter plant–aphid interactions, this is by no means conserved across all plant-virus–aphid interactions or even across all viruses of a particular transmission mode.

Recent research demonstrates that certain plant viruses are able to manipulate plant JA-mediated defense responses and it is suggested that this will have a direct effect on the performance of insect vectors and thus virus transmission (Yang *et al.*, 2008). The βC1 pathogenicity factor encoded by a satellite DNA of *Tomato yellow leaf curl* China *virus* (TYLCCNV) has recently been demonstrated to selectively suppress certain JA-inducible transcripts in *Arabidopsis* by interacting with the plant protein ASMMETRIC LEAVES 1 (AS1), a negative regulator of JA-signaling (Nurmberg *et al.*, 2007; Yang *et al.*, 2008). TYLCCNV is transmitted by the whitefly, *B. tabaci* in a persistent manner. Whiteflies feeding on tobacco plants infected with this virus showed increased fecundity, longevity, and exhibited an enhanced population growth rate (Jiu *et al.*, 2007). Although the suppression of JA-mediated signaling has not yet been shown to be directly responsible for the increased performance of whiteflies on infected plants, it seems likely to be the case given the role of JA in anti-insect defense. Furthermore, recent work which characterized whitefly performance on a range of *Arabidopsis* defense signaling mutants demonstrated that whiteflies show enhanced performance on mutants which activate SA defenses or impair JA defenses. Conversely, whitefly development was delayed on mutants with enhanced JA defenses or impaired SA defenses (Zarate *et al.*, 2007). Research from the same group also found that whitefly feeding activated SA-responsive transcripts but suppressed JA-responsive transcripts (Kempema *et al.*, 2007). Taken together, these papers further confirm a role for JA in anti-insect defense.

III. ENGINEERED RESISTANCE

A. Breeding for resistance

Over the last 60 years modern farming systems have relied on insecticides to control virus vectors. Alternative strategies or integrated pest management approaches are required as insecticides come under increasing pressure both environmentally and in some circumstances politically;

the occurrence and development of resistance mechanisms to a number of different insecticide groups, particularly in *M. persicae* clones is also an important issue. Conventional breeding programs have developed many resistant lines of agriculturally important species such as wheat, barley, and soybean (Bregitzer *et al.*, 2003; Li *et al.*, 2007; Lynch *et al.*, 2003). There are far too many to mention here and to do so would be beyond the scope of this chapter, but some recent examples are highlighted.

The introduction and subsequent spread of the RWA (*D. noxia*) in the United States, which is a serious and perennial pest of both wheat and barley led to the development of new aphid resistant commercial varieties of these crops. Screening of a small grains germplasm collection maintained by the USDA identified several sources of resistance to *D. noxia* in a number of barley accessions (Lynch *et al.*, 2003). Lines that exhibited high levels of resistance were selected for accelerated breeding, resulting in the release of the first RWA-resistant line, STARS-9301B (Mornhinweg *et al.*, 1995) and a second, STARS-9577B in 1999 (Mornhinweg *et al.*, 1999). In both of these lines, RWA resistance is multigenic being conferred by two loci (Mornhinweg *et al.*, 2002) thus making inheritance patterns complex.

The European large raspberry aphid (*Amphorophora idaei*) is known to transmit four viruses, all of which cause severe economic damage to commercially grown red raspberry (*Rubus idaeus*) in Northern Europe (reviewed by McMenemy *et al.*, 2009). Raspberry is a high value and economically important crop in the United Kingdom and across many parts of Europe and North America and there have been extensive and successful efforts to control *A. idaei* through breeding of aphid-resistant raspberry lines over the past 40 years (Sargent *et al.*, 2007). The two most effective genes incorporated into commercially successful raspberry varieties, A_1 and A_{10}, confer monogenic resistance in several different cultivars and the resistance is thought to have antixenotic and antibiotic effects (Birch and Jones, 1988; Jones *et al.*, 2000; Mcmenemy *et al.*, 2009). However, the precise mechanisms of resistance conferred by these genes are as yet unknown (Mcmenemy *et al.*, 2009). Until recently, the use of aphid-resistant raspberry cultivars was successful in controlling *A. idaei* populations and thus limiting virus spread. This placed strong selection pressure on the aphid to overcome this resistance and as a result several resistant-breaking biotypes have emerged (Mcmenemy *et al.*, 2009). These are a serious problem for the raspberry industry and have hastened the need for novel control strategies.

The rapid emergence of resistance-breaking aphid biotypes is currently a huge problem for plant breeders, with resistant biotypes often occurring as quickly as breeders can introduce new resistance traits. Consequently, there is increased pressure on breeders to introduce more durable resistant traits. These are often multigenic and thus have complex inheritance patterns, making it difficult to develop new cultivars quickly. It is likely that new integrated approaches to aphid resistance, combining alternative control methods such as genetic manipulation, chemical and biological control will be needed in order to provide more robust resistance mechanisms in the future.

B. Transgenic resistance

There is huge potential for the use of genetic modification (GM) in the battle against virus vectors as has already been shown for other insect pests. For example, the widespread deployment of *Bt* cotton that has been engineered to express a bacterial protein, confers resistance to the cotton bollworm and other insects (Lynch *et al.*, 2003).

Virus-resistant transgenic crops have been in commercial use for over 20 years, but as yet no transgenic aphid resistance traits have been commercialized. However, as the understanding of molecular and biochemical plant responses to aphid feeding develops, there is increasing potential for the implementation of GM technologies in protecting agriculturally important crop species against aphids and the viruses which they transmit.

As discussed previously, only two aphid *R* genes have been cloned, the tomato *Mi-1* gene and the *Vat* gene from melon. The cloning and identification of more aphid *R* genes and their incorporation into crops should now be an important priority. However, the generation of transgenic crops expressing aphid *R* genes is unlikely to be straightforward as demonstrated by Goggin and colleagues (2006) who transformed eggplant with *Mi-1* but found that while the nematode-resistance phenotype was retained, the aphid resistance conferred by this gene in tomato was lost. However, functional studies of the recently cloned *Vat* gene in transgenic melon, cotton, and tomato showed that this gene retains functionality of its aphid-resistance phenotype when expressed in these plants (Villada *et al.*, 2009). This exciting development paves the way for generation of commercial transgenic, aphid-resistant crop plants that are likely to have a significant impact on the management of aphid pests in the future.

Engineering specific aphid *R* genes into plants is only one way of incorporating aphid resistance into susceptible crop varieties. There is also much potential for the manipulation of plant basal defenses in order to confer aphid resistance, indeed many plant breeding approaches have sought to incorporate elevated basal defenses into new varieties, for example, aphid resistance in certain potato varieties is conferred by the presence of glandular trichomes (Lapointe and Tingey, 1986). Molecular regulation of basal plant defenses against aphids is currently a growing area of research, with the SA and JA pathways under intense investigation. It has been shown that manipulation of these pathways in model plant species such as *Arabidopsis* and *Medicago* either through mutation, transgene expression, or chemical treatments can have significant effects on aphid performance (Bruce *et al.*, 2003; Ellis *et al.*, 2002; Gao *et al.*, 2007; Moran and Thompson, 2001) with constitutive activation of JA-induced defenses significantly reducing aphid infestation. Transgenic manipulation of these pathways in crop species may therefore also have effects in reducing aphid colonization. However, it should also be noted that manipulation of these important defense pathways may also have other, unwanted effects on other aspects of plant defense against other organisms. This was recently highlighted with virus-resistant transgenic squash being more susceptible to cucumber beetles as these plants remained healthy and more attractive but subsequently were exposed to greater incidence of bacterial wilt disease carried by the beetles (Sasu *et al.*, 2009).

For nearly 20 years it has been well documented that the emission of certain volatiles by plants under herbivore attack has the effect of attracting natural enemies of the herbivores (Dicke *et al.*, 1990; Pare and Tumlinson, 1999; Turlings *et al.*, 1990; Unsicker *et al.*, 2009). These volatile compounds are predominantly terpenes and recent advances in our understanding of the metabolic pathways and molecular genetics governing terpene biosynthesis provides a novel opportunity to engineer crop plants that attract natural enemies of herbivores (Degenhardt *et al.*, 2003). Manipulation of plant indirect defense has already successfully been implemented by intercropping maize with an African grass (*Melinis minutiflora*) that is known to release many volatile compounds (Khan *et al.*, 1997, 2000). This intercropping strategy led to a marked reduction in damage of the maize plants by lepidopteran larvae due to an increase in parasitism by parasitic wasps. This indicates that manipulation of the volatile blend of crop plants may be an effective and environmentally friendly strategy in attracting enemies of plant herbivores. Through GM, volatile blends of existing crops could be modified to produce attractant compounds which function as an indirect defense system as well as compounds that would directly deter aphids from feeding. These two strategies have already been shown to be effective in deterring aphids from feeding on transformed *Arabidopsis* expressing a terpene synthase and the aphid alarm pheromone, EBF (Aharoni *et al.*, 2003; Beale *et al.*, 2006; Schnee *et al.*, 2006).

IV. CONCLUDING REMARKS

The two cloned *R* genes conferring aphid resistance clearly have different modes of action, with the *Mi-1* gene conferring resistance by preventing ingestion of sap from the phloem whereas the *Vat* gene initiates an HR in epidermal and mesophyll cells. The initial signaling events in both of these resistant interactions are currently an area of intense research. As yet, very little is known about aphid effector molecules and how they interact with plant *R* genes—there are currently no cloned or identified insect avirulence (*Avr*) genes. However, the recent project to sequence the pea aphid (*A. pisum*) genome as well as the availability of other molecular tools makes it likely that aphid effector proteins and avirulence genes will be identified in the near future (http://www.hgsc.bcm.tmc.edu/projects/aphid).

During aphid feeding, aphids secrete watery and gelling saliva into their plant host. These types of saliva are composed of a cocktail of enzymes and other proteins (reviewed by Miles, 1999) and contain many potential Avr effector candidates. Indeed, aphid salivary proteins/chemicals are the most likely candidates for aphid effector molecules since only these proteins are injected intracellularly. Following further development of the pea aphid molecular toolkit, the next challenge will be to characterize the contents of the aphid saliva and identify putative effector proteins or their targets. This will facilitate a greater understanding of *R* gene-mediated plant–aphid resistance interactions and enable the transformation of crops with *R* genes to produce stable aphid resistance phenotypes (Gao *et al.*, 2008).

RNAi has been used extensively in generating transgenic resistance to plant viruses, however, as yet this technology has not been fully harnessed in providing transgenic resistance to insects (reviewed by Price and Gatehouse, 2008). It has been demonstrated that oral uptake of dsRNA or siRNA by the model nematode species, *Caenorhabditis elegans* can induce systemic silencing (Timmons and Fire, 1998; Timmons *et al.*, 2001) and a gene, *systemic RNA interference deficient (sid-1)*, has been identified which is essential in mediating the uptake and systemic propagation of the silencing signal (Feinberg and Hunter, 2003). The principle of plant-mediated RNAi against nematodes has also been demonstrated in cyst nematode and root-knot nematode species and was thoroughly reviewed by Rosso and colleagues (2009). There are two strategies to enable host plants to produce the required dsRNA or siRNA to silence target genes in the parasite. Plants may either stably transformed to produce hairprin-shaped dsRNA with a coding sequence that corresponds to the nematode target. Alternatively, host plants may be infected with an RNA virus modified to contain a short nematode-derived section of coding sequence. dsRNA is produced as an intermediate of viral replication and is targeted by the plant's silencing machinery. This results in the production of siRNAs which when ingested, will be recruited into the nematode's silencing machinery and result in downregulation of the target gene (Fig. 2). The commonly used VIGS vector, *Tobacco rattle virus* (TRV) has been successfully modified to contain a portion of the nematode glyceraldehyde-3-phosphate gene. This induced systemic RNAi in females feeding on host plants infected with the modified virus and resulted in a significant reduction in their length (Valentine *et al.*, 2007). Stably transformed plants have also successfully been engineered to produce hairpin RNA targeted at nematode housekeeping and parasitism

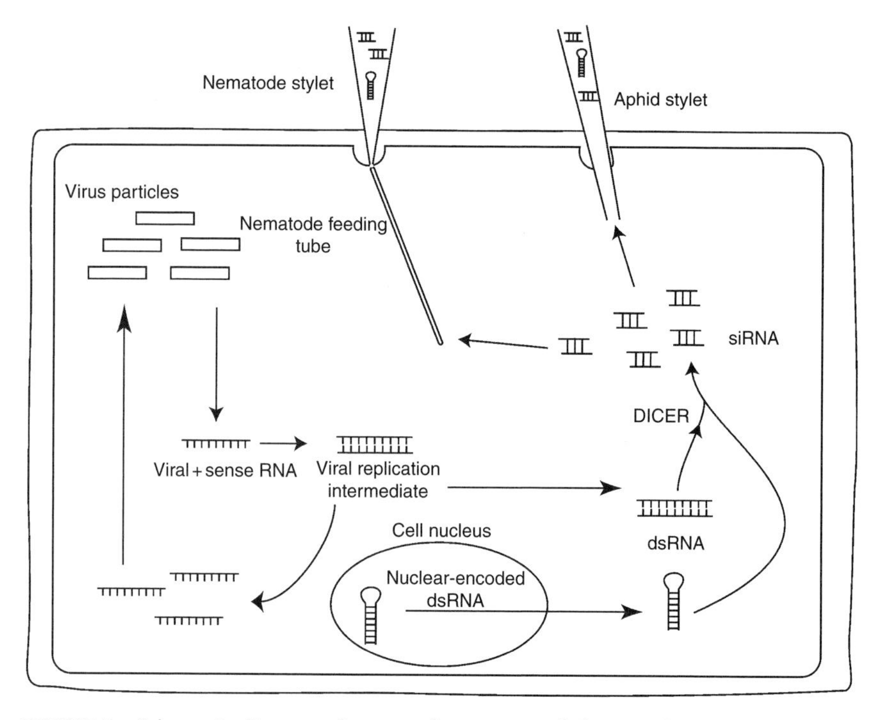

FIGURE 2 Schematic diagram of potential strategies of plant-mediated RNAi targeted against nematodes and aphids. Double-stranded RNA (dsRNA) is produced in feeding sites either transgenically expressed by the plant as hairpin RNA or formed as a replication intermediate by a modified viral vector (e.g., TRV) containing sequence corresponding to the insect gene target. This dsRNA may then be ingested by feeding insects and be recruited into the insect's silencing machinery, targeting genes essential for feeding. Alternatively, dsRNA molecules are processed into siRNA by the plant's silencing machinery, which are then ingested. Based on ideas from Rosso *et al.* (2009).

genes (Rosso *et al.*, 2009). In one instance, root-knot nematodes feeding on transgenic *Arabidopsis thaliana* that produced a dsRNA targeted at the parasitism gene *16D10* resulted in a reduction of up to 90% in the number of galls (Huang *et al.*, 2006).

A homologue to the *C. elegans sid-1* gene has recently been identified in aphids (Xu and Han, 2008), which suggests that plant-mediated RNAi against aphid feeding may be possible. Feeding aphids with dsRNA constructs has been demonstrated to successfully silence aphid genes (Shakesby *et al.*, 2009; Whyard *et al.*, 2009). Whyard *et al.* (2009) showed that feeding pea aphids with a dsRNA which targeted an essential species-specific vATPase significantly increased mortality rates. The authors concluded that dsRNAs have great potential to be developed as species-specific insecticides in the future. The principle has yet to be demonstrated for plants transgenically expressing ds- or siRNAs. Potential targets for developing such an approach include housekeeping genes as well as gene products contained in aphid saliva. There is currently a great deal of interest in understanding the composition and function of compounds contained in saliva. It is thought that effector-like molecules contained in saliva and injected into the plant as part of the feeding process may be responsible for subverting plant defenses and/or serving as recognition points for plant *R* genes. Salivary components from *M. persicae* were recently shown to contain compounds which elicit plant defense responses independent of known SA- and JA-dependent pathways (De Vos and Jander, 2009). Additionally, Mutti *et al.* (2008) showed that a protein produced by the salivary glands of the pea aphid, protein C002, is essential for feeding on a host plant. Downregulating transcription of this gene by injecting siRNA which targeted the c002 gene into the aphid caused the aphid alter its feeding behavior so that it spent very little time feeding from the phloem. This gene may therefore be a good target for plant-mediated RNAi.

There are several examples of plants which have been successfully engineered using this RNAi approach to confer enhanced resistance to economically important pests such as the cotton bollworm and Western corn rootworm, although as yet none have been commercially implemented. (Baum *et al.*, 2007; Mao *et al.*, 2007). Producing plants resistant to phloem-feeding insects in this manner may require expression of dsRNA molecules at feeding sites, including vascular tissue. However, although certain species of RNA (mRNA, viral RNA, small noncoding RNA) are known to be transported systemically via the phloem, there is no evidence that this is that case for dsRNA. This may make it difficult to produce plants resistant to whitefly, which feed exclusively from the phloem without making symplastic punctures (Janssen *et al.*, 1989). However, during aphid feeding, epidermal and mesophyll cells are frequently punctured and ingested from. Because of the similarities between aphid and nematode feeding methods, it may therefore be possible for aphids to acquire plant-derived dsRNA molecules in this manner providing new tools to tackle plant viruses in the future.

ACKNOWLEDGMENTS

Work in authors' laboratories is funded by a Grant from the Leverhulme Trust (F/09741/F).

REFERENCES

Agrawal, A. A. (1999). Induced responses to herbivory in wild radish: Effects on several herbivores and plant fitness. *Ecology* **80:**1713–1723.

Aharoni, A., Giri, A. P., Deuerlein, S., Griepink, F., de Kogel, W. J., Verstappen, F. W. A., Verhoeven, H. A., Jongsma, M. A., Schwab, W., and Bouwmeester, H. J. (2003). Terpenoid metabolism in wild-type and transgenic *Arabidopsis* plants. *Plant Cell* **15:**2866–2884.

Alvarez, A. E., Garzo, E., Verbeek, M., Vosman, B., Dicke, M., and Tjallingii, W. F. (2007). Infection of potato plants with potato leafroll virus changes attraction and feeding behaviour of *Myzus persicae*. *Entomol. Exp. Appl.* **125:**135–144.

Ameline-Torregrosa, C., Wang, B. B., O'Bleness, M. S., Deshpande, S., Zhu, H. Y., Roe, B., Young, N. D., and Cannon, S. B. (2008). Identification and characterization of nucleotide-binding site-Leucine-rich repeat genes in the model plant *Medicago truncatula*. *Plant Physiol.* **146:**5–21.

Arimura, G., Kost, C., and Boland, W. (2005). Herbivore-induced, indirect plant defences. *Biochim. Biophys. Acta-Mol. Cell Biol. Lipids* **1734:**91–111.

Baum, J. A., Bogaert, T., Clinton, W., Heck, G. R., Feldmann, P., Ilagan, O., Johnson, S., Plaetinck, G., Munyikwa, T., Pleau, M., Vaughn, T., and Roberts, J. (2007). Control of coleopteran insect pests through RNA interference. *Nat. Biotechnol.* **25:**1322–1326.

Beale, M. H., Birkett, M. A., Bruce, T. J. A., Chamberlain, K., Field, L. M., Huttly, A. K., Martin, J. L., Parker, R., Phillips, A. L., Pickett, J. A., Prosser, I. M., Shewry, P. R., *et al.* (2006). Aphid alarm pheromone produced by transgenic plants affects aphid and parasitoid behavior. *Proc. Natl. Acad. Sci. USA* **103:**10509–10513.

Belefantmiller, H., Porter, D. R., Pierce, M. L., and Mort, A. J. (1994). An early indicator of resistance in barley to Russian wheat aphid. *Plant Physiol.* **105:**1289–1294.

Birch, A. N. E., and Jones, A. T. (1988). Levels and components of resistance to *Amphorophora idaei* in raspberry cultivars containing different resistance genes. *Ann. Appl. Biol.* **113:**567–578.

Blande, J. D., Pickett, J. A., and Poppy, G. M. (2007). A comparison of semiochemically mediated interactions involving specialist and generalist *Brassica*-feeding aphids and the braconid parasitoid *Diaeretiella rapae*. *J. Chem. Ecol.* **33:**767–779.

Boughton, A. J., Hoover, K., and Felton, G. W. (2005). Methyl jasmonate application induces increased densities of glandular trichomes on tomato, *Lycopersicon esculentum*. *J. Chem. Ecol.* **31:**2211–2216.

Bregitzer, P., Mornhinweg, D. W., and Jones, B. L. (2003). Resistance to Russian wheat aphid damage derived from STARS 9301B protects agronomic performance and malting quality when transferred to adapted barley germplasm. *Crop Sci.* **43:**2050–2057.

Bruce, T. J. A., Martin, J. L., Pickett, J. A., Pye, B. J., Smart, L. E., and Wadhams, L. J. (2003). *cis*-Jasmone treatment induces resistance in wheat plants against the grain aphid, *Sitobion avenae* (Fabricius) (Homoptera: Aphididae). *Pest Manage. Sci.* **59:**1031–1036.

Bruce, T. J. A., Matthes, M. C., Chamberlain, K., Woodcock, C. M., Mohib, A., Webster, B., Smart, L. E., Birkett, M. A., Pickett, J. A., and Napier, J. A. (2008). *cis*-Jasmone induces *Arabidopsis* genes that affect the chemical ecology of multitrophic interactions with aphids and their parasitoids. *Proc. Natl. Acad. Sci. USA* **105:**4553–4558.

Cai, Q. N., Ma, X. M., Zhao, X., Cao, Y. Z., and Yang, X. Q. (2009). Effects of host plant resistance on insect pests and its parasitoid: A case study of wheat-aphid-parasitoid system. *Biol. Control* **49:**134–138.

Casteel, C. L., Walling, L. L., and Paine, T. D. (2006). Behavior and biology of the tomato psyllid, *Bactericerca cockerelli*, in response to the *Mi-1.2* gene. *Entomol. Exp. Appl.* **121:**67–72.

Castle, S. J., and Berger, P. H. (1993). Rates of growth and increase of *Myzus persicae* on virus-infected potatoes according to type of virus-vector relationship. *Entomol. Exp. Appl.* **69:**51–60.

Chini, A., Fonseca, S., Fernandez, G., Adie, B., Chico, J. M., Lorenzo, O., Garcia-Casado, G., Lopez-Vidriero, I., Lozano, F. M., Ponce, M. R., Micol, J. L., and Solano, R. (2007). The JAZ family of repressors is the missing link in jasmonate signalling. *Nature* **448:**666–671.

Colvin, J., Omongo, C. A., Govindappa, M. R., Stevenson, P. C., Maruthi, M. N., Gibson, G., Seal, S. E., and Muniyappa, V. (2006). Host-plant viral infection effects on arthropod-vector population growth, development and behaviour: Management and epidemiological implications. *Adv. Virus Res.* **67**:419–452.

Cooper, W. R., and Goggin, F. L. (2005). Effects of jasmonate-induced defenses in tomato on the potato aphid, *Macrosiphum euphorbiae*. *Entomol. Exp. Appl.* **115**:107–115.

Cooper, W. C., Jia, L., and Goggin, F. L. (2004). Acquired and *R*-gene-mediated resistance against the potato aphid in tomato. *J. Chem. Ecol.* **30**:2527–2542.

Dangl, J. L., and Jones, J. D. G. (2001). Plant pathogens and integrated defence responses to infection. *Nature* **411**:826–833.

de Ilarduya, O. M., Xie, Q. G., and Kaloshian, I. (2003). Aphid-induced defense responses in *Mi-1*-mediated compatible and incompatible tomato interactions. *Mol. Plant–Microbe Interact.* **16**:699–708.

De Vos, M., and Jander, G. (2009). *Myzus persicae* (green peach aphid) salivary components induce defence responses in *Arabidopsis thaliana*. *Plant Cell Environ.* **32**:1548–1560.

de Vos, M., Kim, J. H., and Jander, G. (2007). Biochemistry and molecular biology of *Arabidopsis*–aphid interactions. *Bioessays* **29**:871–883.

Degenhardt, J., Gershenzon, J., Baldwin, I. T., and Kessler, A. (2003). Attracting friends to feast on foes: Engineering terpene emission to make crop plants more attractive to herbivore enemies. *Curr. Opin. Biotechnol.* **14**:169–176.

Delp, G., Gradin, T., Ahman, I., and Jonsson, L. M. V. (2009). Microarray analysis of the interaction between the aphid *Rhopalosiphum padi* and host plants reveals both differences and similarities between susceptible and partially resistant barley lines. *Mol. Genet. Genomics* **281**:233–248.

Devine, G. J., Harling, Z. K., Scarr, A. W., and Devonshire, A. L. (1996). Lethal and sublethal effects of imidacloprid on nicotine-tolerant *Myzus nicotianae* and *Myzus persicae*. *Pestic. Sci.* **48**:57–62.

Dicke, M., Van Beek, T. A., Posthumus, M. A., Bendom, N., Vanbokhoven, H., and Degroot, A. E. (1990). Isolation and identification of volatile kairomone that affects acarine predatorprey interactions involvement of host plant in its production. *J. Chem. Ecol.* **16**:381–396.

Divol, F., Vilaine, F., Thibivilliers, S., Amselem, J., Palauqui, J. C., Kusiak, C., and Dinant, S. (2005). Systemic response to aphid infestation by *Myzus persicae* in the phloem of *Apium graveolens*. *Plant Mol. Biol.* **57**:517–540.

Divol, F., Vilaine, F., Thibivilliers, S., Kusiak, C., Sauge, M. H., and Dinant, S. (2007). Involvement of the xyloglucan endotransglycosylase/hydrolases encoded by celery *XTH1* and Arabidopsis *XTH33* in the phloem response to aphids. *Plant Cell Environ.* **30**:187–201.

Dixon, R. A., Achnine, L., Kota, P., Liu, C. J., Reddy, M. S. S., and Wang, L. J. (2002). The phenylpropanoid pathway and plant defence—A genomics perspective. *Mol. Plant Pathol.* **3**:371–390.

Eigenbrode, S. D., Ding, H. J., Shiel, P., and Berger, P. H. (2002). Volatiles from potato plants infected with potato leafroll virus attract and arrest the virus vector, *Myzus persicae* (Homoptera: Aphididae). *Proc. R. Soc. Lond. Ser. B. Biol. Sci.* **269**:455–460.

Ellis, P. R., Singh, R., Pink, D. A. C., Lynn, J. R., and Saw, P. L. (1996). Resistance to *Brevicoryne brassicae* in horticultural brassicas. *Euphytica* **88**:85–96.

Ellis, C., Karafyllidis, L., and Turner, J. G. (2002). Constitutive activation of jasmonate signaling in an *Arabidopsis* mutant correlates with enhanced resistance to *Erysiphe cichoracearum, Pseudomonas syringae,* and *Myzus persicae*. *Mol. Plant–Microbe Interact.* **15**:1025–1030.

Fahey, J. W., Zalcmann, A. T., and Talalay, P. (2001). The chemical diversity and distribution of glucosinolates and isothiocyanates among plants. *Phytochemistry* **56**:5–51.

Farmer, E. E., Liechti, R., and Gfeller, A. (2009). Arabidopsis jasmonate signaling pathway. *Sci. Signal.*(Connections Map in the Database of Cell Signaling, as seen on 11 November 2009).

Feinberg, E. H., and Hunter, C. P. (2003). Transport of dsRNA into cells by the transmembrane protein SID-1. *Science* **301:**1545–1547.

Fidantsef, A. L., Stout, M. J., Thaler, J. S., Duffey, S. S., and Bostock, R. M. (1999). Signal interactions in pathogen and insect attack: Expression of lipoxygenase, proteinase inhibitor II, and pathogenesis-related protein P4 in the tomato, *Lycopersicon esculentum*. *Physiol. Mol. Plant Pathol.* **54:**97–114.

Flor, H. H. (1955). Host–parasite interaction in flax rust—Its genetics and other implications. *Phytopathology* **45:**680–685.

Flor, H. H. (1971). Current status of gene-for-gene concept. *Annu. Rev. Phytopathol.* **9:**275–296.

Foster, S. P., Denholm, I., and Thompson, R. (2003). Variation in response to neonicotinoid insecticides in peach-potato aphids, *Myzus persicae* (Hemiptera: Aphididae). *Pest Manage. Sci.* **59:**166–173.

Foster, S. P., Devine, G. J., and Devonshire, A. L. (2007). Insecticide resistance. *In* "Aphids as Crop Pests" (H. F. van Emden and R. Harrington, eds.), pp. 261–286. CABI, Wallingford, Oxfordshire.

Gao, L. L., Anderson, J. P., Klingler, J. P., Nair, R. M., Edwards, O. R., and Singh, K. B. (2007). Involvement of the octadecanoid pathway in bluegreen aphid resistance in *Medicago truncatula*. *Mol. Plant–Microbe Interact.* **20:**82–93.

Gao, L. L., Klingler, J. P., Anderson, J. P., Edwards, O. R., and Singh, K. B. (2008). Characterization of pea aphid resistance in *Medicago truncatula*. *Plant Physiol.* **146:**996–1009.

Garcia-Brugger, A., Lamotte, O., Vandelle, E., Bourque, S., Lecourieux, D., Poinssot, B., Wendehenne, D., and Pugin, A. (2006). Early signaling events induced by elicitors of plant defenses. *Mol. Plant–Microbe Interact.* **19:**711–724.

Gibson, R. W. (1976). Glandular hairs are a possible means of limiting aphid damage to the potato crop. *Ann. Appl. Biol.* **82:**143–146.

Gibson, R. W., and Pickett, J. A. (1983). Wild potato repels aphids by release of aphid alarm pheromone. *Nature* **302:**608–609.

Girling, R. D., Hassall, M., Turner, J. G., and Poppy, G. M. (2006). Behavioural responses of the aphid parasitoid *Diaeretiella rapae* to volatiles from *Arabidopsis thaliana* induced by *Myzus persicae*. *Entomol. Exp. Appl.* **120:**1–9.

Glazebrook, J., Zook, M., Mert, F., Kagan, I., Rogers, E. E., Crute, I. R., Holub, E. B., Hammerschmidt, R., and Ausubel, F. M. (1997). Phytoalexin-deficient mutants of *Arabidopsis* reveal that *PAD4* encodes a regulatory factor and that four *PAD* genes contribute to downy mildew resistance. *Genetics* **146:**381–392.

Goggin, F. L. (2007). Plant–aphid interactions: Molecular and ecological perspectives. *Curr. Opin. Plant Biol.* **10:**399–408.

Goggin, F. L., Jia, L. L., Shah, G., Hebert, S., Williamson, V. M., and Ullman, D. E. (2006). Heterologous expression of the *Mi-1.2* gene from tomato confers resistance against nematodes but not aphids in eggplant. *Mol. Plant–Microbe Interact.* **19:**383–388.

Guerrieri, E., Poppy, G. M., Powell, W., Tremblay, E., and Pennacchio, F. (1999). Induction and systemic release of herbivore-induced plant volatiles mediating in-flight orientation of *Aphidius ervi*. *J. Chem. Ecol.* **25:**1247–1261.

Guo, S., Kamphuis, L. G., Gao, L., Edwards, O. R., and Singh, K. B. (2009). Two independent resistance genes in the *Medicago truncatula* cultivar jester confer resistance to two different aphid species of the genus *Acyrthosiphon*. *Plant Signal. Behav.* **4:**328–331.

Hahlbrock, K., and Scheel, D. (1989). Physiology and molecular biology of phenylpropanoid metabolism. *Annu. Rev. Plant Physiol. Plant Mol. Biol.* **40:**347–369.

Halkier, B. A., and Gershenzon, J. (2006). Biology and biochemistry of glucosinolates. *Annu. Rev. Plant Biol.* **57:**303–333.

Heidel, A. J., and Baldwin, I. T. (2004). Microarray analysis of salicylic acid- and jasmonic acid-signalling in responses of *Nicotiana attenuata* to attack by insects from multiple feeding guilds. *Plant Cell Environ.* **27:**1362–1373.

Hodge, S., and Powell, G. (2008). Complex interactions between a plant pathogen and insect parasitoid via the shared vector-host: Consequences for host plant infection. *Oecologia* **157:**387–397.

Hodge, S., and Powell, G. (2008). Do plant viruses facilitate their aphid vectors by inducing symptoms that alter behavior and performance? *Environ. Entomol.* **37:**1573–1581.

Huang, G. Z., Allen, R., Davis, E. L., Baum, T. J., and Hussey, R. S. (2006). Engineering broad root-knot resistance in transgenic plants by RNAi silencing of a conserved and essential root-knot nematode parasitism gene. *Proc. Natl. Acad. Sci. USA* **103:**14302–14306.

Iglesias, V. A., and Meins, F. (2000). Movement of plant viruses is delayed in a beta-1, 3-glucanase-deficient mutant showing a reduced plasmodesmatal size exclusion limit and enhanced callose deposition. *Plant J.* **21:**157–166.

Janssen, J. A. M., Tjallingii, W. F., and Vanlenteren, J. C. (1989). Electrical recording and ultrastructure of stylet penetration by the greenhouse whitefly. *Entomol. Exp. Appl.* **52:**69–81.

Jimenez-Martinez, E. S., Bosque-Perez, N. A., Berger, P. H., Zemetra, R., Ding, H. J., and Eigenbrode, S. D. (2004). Volatile cues influence the response of *Rhopalosiphum padi* (Homoptera: Aphididae) to Barley yellow dwarf virus-infected transgenic and untransformed wheat. *Environ. Entomol.* **33:**1207–1216.

Jiu, M., Zhou, X.-P., Tong, L., Xu, J., Yang, X., Wan, F.-H., and Liu, S.-S. (2007). Vector-virus mutualism accelerates population increase of an invasive whitefly. *PLoS One* **2:**e182.

Johnson, B. (1953). The injurious effects of the hooked epidermal hairs of French beans (*Phaseolus vulgaris* L.) on *Aphis cracctvora* Koch. *Bull. Entomol. Res.* **44:**779–788.

Jones, J. D. G., and Dangl, J. L. (2006). The plant immune system. *Nature* **444:**323–329.

Jones, A. T., McGavin, W. J., and Birch, A. N. E. (2000). Effectiveness of resistance genes to the large raspberry aphid, *Amphorophora idaei* Borner, in different raspberry (*Rubus idaeus* L.) genotypes and under different environmental conditions. *Ann. Appl. Biol.* **136:**107–113.

Kaloshian, I. (2004). Gene-for-gene disease resistance: Bridging insect pest and pathogen defense. *J. Chem. Ecol.* **30:**2419–2438.

Kaloshian, I., Lange, W. H., and Williamson, V. M. (1995). An aphid-resistance locus is tightly linked to the nematode-resistance gene, *Mi*, in tomato. *Proc. Natl. Acad. Sci. USA* **92:**622–625.

Kaloshian, I., Kinsey, M. G., Ullman, D. E., and Williamson, V. M. (1997). The impact of *Meu1*-mediated resistance in tomato on longevity, fecundity and behavior of the potato aphid, *Macrosiphum euphorbiae. Entomol. Exp. Appl.* **83:**181–187.

Kaloshian, I., Kinsey, M. G., Williamson, V. M., and Ullman, D. E. (2000). *Mi*-mediated resistance against the potato aphid *Macrosiphum euphorbiae* (Hemiptera: Aphididae) limits sieve element ingestion. *Environ. Entomol.* **29:**690–695.

Katis, N. I., Tsitsipis, J. A., Stevens, M., and Powell, G. (2007). Transmission of plant viruses. *In* "Aphids as Crop Pests" (H. F. van Emden and R. Harrington, eds.), pp. 353–390. CABI International, Wallingford.

Kempema, L. A., Cui, X. P., Holzer, F. M., and Walling, L. L. (2007). *Arabidopsis* transcriptome changes in response to phloem-feeding silverleaf whitefly nymphs. Similarities and distinctions in responses to aphids. *Plant Physiol.* **143:**849–865.

Kessler, A., and Baldwin, I. T. (2001). Defensive function of herbivore-induced plant volatile emissions in nature. *Science* **291:**2141–2144.

Kessler, A., and Baldwin, I. T. (2002). Plant responses to insect herbivory: The emerging molecular analysis. *Annu. Rev. Plant Biol.* **53:**299–328.

Khan, Z. R., AmpongNyarko, K., Chiliswa, P., Hassanali, A., Kimani, S., Lwande, W., Overholt, W. A., Pickett, J. A., Smart, L. E., Wadhams, L. J., and Woodcock, C. M. (1997). Intercropping increases parasitism of pests. *Nature* **388:**631–632.

Khan, Z. R., Pickett, J. A., van den Berg, J., Wadhams, L. J., and Woodcock, C. M. (2000). Exploiting chemical ecology and species diversity: Stem borer and striga control for maize and sorghum in Africa. *Pest Manage. Sci.* **56:**957–962.

Kim, J. H., and Jander, G. (2007). *Myzus persicae* (green peach aphid) feeding on *Arabidopsis* induces the formation of a deterrent indole glucosinolate. *Plant J.* **49:**1008–1019.

Klingler, J., Creasy, R., Gao, L. L., Nair, R. M., Calix, A. S., Jacob, H. S., Edwards, O. R., and Singh, K. B. (2005). Aphid resistance in *Medicago truncatula* involves antixenosis and phloem-specific, inducible antibiosis, and maps to a single locus flanked by NBS-LRR resistance gene analogs. *Plant Physiol.* **137:**1445–1455.

Klingler, J. P., Edwards, O. R., and Singh, K. B. (2007). Independent action and contrasting phenotypes of resistance genes against spotted alfalfa aphid and bluegreen aphid in *Medicago truncatula*. *New Phytol.* **173:**630–640.

Klingler, J. P., Nair, M. N., Edwards, O. R., and Singh, K. B. (2009). A single gene, *AIN*, in *Medicago truncatula* mediates a hypersensitive response to both bluegreen aphid and pea aphid, but confers resistance only to bluegreen aphid. *J. Exp. Bot.* **60:**4115–4127.

Knoblauch, M., and Peters, W. S. (2004). Forisomes, a novel type of Ca^{2+}-dependent contractile protein motor. *Cell Motil. Cytoskeleton* **58:**137–142.

Knoblauch, M., and van Bel, A. J. E. (1998). Sieve tubes in action. *Plant Cell* **10:**35–50.

Knoblauch, M., Peters, W. S., Ehlers, K., and van Bel, A. J. E. (2001). Reversible calcium-regulated stopcocks in legume sieve tubes. *Plant Cell* **13:**1221–1230.

Knoblauch, M., Noll, G. A., Muller, T., Prufer, D., Schneider-Huther, I., Scharner, D., Van Bel, A. J. E., and Peters, W. S. (2003). ATP-independent contractile proteins from plants. *Nat. Mater.* **2:**600–603.

Koornneef, A., and Pieterse, C. M. J. (2008). Cross talk in defense signaling. *Plant Physiol.* **146:**839–844.

Lapointe, S. L., and Tingey, W. M. (1986). Glandular trichomes of *Solanum neocardenasii* confer resistance to green peach aphid (Homoptera: Aphididae). *J. Econ. Entomol.* **79:**1264–1268.

Levin, D. A. (1973). The role of trichomes in plant defense. *Q. Rev. Biol.* **48:**3–15.

Li, Q., Xie, Q. G., Smith-Becker, J., Navarre, D. A., and Kaloshian, I. (2006). *Mi-1*-mediated aphid resistance involves salicylic acid and mitogen-activated protein kinase signaling cascades. *Mol. Plant–Microbe Interact.* **19:**655–664.

Li, Y., Hill, C. B., Carlson, S. R., Diers, B. W., and Hartman, G. L. (2007). Soybean aphid resistance genes in the soybean cultivars Dowling and Jackson map to linkage group M. *Mol. Breed.* **19:**25–34.

Lowe, H. J. B., Murphy, G. J. P., and Parker, M. L. (1985). Non-glaucousness, a probable aphid-resistance character of wheat. *Ann. Appl. Biol.* **106:**555–560.

Lynch, R. E., Guo, B. Z., Timper, P., and Wilson, J. P. (2003). United States department of agriculture agricultural research service research on improving host-plant resistance to pest. *Pest Manage. Sci.* **59:**718–727.

Mantelin, S., Bhattarai, K. K., and Kaloshian, I. (2009). Ethylene contributes to potato aphid susceptibility in a compatible tomato host. *New Phytol.* **183:**444–456.

Mao, Y. B., Cai, W. J., Wang, J. W., Hong, G. J., Tao, X. Y., Wang, L. J., Huang, Y. P., and Chen, X. Y. (2007). Silencing a cotton bollworm P450 monooxygenase gene by plant-mediated RNAi impairs larval tolerance of gossypol. *Nat. Biotechnol.* **25:**1307–1313.

Martin, B., Rahbe, Y., and Fereres, A. (2003). Blockage of stylet tips as the mechanism of resistance to virus transmission by *Aphis gossypii* in melon lines bearing the *Vat* gene. *Ann. Appl. Biol.* **142:**245–250.

McMenemy, L. S., Mitchell, C., and Johnson, S. N. (2009). Biology of the European large raspberry aphid (*Amphorophora idaei*): Its role in virus transmission and resistance breakdown in red raspberry. *Agric. For. Entomol.* **11:**61–71.

Mewis, I., Appel, H. M., Hom, A., Raina, R., and Schultz, J. C. (2005). Major signaling pathways modulate *Arabidopsis* glucosinolate accumulation and response to both phloem-feeding and chewing insects. *Plant Physiol.* **138:**1149–1162.

Mewis, I., Tokuhisa, J. G., Schultz, J. C., Appel, H. M., Ulrichs, C., and Gershenzon, J. (2006). Gene expression and glucosinolate accumulation in *Arabidopsis thaliana* in response to generalist and specialist herbivores of different feeding guilds and the role of defense signaling pathways. *Phytochemistry* **67:**2450–2462.

Miles, P. W. (1999). Aphid saliva. *Biological Reviews* **74:**41-85.

Milligan, S. B., Bodeau, J., Yaghoobi, J., Kaloshian, I., Zabel, P., and Williamson, V. M. (1998). The root knot nematode resistance gene *Mi* from tomato is a member of the leucine zipper, nucleotide binding, leucine-rich repeat family of plant genes. *Plant Cell* **10:**1307–1319.

Moloi, M. J., and van der Westhuizen, A. J. (2006). The reactive oxygen species are involved in resistance responses of wheat to the Russian wheat aphid. *J. Plant Physiol.* **163:**1118–1125.

Moran, P. J., and Thompson, G. A. (2001). Molecular responses to aphid feeding in *Arabidopsis* in relation to plant defense pathways. *Plant Physiol.* **125:**1074–1085.

Mornhinweg, D. W., Porter, D. R., and Webster, J. S. (1995). Registration of STARS-9301B barley germplasm resistant to Russian wheat aphid. *Crop Sci.* **35:**602.

Mornhinweg, D. W., Porter, D. R., and Webster, J. A. (1999). Registration of STARS-9577B Russian wheat aphid resistant barley germplasm. *Crop Sci.* **39:**882–883.

Mornhinweg, D. W., Porter, D. R., and Webster, J. A. (2002). Inheritance of Russian wheat aphid resistance in spring barley germplasm line STARS-9577B. *Crop Sci.* **42:**1891–1893.

Mowry, T. M., and Ophus, J. D. (2006). Influence of the Potato leafroll virus and virus-infected plants on the arrestment of the aphid, *Myzus persicae*. *J. Insect Sci.* **6:**22.

Mutti, N. S., Louis, J., Pappan, L. K., Pappan, K., Begum, K., Chen, M. S., Park, Y., Dittmer, N., Marshall, J., Reese, J. C., and Reeck, G. R. (2008). A protein from the salivary glands of the pea aphid, *Acyrthosiphon pisum*, is essential in feeding on a host plant. *Proc. Natl. Acad. Sci. USA* **105:**9965–9969.

Nair, R. M., Craig, A. D., Auricht, G. C., Edwards, O. R., Robinson, S. S., Otterspoor, M. J., and Jones, J. A. (2003). Evaluating pasture legumes for resistance to aphids. *Aust. J. Exp. Agric.* **43:**1345–1349.

Nauen, R., and Denholm, I. (2005). Resistance of insect pests to neonicotinoid insecticides: Current status and future prospects. *Arch. Insect Biochem. Physiol.* **58:**200–215.

Ng, J. C. K., and Perry, K. L. (2004). Transmission of plant viruses by aphid vectors. *Mol. Plant Pathol.* **5:**505–511.

Nombela, G., Williamson, V. M., and Muniz, M. (2003). The root-knot nematode resistance gene *Mi-1.2* of tomato is responsible for resistance against the whitefly *Bemisia tabaci*. *Mol. Plant–Microbe Interact.* **16:**645–649.

Nurmberg, P. L., Knox, K. A., Yun, B. W., Morris, P. C., Shafiei, R., Hudson, A., and Loake, G. J. (2007). The developmental selector *AS1* is an evolutionarily conserved regulator of the plant immune response. *Proc. Natl. Acad. Sci. USA* **104:**18795–18800.

Pare, P. W., and Tumlinson, J. H. (1999). Plant volatiles as a defense against insect herbivores. *Plant Physiol.* **121:**325–331.

Pauwels, L., Morreel, K., De Witte, E., Lammertyn, F., Van Montagu, M., Boerjan, W., Inze, D., and Goossens, A. (2008). Mapping methyl jasmonate-mediated transcriptional reprogramming of metabolism and cell cycle progression in cultured *Arabidopsis* cells. *Proc. Natl. Acad. Sci. USA* **105:**1380–1385.

Pegadaraju, V., Knepper, C., Reese, J., and Shah, J. (2005). Premature leaf senescence modulated by the *Arabidopsis PHYTOALEXIN DEFICIENT4* gene is associated with defense against the phloem-feeding green peach aphid. *Plant Physiol.* **139:**1927–1934.

Pegadaraju, V., Louis, J., Singh, V., Reese, J. C., Bautor, J., Feys, B. J., Cook, G., Parker, J. E., and Shah, J. (2007). Phloem-based resistance to green peach aphid is controlled by *Arabidopsis PHYTOALEXIN DEFICIENT4* without its signaling partner *ENHANCED DISEASE SUSCEPTIBILITY1*. *Plant J.* **52:**332–341.

Pennazio, S., Redolfi, P., and Sapetti, C. (1981). Callose formation and permeability changes during the partly localized reaction of *Gomphrena globosa* to Potato virus X. *J. Phytopathol.* **100:**172–181.

Pope, T. W., Kissen, R., Grant, M., Pickett, J. A., Rossiter, J. T., and Powell, G. (2008). Comparative innate responses of the aphid parasitoid *Diaeretiella rapae* to alkenyl glucosinolate derived isothiocyanates, nitriles, and epithionitriles. *J. Chem. Ecol.* **34:**1302–1310.

Price, D. R. G., and Gatehouse, J. A. (2008). RNAi-mediated crop protection against insects. *Trends Biotechnol.* **26:**393–400.

Ramirez, C. C., and Niemeyer, H. M. (2000). The influence of previous experience and starvation on aphid feeding behavior. *J. Insect Behav.* **13:**699–709.

Read, D. P., Feeny, P. P., and Root, R. B. (1970). Habitat selection by the aphid parasite *Diaeretiella rapae* (Hymenoptera: Braconidae) and hyperparasite *Charips brassicae* (Hymenoptera: Cynipidae). *Can. Entomol.* **102:**1567–1578.

Reed, H. C., Tan, S. H., Haapanen, K., Killmon, M., Reed, D. K., and Elliott, N. C. (1995). Olfactory responses of the parasitoid *Diaeretiella rapae* (Hymenoptera: Aphidiidae) to odor of plants, aphids, and plant–aphid complexes. *J. Chem. Ecol.* **21:**407–418.

Rossi, M., Goggin, F. L., Milligan, S. B., Kaloshian, I., Ullman, D. E., and Williamson, V. M. (1998). The nematode resistance gene *Mi* of tomato confers resistance against the potato aphid. *Proc. Natl. Acad. Sci. USA* **95:**9750–9754.

Rosso, M. N., Jones, J. T., and Abad, P. (2009). RNAi and functional genomics in plant parasitic nematodes. *Annu. Rev. Phytopathol.* **47:**207–232.

Salzman, R. A., Brady, J. A., Finlayson, S. A., Buchanan, C. D., Summer, E. J., Sun, F., Klein, P. E., Klein, R. R., Pratt, L. H., Cordonnier-Pratt, M. M., and Mullet, J. E. (2005). Transcriptional profiling of sorghum induced by methyl jasmonate, salicylic acid, and aminocyclopropane carboxylic acid reveals cooperative regulation and novel gene responses. *Plant Physiol.* **138:**352–368.

Sargent, D. J., Fernandez-Fernandez, F., Rys, A., Knight, V. H., Simpson, D. W., and Tobutt, K. R. (2007). Mapping of A_1 conferring resistance to the aphid *Amphorophora idaei* and dw (dwarfing habit) in red raspberry (*Rubus idaeus* L.) using AFLP and microsatellite markers. *BMC Plant Biol.* **7:**15.

Sasu, M. A., Ferrari, M. J., Du, D., Winsor, J. A., and Stephenson, A. G. (2009). Indirect costs of a nontarget pathogen mitigate the direct benefits of a virus-resistant transgene in wild *Cucurbita*. *Proc. Natl. Acad. Sci. USA* **106**(45)**:**19067–19071.

Schnee, C., Köllner, T. G., Held, M., Turlings, T. C. J., Gershenzon, J., and Degenhardt, J. (2006). The products of a single maize sesquiterpene synthase form a volatile defense signal that attracts natural enemies of maize herbivores. *Proc. Natl. Acad. Sci. USA* **103:**1129–1134.

Schommer, C., Palatnik, J. F., Aggarwal, P., Chetelat, A., Cubas, P., Farmer, E. E., Nath, U., and Weigel, D. (2008). Control of jasmonate biosynthesis and senescence by miR319 targets. *PLoS Biol.* **6:**1991–2001.

Shakesby, A. J., Wallace, I. S., Isaacs, H. V., Pritchard, J., Roberts, D. M., and Douglas, A. E. (2009). A water-specific aquaporin involved in aphid osmoregulation. *Insect Biochem. Mol. Biol.* **39:**1–10.

Simmons, A. T., Gurr, G. M., McGrath, D., Nicol, H. I., and Martin, P. M. (2003). Trichomes of *Lycopersicon* spp. and their effect on *Myzus persicae* (Sulzer) (Hemiptera: Aphididae). *Aust. J. Entomol.* **42:**373–378.

Simmons, A. T., McGrath, D., and Gurr, G. M. (2005). Trichome characteristics of F_1 *Lycopersicon esculentum* x *L. cheesmanii* f. *minor* and *L. esculentum* x *L. pennellii* hybrids and effects on *Myzus persicae*. *Euphytica* **144:**313–320.
Smith, S. D. J., and Furk, C. (1989). The spread of the resistant aphid. *Sugar Beet Review*:4-6.
Spoel, S. H., and Dong, X. (2008). Making sense of hormone crosstalk during plant immune responses. *Cell Host Microbe* **3:**348–351.
Spoel, S. H., Koornneef, A., Claessens, S. M. C., Korzelius, J. P., Van Pelt, J. A., Mueller, M. J., Buchala, A. J., Metraux, J. P., Brown, R., Kazan, K., Van Loon, L. C., Dong, X. N., *et al.* (2003). NPR1 modulates cross-talk between salicylate- and jasmonate-dependent defense pathways through a novel function in the cytosol. *Plant Cell* **15:**760–770.
Stewart, S. A., Hodge, S., Ismail, N., Mansfield, J. W., Feys, B. J., Prospéri, J.-M., Huguet, T., Ben, C., Gentzbittel, L., and Powell, G. (2009). The *RAP1* gene confers effective, race-specific resistance to the Pea aphid in *Medicago truncatula* independent of the hypersensitive reaction. *Mol. Plant–Microbe Interact.* **22:**1645–1655.
Stoner, K. A. (1992). Density of imported cabbageworms (Lepidoptera: Pieridae), cabbage aphids (Homoptera: Aphididae), and flea beetles (Coleoptera: Chrysomelidae) on glossy and trichome-bearing lines of *Brassica oleracea*. *J. Econ. Entomol.* **85:**1023–1030.
Stuhlfelder, C., Mueller, M. J., and Warzecha, H. (2004). Cloning and expression of a tomato cDNA encoding a methyl jasmonate cleaving esterase. *Eur. J. Biochem.* **271:**2976–2983.
Takken, F. L. W., Albrecht, M., and Tameling, W. I. L. (2006). Resistance proteins: Molecular switches of plant defence. *Curr. Opin. Plant Biol.* **9:**383–390.
Thines, B., Katsir, L., Melotto, M., Niu, Y., Mandaokar, A., Liu, G. H., Nomura, K., He, S. Y., Howe, G. A., and Browse, J. (2007). JAZ repressor proteins are targets of the SCF CO11 complex during jasmonate signalling. *Nature* **448:**661–665.
Thompson, G. A., and Goggin, F. L. (2006). Transcriptomics and functional genomics of plant defence induction by phloem-feeding insects. *J. Exp. Bot.* **57:**755–766.
Timmons, L., and Fire, A. (1998). Specific interference by ingested dsRNA. *Nature* **395:**854.
Timmons, L., Court, D. L., and Fire, A. (2001). Ingestion of bacterially expressed dsRNAs can produce specific and potent genetic interference in *Caenorhabditis elegans*. *Gene* **263:**103–112.
Traw, M. B., and Bergelson, J. (2003). Interactive effects of jasmonic acid, salicylic acid, and gibberellin on induction of trichomes in *Arabidopsis*. *Plant Physiol.* **133:**1367–1375.
Traw, M. B., and Dawson, T. E. (2002). Differential induction of trichomes by three herbivores of black mustard. *Oecologia* **131:**526–532.
Turlings, T. C. J., Tumlinson, J. H., and Lewis, W. J. (1990). Exploitation of herbivore-induced plant odors by host-seeking parasitic wasps. *Science* **250:**1251–1253.
Unsicker, S. B., Kunert, G., and Gershenzon, J. (2009). Protective perfumes: The role of vegetative volatiles in plant defense against herbivores. *Curr. Opin. Plant Biol.* **12:**479–485.
Valentine, T. A., Randall, E., Wypijewski, K., Chapman, S., Jones, J., and Oparka, K. J. (2007). Delivery of macromolecules to plant parasitic nematodes using a tobacco rattle virus vector. *Plant Biotechnol. J.* **5:**827–834.
van der Westhuizen, A. J., Qian, X. M., and Botha, A. M. (1998). beta-1, 3-glucanases in wheat and resistance to the Russian wheat aphid. *Physiol. Plant.* **103:**125–131.
van der Westhuizen, A. J., Qian, X. M., and Botha, A. M. (1998). Differential induction of apoplastic peroxidase and chitinase activities in susceptible and resistant wheat cultivars by Russian wheat aphid infestation. *Plant Cell Rep.* **18:**132–137.
van Emden, H. F. (2007). Host-plant resistance. *In* "Aphids as Crop Pests" (H. F. van Emden and R. Harrington, eds.), pp. 447–468. CABI International, Wallingford.
Villada, E. S., Gonzalez, E. G., Lopez-Sese, A. I., Castiel, A. F., and Gomez-Guillamon, M. L. (2009). Hypersensitive response to *Aphis gossypii* Glover in melon genotypes carrying the *Vat* gene. *J. Exp. Bot.* **60:**3269–3277.

Voelckel, C., Weisser, W. W., and Baldwin, I. T. (2004). An analysis of plant–aphid interactions by different microarray hybridization strategies. *Mol. Ecol.* **13:**3187–3195.

Walling, L. L. (2008). Avoiding effective defenses: Strategies employed by phloem-feeding insects. *Plant Physiol.* **146:**859–866.

Wasternack, C. (2007). Jasmonates: An update on biosynthesis, signal transduction and action in plant stress response, growth and development. *Ann. Bot.* **100:**681–697.

Weathersbee, A. A., and Hardee, D. D. (1994). Abundance of cotton aphids (Homoptera: Aphididae) and associated biological control agents on six cotton cultivars. *J. Econ. Entomol.* **87:**258–265.

Weathersbee, A. A., Hardee, D. D., and Meredith, W. R. (1994). Effects of cotton genotype on seasonal abundance of cotton aphid (Homoptera: Aphididae). *J. Agric. Entomol.* **11:**29–37.

Weathersbee, A. A., Hardee, D. D., and Meredith, W. R. (1995). Differences in yield response to cotton aphids (Homoptera: Aphididae) between smooth-leaf and hairy-leaf isogenic cotton lines. *J. Econ. Entomol.* **88:**749–754.

Webster, J. A., Inayatullah, C., Hamissou, M., and Mirkes, K. A. (1994). Leaf pubescence effects in wheat on yellow sugarcane aphids and greenbugs (Homoptera: Aphididae). *J. Econ. Entomol.* **87:**231–240.

White, C., and Eigenbrode, S. D. (2000). Effects of surface wax variation in *Pisum sativum* on herbivorous and entomophagous insects in the field. *Environ. Entomol.* **29:**773–780.

Whyard, S., Singh, A. D., and Wong, S. (2009). Ingested double-stranded RNAs can act as species-specific insecticides. *Insect Biochem. Mol. Biol.* **39:**824–832.

Wiermer, M., Feys, B. J., and Parker, J. E. (2005). Plant immunity: The EDS1 regulatory node. *Curr. Opin. Plant Biol.* **8:**383–389.

Will, T., and van Bel, A. J. E. (2006). Physical and chemical interactions between aphids and plants. *J. Exp. Bot.* **57:**729–737.

Will, T., Tjallingii, W. F., Thonnessen, A., and van Bel, A. J. E. (2007). Molecular sabotage of plant defense by aphid saliva. *Proc. Natl. Acad. Sci. USA* **104:**10536–10541.

Wright, K. M., Duncan, G. H., Pradel, K. S., Carr, F., Wood, S., Oparka, K. J., and Cruz, S. S. (2000). Analysis of the *N* gene hypersensitive response induced by a fluorescently tagged tobacco mosaic virus. *Plant Physiol.* **123:**1375–1385.

Wroblewski, T., Piskurewicz, U., Tomczak, A., Ochoa, O., and Michelmore, R. W. (2007). Silencing of the major family of NBS-LRR-encoding genes in lettuce results in the loss of multiple resistance specificities. *Plant J.* **51:**803–818.

Wu, J. S., Wang, L., and Baldwin, I. T. (2008). Methyl jasmonate-elicited herbivore resistance: Does MeJA function as a signal without being hydrolyzed to JA? *Planta* **227:**1161–1168.

Xu, W. N., and Han, Z. J. (2008). Cloning and phylogenetic analysis of *sid-1*-like genes from aphids. *J. Insect Sci.* **8:**30–35.

Yang, J. Y., Iwasaki, M., Machida, C., Machida, Y., Zhou, X. P., and Chua, N. H. (2008). beta C1, the pathogenicity factor of TYLCCNV, interacts with AS1 to alter leaf development and suppress selective jasmonic acid responses. *Genes Dev.* **22:**2564–2577.

Yoshida, Y., Sano, R., Wada, T., Takabayashi, J., and Okada, K. (2009). Jasmonic acid control of GLABRA3 links inducible defense and trichome patterning in *Arabidopsis*. *Development* **136:**1039–1048.

Zarate, S. I., Kempema, L. A., and Walling, L. L. (2007). Silverleaf whitefly induces salicylic acid defenses and suppresses effectual jasmonic acid defenses. *Plant Physiol.* **143:**866–875.

Zhu-Salzman, K., Salzman, R. A., Ahn, J. E., and Koiwa, H. (2004). Transcriptional regulation of sorghum defense determinants against a phloem-feeding aphid. *Plant Physiol.* **134:**420–431.

CHAPTER 6

Cross-Protection: A Century of Mystery

Heiko Ziebell[1] and **John Peter Carr**

Contents

Abstract

Cross-protection is a phenomenon in which infection of a plant with a mild virus or viroid strain protects it from disease resulting from a subsequent encounter with a severe strain of the same virus

Department of Plant Sciences, University of Cambridge, Cambridge CB2 3EA, United Kingdom
[1] Present address: Department of Plant Pathology and Plant–Microbe Biology, Cornell University, Ithaca, New York, USA, E-mail: heiko.ziebell@cantab.net/hz237@cornell.edu

Advances in Virus Research, Volume 76
ISSN 0065-3527, DOI: 10.1016/S0065-3527(10)76006-1

or viroid. In this chapter, we review the history of cross-protection with regard to the development of ideas concerning its likely mechanisms, including RNA silencing and exclusion, and its influence on the early development of genetically engineered virus resistance. We also examine examples of the practical use of cross-protection in averting crop losses due to viruses, as well as the use of satellite RNAs to ameliorate the impact of virus-induced diseases. We also discuss the potential of cross-protection to contribute in future to the maintenance of crop health in the face of emerging virus diseases and related threats to agricultural production.

I. INTRODUCTION

Food production faces serious challenges. Foremost among these is increasing population (Anonymous, 2004; Strange and Scott, 2005). But an additional challenge arises from global warming, which may decrease the land available for cultivation (Canto *et al.*, 2009). Further problems, which are likely to be exacerbated by climate change, result from the action of pests and diseases that can cause severe crop losses (Anderson *et al.*, 2004; Canto *et al.*, 2009). Environmental change, combined with changing patterns in world trade and the opening up of new land to agriculture appear to be promoting the emergence of "new" diseases. Viruses cause around half of the emerging diseases, possibly because of changes in the distribution of the viruses' invertebrate vectors and the effect of temperature on plant-resistance mechanisms, among other causes (Anderson *et al.*, 2004; Canto *et al.*, 2009). Some authors have also raised the specter of agricultural bioterrorism, in which the intentional application of plant pathogens may in the future become an additional threat to regional, national, or international food and crop biosecurity (Madden and Wheelis, 2003; Pimentel *et al.*, 2000; Rodini, 2009).

What is to be done? All major food crops (e.g., banana, barley, cassava, maize, potatoes, sweet potatoes, rice, and wheat) already suffer economically significant losses caused by viruses (Strange and Scott, 2005). Considerable efforts have been expended in producing virus-free propagative stocks, in preventing the spread of viral diseases by vectors (insects, mites, fungi, and even humans), in breeding resistant varieties, in development of genetically engineered virus-resistant plants, and the investigation of resistance-inducing agents (Fraile and García-Arenal, 2010; Gottula and Fuchs, 2009; Hull, 2002; Loebenstein, 2009). The approaches are not without difficulties, though. For example, pesticides used against viral vectors may not efficiently prevent transmission, or might have negative impacts on the environment (Castle *et al.*, 2009;

Gallitelli, 2000). Conventional breeding of virus-resistant crop plants may be impossible when resistance genes are not present in genetically compatible wild relatives and although genetic engineering can solve many of these problems it has not always proved to be a popular approach (Reddy *et al.*, 2009; Thompson and Tepfer, 2010).

Perhaps, therefore, it is time to revisit effective but relatively underused plant protection methods. An example of such a method is cross-protection, in which plants are infected with one strain of a virus or viroid to protect against subsequent infection with another strain or strains of the virus or viroid (Hull, 2002). The phenomenon was described almost one hundred years ago and over that time has been used successfully to protect various crops, yet an understanding of its mechanisms remains elusive. This review aims to summarize the history of cross-protection, its appliance and its potential mechanisms.

II. GENERAL REMARKS

A. Definition of cross-protection

In this review, the term cross-protection describes the phenomenon in which infection with mild or attenuated virus strains protects plants against subsequent (or "challenge") inoculation or infection with more severe strains of the same virus (Hull, 2002; Ziebell, 2008). Where appropriate, we will also use the term to describe protection of plants from viroid-induced disease by prior inoculation with viroid strains that are asymptomatic or causing mild symptoms only. We shall also deal with some examples of protection provided by satellite RNAs, although this is not cross-protection *sensu stricto*.

Mild, attenuated, or nonsymptomatic virus strains and mutants have been used to cross-protect agricultural and horticultural crop plants from subsequent infection by severe strains. This practical aspect is reflected in the more tightly focused definition used by Gonsalves and Garnsey (1989), who described cross-protection as "the use of a mild virus isolate to protect plants against economic damage caused by infection with a severe challenge strain(s) of the same virus." However, it is also possible to preinoculate plants with a severe strain that prevents subsequent infection by a milder strain (Aapola and Rochow, 1971; Bodaghi *et al.*, 2004; Dodds, 1982; Tian *et al.*, 2009; Ziebell and Carr, 2009). Thus, although cross-protection can be exploited to protect plants against virus-induced disease, there is nothing intrinsically "protective" about the mechanism or mechanisms underlying this phenomenon.

Over the last century, and as the field has developed, numerous terms have been used to describe the phenomenon of cross-protection. These have included acquired immunity, antagonism, cross-immunization, dominance, induced immunity, induced resistance, interference, premunity, prophylactic inoculation, protective inoculation, and so on (Bennet, 1951; Bozarth and Ford, 1989; Fulton, 1986; Pennazio *et al.*, 2001; Sherwood, 1987b; Urban *et al.*, 1990). Many of these terms have since gained very specific meanings within plant pathology and are rarely if ever used to describe cross-protection nowadays. The term cross-protection has occasionally been used to describe resistance phenomena in transgenic plants expressing virus-derived transgenes (Culver, 1996; Goregaoker *et al.*, 2000; Lu *et al.*, 1998; Ratcliff *et al.*, 1999; Tumer *et al.*, 1987; Yamaya *et al.*, 1988).

Crop protection is not the only practical use to which cross-protection has been put. Because cross-protection works only between strains of the same virus, cross-protection was once used in diagnosis and in the characterization of virus strains, but the advent of serological- and nucleic acid-based methods make this ingenious use of the phenomenon rare, if not defunct, in contemporary studies (Aapola and Rochow, 1971; Latorre and Flores, 1985; Matthews, 1949; McKinney, 1941; Nelson and Wheeler, 1978; Olson, 1956, 1958; Price, 1936, 1940; Stubbs, 1964; Valenta, 1959a,b; Webb *et al.*, 1952).

B. Properties of mild strains

Fulton (1986) and Lecoq (1998) laid out the requirements for mild or attenuated virus strains for field application of cross-protection. Thus, a protective strain should only induce mild symptoms not only on the target crops but also on other hosts in the vicinity of the protected field and it should protect against a broad range of severe strains. Additionally, crop quality and yield should not be significantly affected by the mild strain. Furthermore, the protective strain needs to be able to infect the plant systemically and be genetically stable within the plant so that it does not mutate into a severe strain. If possible, the protective strain should not be capable of dissemination to other nontarget plants by vectors. Lastly, easy propagation, storage, and application of protective strain inocula to crop plants are requirements for the practical cross-protection in the field or glasshouse.

C. Disadvantages of cross-protection

Some practical difficulties have been reported with cross-protection. Incomplete infection of a crop with the protective strain enables the challenging strain to invade some of the plants and if insufficient time is allowed for the establishment of systemic infection by the protective strain protection will not occur (Costa and Müller, 1980; Lecoq and

Raccah, 2001). Breakdown of cross-protection was observed in some cases due to high disease pressure, the advent of new virus vectors, physiological changes of host plants over time, or other reasons (Bar-Joseph, 1978; Fulton, 1986; Lecoq *et al.*, 1991; Powell *et al.*, 1992, 2003; Rezende and Sherwood, 1991; Wang *et al.*, 1987; Yokomi *et al.*, 1991). Virus-induced symptoms are often host-dependent so that a spread of the protective strain onto nontarget plants may induce severe symptoms in a new host or even in different varieties of the target host (Fulton, 1986; Gonsalves, 1998; Lecoq, 1998). Concerns have been voiced regarding heteroencapsidation [i.e., encapsidation of one virus by the coat protein (CP) of another, which may affect virus–vector interactions] or mutations of mild strains into severe ones (Fulton, 1986; Gal-On and Shiboleth, 2006; Hammond *et al.*, 1999; Lecoq, 1998). Similarly, synergism between a mild virus strain and an unrelated virus strain may lead to severe symptoms though it appears to be of little practical concern (Lecoq and Raccah, 2001). Growers' hesitation to deliberately infect their crops with viruses may also adversely affect the acceptance of cross-protection in agricultural practice (Gonsalves, 1998).

III. HISTORY OF CROSS-PROTECTION

Cross-protection is often discussed in relation to the phenomenon of "recovery." As early as 1928, Wingard observed that plants inoculated with a ringspot virus (TRSV; see Table I for a list of all abbreviations of names of viruses mentioned in this review) would exhibit typical ringspot symptoms on infected tobacco plants but those symptoms would disappear after a few days leaving the newly developing growth in a healthy appearance (Wingard, 1928). He was unable to recreate symptoms on those asymptomatic leaves by reinoculation with infectious plant sap; an observation that was further explored by Price (1932) for TRSV and Wallace and coworkers for "curly top" disease in tomato and tobacco caused by BCTV (Lesley and Wallace, 1938; Wallace, 1939, 1940, 1944). What we know today is that this phenomenon of recovery is based on RNA silencing (Baulcombe, 2004; Covey *et al.*, 1997; Csorba *et al.*, 2009; Ratcliff *et al.*, 1997). The year after the publication of Wingard's study, the phenomenon of cross-protection was discovered by McKinney (1929). While working with TMV, he found that tobacco plants systemically infected with a strain of TMV inducing "greenish" mosaic symptoms were protected against a challenge inoculation with a TMV strain inducing yellow mosaic symptoms. Soon after, Thung (1931) found that pre-infection of tobacco with the "common" TMV strain would prevent secondary infection with a "white mosaic" strain.

TABLE I Virus and viroid acronyms used in this chapter

Acronym	Virus name	Acronym	Virus name
ACLSV	*Apple chlorotic leaf spot virus*	PMMoV	*Pepper mild mottle virus*
AIMV	*Alfalfa mosaic virus*		
ALSV	*Apple latent spherical virus*	PPV	*Plum pox virus*
ApMV	*Apple mosaic virus*	PRSV	*Papaya ringspot virus*
ArMV	*Arabis mosaic virus*	PSV	*Peanut stunt virus*
BCTV	*Beet curly top virus*	PSTVd	*Potato spindle tuber viroid*
BMV	*Brome mosaic virus*	PVA	*Potato virus A*
BNYVV	*Beet necrotic yellow vein virus*	PVX	*Potato virus X*
BSBMV	*Beet soilborne mosaic virus*	PVY	*Potato virus Y*
BYDV	*Barley yellow dwarf virus*	PWV	*Passion fruit woodiness virus*
BYMV	*Bean yellow mosaic virus*	SHMV	*Sunn-hemp mosaic virus*
CaMV	*Cauliflower mosaic virus*	SMV	*Soybean mosaic virus*
CChMVd	*Chrysanthemum chlorotic mottle viroid*	SPMV	*Satellite panicum tobacco mosaic virus*
CEVd	*Citrus exocortis viroid*	STMV	*Satellite tobacco mosaic virus*
CGMMV	*Cucumber green mottle mosaic virus*	STNV	*Satellite tobacco necrosis virus*
CLVd-N	*Columnea latent viroid strain Nematanthus*	TAV	*Tomato aspermy virus*
CMD	*Cassava mosaic disease*	TCV	*Turnip crinkle virus*
CMGs	Cassava mosaic geminiviruses	TMGMV	*Tobacco mild green mosaic virus*
CMV	*Cucumber mosaic virus*	ToMV	*Tomato mosaic virus*
CSSV	*Cocoa swollen shoot virus*	TMV	*Tobacco mosaic virus*
CTV	*Citrus tristeza virus*	TNV	*Tobacco necrosis virus*
CYNMV	*Chinese yam necrotic mosaic virus*	ToRSV	*Tomato ringspot virus*

(continued)

TABLE I (*continued*)

Acronym	Virus name	Acronym	Virus name
GFLV	*Grapevine fanleaf virus*	TRSV	*Tobacco ringspot virus*
HSVd	*Hop stunt viroid*	TSWV	*Tomato spotted wilt virus*
JYMV	*Japanese yam mosaic virus*	TVMV	*Tobacco vein mottling virus*
ORMV	*Oilseed rape mosaic virus*	TYLCCV	*Tomato yellow leaf curl China virus*
PcMV	*Peach mosaic virus*	VNPV	*Vanilla necrosis virus*
PLMVd	*Peach latent mosaic viroid*	WMV-2	*Watermelon mosaic virus 2*
PLRV	*Potato leaf roll virus*	WSMV	*Wheat streak mosaic virus*
PMV	*Panicum mosaic virus*	ZYMV	*Zucchini yellow mosaic virus*

Salaman described the use of a mild PVX strain that protected tobacco and *Datura stramonium* plants from subsequent infection with a more severe PVX strain (Salaman, 1933; Salaman *et al.*, 1938). In the following years, many more reports were published describing virus strains that protected against closely related severe strains (Bodine, 1942; Caldwell, 1935; Crowdy and Posnette, 1947; Matthews, 1949; Price, 1935; Thung, 1935). In his review of this emerging field, Price (1940) used the term "cross immunity" for the protection of one inoculated virus strain against challenge inoculation of a different strain.

Cross-protection was for a while a useful tool for establishing the relatedness of viruses, as exemplified by an extensive study of the relationships of different PLRV strains carried out by Webb and colleagues (1952). The 1950s also saw the first use of cross-protection as a method to prevent viral infections in commercially grown plants. The protection of cocoa in Africa against CSSV (Posnette and Todd, 1951, 1955) and citrus plants in South America against CTV (Grant and Costa, 1951) were the first examples of the commercial exploitation of cross-protection. Later, tomato crops were protected by cross-protection against tobamoviruses under both glasshouse and field conditions (Rast, 1972, 1975), although this has now been mostly superseded by the deployment of lines possessing resistance genes (Ahoonmanesh and Shalla, 1981; Fletcher and Rowe, 1975).

IV. APPLICATIONS

For many plant viruses, mild or asymptomatic strains have been described that can protect against challenge with other strains of the same virus. This includes PMMoV (Ichiki *et al.*, 2005; Tsuda *et al.*, 2007; Yoon *et al.*, 2006), ToMV (Ahoonmanesh and Shalla, 1981; Goto *et al.*, 1966; Kurihara and Watanabe, 2003; Oshima, 1981 and review by Broadbent, 1975), TSWV (Wang and Gonsalves, 1992), JYMV (Fuji *et al.*, 2000; Kajihara *et al.*, 2000), ApMV (Chamberlain *et al.*, 1969; Posnette and Cropley, 1956), PPV (Rancovic and Paunovic, 1988; Ravelonandro *et al.*, 2008), TMV (Boyle and Bergman, 1969; Burgyán and Gáborjányi, 1984; Cassells and Herrick, 1977; Fletcher and Rowe, 1975; Lu *et al.*, 1998; Rast, 1972, 1975; Rezende *et al.*, 1992 and review by Fletcher, 1978), between the *Tobamoviruses* TMV and SHMV (Zinnen and Fulton, 1986), ToRSV (Bitterlin and Gonsalves, 1988), PcMV (later shown to be the viroid PLMVd; Cochran, 1954), CaMV (Tomlinson and Shepherd, 1978), ACLSV (Marenaud *et al.*, 1976), PWV (Novaes and Rezende, 2005; Simmonds, 1959), BYDV (Aapola and Rochow, 1971; Jedlinski and Brown, 1965; Smith, 1963; Wen *et al.*, 1991), CSSV (Hughes and Ollennu, 1994; Posnette and Todd, 1951, 1955), TMGMV (Bodaghi *et al.*, 2004), CMV (Dodds, 1982; Dodds *et al.*, 1985; Kosaka and Fukunishi, 1997; Loebenstein *et al.*, 1977; Nakazono-Nagaoka *et al.*, 2005; Price, 1935; Rodríguez-Alvarado *et al.*, 2001; Sclavounos *et al.*, 2006; Tomaru *et al.*, 1967; Ziebell and Carr, 2009; Ziebell *et al.*, 2007), CGMMV (Nishiguchi, 2007; Tan *et al.*, 1997), SMV (Kosaka and Fukunishi, 1993), BYMV (Nakazono-Nagaoka *et al.*, 2009; Nakazono-Nagaoka *et al.*, 2004; Uga *et al.*, 2004), CYNMV (Kondo *et al.*, 2007), TYLCCNV (Ye *et al.*, 2006), VNPV, now known as WMV-2 (Liefting *et al.*, 1992), WMV-2 (Kameya-Iwaki *et al.*, 1992), TAV (Kuti and Moline, 1986), between the *Nepoviruses* ArMV and GFLV (Huss *et al.*, 1989; Komar *et al.*, 2008; Vigne *et al.*, 2009), WSMV (Hall *et al.*, 2001), between the *Benyviruses* BSBMV and BNYVV (Mahmood and Rush, 1999), between the *Tobamoviruses* ORMV and TMGMV (Aguilar *et al.*, 2000), PVY (Latorre and Flores, 1985; Singh and Singh, 1995) and cassava mosaic disease caused by CMGs (Legg *et al.*, 2006; Owor *et al.*, 2004; Storey and Nichols, 1938).

Case studies of three viruses that have been controlled by cross-protection under commercial growing conditions will be discussed in more detail in the following sections.

A. ZYMV

ZYMV is one of the few examples in which cross-protection has been used to protect annual crops from infection with severe virus strains and consequent crop losses. ZYMV was first described in Europe in the early 1980s but occurs worldwide, predominantly in regions of cucurbit production (Desbiez and Lecoq, 1997). ZYMV belongs to the genus *Potyvirus* and is aphid-transmitted in a nonpersistent manner but may also be transmitted via seeds (Desbiez and Lecoq, 1997; Gal-On, 2007).

ZYMV is one of the major viruses of cucurbit crops, causing yield losses of up to 100% (Desbiez and Lecoq, 1997). Development of ZYMV-resistant cucurbit varieties has been problematic because of a limited pool of resistance genes and in some cases a requirement for multiple genes to provide effective resistance (Gaba *et al.*, 2004). Numerous strains of ZYMV have been found which allowed for the selection of mild strains for protection (Desbiez and Lecoq, 1997). Lecoq and colleagues (1991) described a mild strain, ZYMV-WK, which they used for cross-protection of zucchini (courgette) squash under field conditions in France. Meanwhile, Wang *et al.* (1991) used ZYMV-WK to preinoculate cucumber, melon, and zucchini squash plants that were challenged under greenhouse conditions and exposed to natural infection pressures in field trials in southern Taiwan. In both cases, protected plants produced a higher yield of marketable fruits compared to unprotected control plants that were subsequently infected with severe strains of ZYMV. Similar results with ZYMV-WK were obtained in the United Kingdom (Walkey, 1992; Walkey *et al.*, 1992) although preinoculation with the mild strain caused yield reductions on courgette and marrow (Spence *et al.*, 1996). Cross-protection was also successfully employed in Hawaii, the United States, Italy, and Turkey (Cho *et al.*, 1992; Desbiez and Lecoq, 1997; Perring *et al.*, 1995). Yarden *et al.* (2000) reported that in the 1990s cross-protection was successfully used to protect watermelon, squash, melon, pumpkin, cucumber, and gourds grown on commercial scales in Israel. Under field conditions they found no differences in yield between unprotected and mild strain-infected plants in the absence of any challenge virus but cross-protection prevented significant losses from occurring when crops were exposed to severe ZYMV strains. These workers also reported on the deployment of efficient inoculation machines in nurseries that insured high inoculation rates with the mild virus strain (Yarden *et al.*, 2000).

In 2000, Gal-On and coworkers reported the engineering of a mild ZYMV strain designated ZYMV-AG using site-directed mutagenesis to replace an arginine with an isoleucine residue in a highly conserved domain within the virus' helper component-protease (HC-Pro) (Gal-On, 2000; Gal-On *et al.*, 2000). Infection with ZYMV-AG induced mild symptoms in squash and was asymptomatic in several other cucurbits (cucumber, melon, and watermelon). Furthermore, this engineered mutant protected against a range of more severe strains including the wild-type ZYMV-AT strain, from which ZYMV-AG was derived (Gal-On, 2000). Interestingly, HC-Pro is a multifunctional protein involved in aphid transmission, genome amplification, viral long-distance movement, synergism, symptom expression and, perhaps most significantly, suppression of RNA silencing (see reviews by Gal-On, 2007; Kasschau and Carrington, 1998; Maia *et al.*, 1996; Urcuqui-Inchima *et al.*, 2001; also see Section V.C).

In Japan, cold treatment of infected plants was used as a method to randomly generate mutant, attenuated strains of ZYMV, SMV, and WMV-2 (Kameya-Iwaki *et al.*, 1992; Kosaka and Fukunishi, 1993, 1997; Kosaka *et al.*, 2006). Under field conditions, ZYMV-2002 successfully protected cucumber plants (Kosaka *et al.*, 2006). Further investigations determined seven amino acid substitutions in this strain compared to the parental strain, four of which were mutations affecting the HC-Pro sequence (Wang *et al.*, 2006). By site-specifically mutating the HC-Pro region of other ZYMV strains, more mild ZYMV strains were generated (Lin *et al.*, 2007) (see Section V.C).

B. PRSV

Not least due to the pioneering efforts of Gonsalves and colleagues, one of the best characterized cross-protection systems involved the control of PRSV on papaya (*Carica papaya*) (see reviews by Fuchs *et al.*, 1997; Gonsalves, 1998, 2006; Gottula and Fuchs, 2009; Tripathi *et al.*, 2008; Yeh and Gonsalves, 1994; Yeh *et al.*, 1988). Like ZYMV (Section IV.A), PRSV belongs to the genus *Potyvirus* and is readily aphid-transmitted in a nonpersistent manner (Purcifull *et al.*, 1984). Strains are classified into two groups, PRSV-W (watermelon mosaic virus 1, WMV1) isolates, which do not infect papaya but cause disease in watermelon and other cucurbits, and PRSV-P isolates which infect both papayas and cucurbits (Yeh and Gonsalves, 1994). PRSV is found in all papaya growing/producing regions in the world where it causes great economic losses to papaya growers (Fuchs *et al.*, 1997; Gonsalves, 1998, 2006; Herold and Weibel, 1962; Yeh and Gonsalves, 1984, 1994).

Disease on papaya is expressed in leaves (symptoms including mosaic, chlorosis, and distortion of young leaves), on the petioles and upper parts of the trunk as water-soaked oily streaks, and the virus causes stunting of whole plants and flower abortion and consequently decreased fruit production (Tripathi *et al.*, 2008). The fruit that develop on PRSV-infected papaya is of reduced quality, displaying virus-induced ringspots and greatly decreased sugar levels, making them unmarketable (Tripathi *et al.*, 2008). The papaya industry was especially severely affected by PRSV in Hawaii, where the impact of the disease led almost to the abandonment of cultivation. Due to the limitation of natural resistance genes available for papaya and related species, conventional breeding for resistance or tolerance to PRSV did not appear to be an option (Ferreira *et al.*, 2002; Gonsalves, 1998, 2006; Tripathi *et al.*, 2008).

In Hawaii, vigorous surveillance and scrupulous removal of PRSV-infected trees initially slowed down the spread of the disease but did not contain the virus long term (Gonsalves, 1998). Cross-protection offered the potential to "immunize" plantations ahead of the arrival of the virus in previously unaffected areas. No naturally occurring mild strains of PRSV were found but two mild strains, PRSV-HA 5-1 and PRSV-HA 6-1, were obtained by random chemical mutagenesis of a severe strain (PRSV-HA) using nitrous acid treatment of infected sap (Gonsalves and Ishii, 1980; Yeh and Gonsalves, 1984). This approach was modeled on that previously used to considerable success in the development of strains for cross-protection of tomato against TMV (Rast, 1972).

Under greenhouse conditions, PRSV-HA 5-1 gave promising protection against the parental strain, encouraging further evaluation (Yeh and Gonsalves, 1984). Both strains were used in greenhouse and field trials in Taiwan, where PRSV had been a significant problem since the mid-1970s (Wang *et al.*, 1987; Yeh *et al.*, 1988). However, the effectiveness of the protective strains varied between greenhouse and field trials and a special design of planting blocks and continuous removal of symptomatic plants were necessary to effectively protect the crop (Sheen *et al.*, 1998; Wang *et al.*, 1987). However, Taiwan government funding supported further large-scale field trials and encouraging yield increases led to mass-inoculation of papaya seedlings with the mild strains (Yeh *et al.*, 1988). However, the protection afforded by PRSV-HA 5-1 was highly strain-specific and susceptible to breakdown, particularly in areas of Taiwan with high disease pressure, or after periods of severe cold, which also led to breakdown of cross-protection (Yeh and Gonsalves, 1994; Yeh and Kung, 2007).

Similar results were obtained during field trials in Hawaii with mild PRSV strains (Yeh and Gonsalves, 1994). Initially, PRSV infection and disease were confined to specific geographical locations in Hawaii, but there were concerns that the infection would spread into the main area of papaya production in the Puna district of that island (Gonsalves, 1998, 2006). Since the mild PRSV strains such as PRSV-HA 5-1 used for cross-protection in Taiwan were developed from Hawaiian strains, it was thought that they had a great potential for the application in Hawaii (Yeh and Gonsalves, 1984, 1994). Greenhouse and field tests were successful but different varieties of papaya reacted differently to the infection with "protective" strains. Some varieties produced fruits displaying mild symptoms but others had severe, sunken ringspots making them unmarketable (Yeh and Gonsalves, 1994; review by Gonsalves, 1998). Furthermore, the propensity of the protective strains to induce damaging symptoms was dependent upon environmental conditions. Thus, if temperatures dropped below 20 °C, the protective strains induced chlorotic leaf spots and ringspots on fruit although there was no gross deformation of fruit or reduction in size (Yeh and Gonsalves, 1994). Additional fertilization of crops repressed those mild symptoms. However, the potential adverse effects of the protective strains, the extra labor required for inoculation and cultivation, and growers' reluctance to deliberately infect their crops proved to be significant barriers to the adoption of routine cross-protection (Gonsalves, 1998).

The tight strain-specificity of protection afforded by PRSV-HA 5-1 made it difficult to employ this strain in parts of the world other than Hawaii (Tennant *et al.*, 1994; Wang *et al.*, 1987; Yeh and Gonsalves, 1994). PRSV populations within and between countries can display great diversity making it hard to find mild strains that protect against a sufficiently wide range of potential challenge strains (Bateson *et al.*, 2002; Fernández-Rodríguez *et al.*, 2008; Inoue-Nagata *et al.*, 2007). Trials using cross-protection against PRSV were carried out in Florida, Mexico, and Thailand, though with variable degrees of success (McMillan and Gonsalves, 1987; Yeh and Gonsalves, 1994). In any locality, it is preferable to obtain mild, potentially protective isolates derived from endemic strains. But screening for such strains does not always yield practically useful cross-protecting viruses. A search for mild isolates from among naturally occurring PRSV populations was carried out in Thailand but unfortunately cross-protection provided by them was also very specific, acting against only certain severe strains (Chatchawankanphanich *et al.*, 2000). In contrast, in India, local mild PRSV strains yielded more promising results in glasshouse and field experiments (Ram *et al.*, 2006).

In Brazil, Rezende and Pacheco (1998) investigated mild strains derived from natural PRSV populations for their potential to cross-protect cucurbits from infection with PRSV-W isolates. These mild strains displayed great cross-protection potential in greenhouse and field trials for protection of zucchini squash from infection with severe strains (Rezende and Pacheco, 1998). Although mild strains were also able to protect watermelons from infection with severe strains, cross-protection was not as satisfactory as that seen in zucchini, and fruit yield losses of up to 50% in comparison to healthy plants were reported for some varieties, despite the presence of protective strain (Dias and Rezende, 2001).

Through recombination experiments involving both PRSV-P and PRSV-W strains, mild strains were generated that protected against a broader range of isolates in cucurbits (You *et al.*, 2005). Recently, the mild symptoms induced by PRSV-HA 5-1 have been attributed to mutations in the P1 and HC-Pro genes (Chiang *et al.*, 2007). This lends further credence to the idea that site-directed mutagenesis of specific viral genes has the potential to produce mild and broad-acting protective strains in the future (see Sections V.C and VII). However, most recent research aimed at combating PRSV has focused on transgenic approaches (as described by Gottula and Fuchs, 2009).

C. CTV

To date, the greatest success story for cross-protection has been in combating CTV. CTV, a member of the *Closteroviridae*, is responsible for devastating losses to citrus production all over the world (for more information, see reviews by Bar-Joseph *et al.*, 1981, 1989; Garnsey *et al.*, 1998; Moreno *et al.*, 2008). Bar-Joseph and colleagues (1989) estimated that by 1983 40 million trees had fallen victim to CTV, rising to 60 million by the end of the decade. Aphids transmit CTV in a semipersistent manner and are important vectors for spread of the virus between neighboring trees. However, spread via infected buds and grafts by exchange of planting material between growing regions facilitated the dispersal of CTV around the world (Moreno *et al.*, 2008).

CTV causes three different disease syndromes, depending on the virus strain and host or rootstock/scion combination. The first syndrome, tristeza (named for the Portuguese/Spanish word for "sadness") materializes as a quick decline of different citrus species such as sweet oranges [*Citrus x sinensis* (L.) Osb.], mandarins [*C. reticulata* Blanco], grapefruit [*C. x paradisi* Macf.], kumquats [*Fortunella japonica* (Thunb.) Swingle], or Mexican limes [*C. aurantifolia* (Christm.) Swingle] propagated on sour orange [*C. aurantium* L.] or lemon [*C. x limon* (L.) Burm. f.] rootstocks (Moreno *et al.*, 2008). Infected trees display dull green or yellowish, thin foliage, leaf shedding and twig dieback or small chlorotic leaves and unmarketable small and pale-colored fruits (Moreno *et al.*, 2008). Virus infection causes obliteration, collapse, and necrosis of sieve tubes and companion cells near the graft bud union. Excessive proliferation of phloem cells, which are nonfunctional, causes water and nutrient deficiency and death of rootstocks (Moreno *et al.*, 2008). However, the use of tristeza decline-tolerant rootstocks can prevent this syndrome.

Another syndrome caused by CTV is stem pitting. Unlike tristeza, stem pitting does not kill infected trees; however, they display an unthrifty growth with chronic yield reduction leading to high economic losses (Garnsey *et al.*, 1998; Moreno *et al.*, 2008). Extensive pitting of stems may limit radial growth, and stunted trees with thin, chlorotic leaves may produce fewer, smaller fruits with low juice content. Sensitive cultivars include acid limes, grapefruits and sweet oranges, which can be affected as seedlings, as rootstock or as scion buds. Once a district has been invaded by stem pitting CTV isolates, the production of these citrus crops may be permanently limited (Moreno *et al.*, 2008).

The third syndrome, seedling yellows, affects the seedlings of sour oranges, grapefruits, or lemons (Moreno *et al.*, 2008). The plants remain stunted with small pale or yellowish leaves and a reduced root system. Sometimes, recovery from symptoms can occur leading to new leaves with normal growth and leaf color (Moreno *et al.*, 2008).

Why did tristeza become such a devastating disease? Sour orange rootstocks were popular for bud propagation of many citrus types, as they were tolerant of the causative agent of citrus blight and resistant to the oomycete responsible for citrus foot rot (Brlansky *et al.*, 1986; Moreno *et al.*, 2008; Rocha-Peña *et al.*, 1991). Furthermore, sour orange rootstocks are adaptable to a variety of soil conditions and climates and support scions that produce high-quality fruit and high yields. The advantages of sour orange rootstocks enabled the expansion of citrus production in many parts of the world during the twentieth century and sour orange became almost the exclusive rootstock type in the Mediterranean region and Americas (Garnsey *et al.*, 1998; Moreno *et al.*, 2008). While many citrus varieties grown on their own rootstocks are tolerant to tristeza, the grafted varieties are not. Subsequently, tristeza disease soon destroyed millions of trees around the world, in particular in South America, Southern US, and Mediterranean countries (see review by Moreno *et al.*, 2008).

However, although the use of tristeza-resistant rootstocks avoided tristeza disease, their use created new agronomic problems in terms of fruit quality and susceptibility to diseases caused by other viruses, as well as fungi and bacteria (Moreno *et al.*, 2008; Román *et al.*, 2004). Furthermore, stem pitting CTV isolates were discovered in areas where tristeza-resistant rootstocks were used after previous tristeza outbreaks (Moreno *et al.*, 2008). In areas with low occurrence of CTV, eradication programs with removal of infected trees have been used, for example, in certain areas of California, Israel, Spain, Cuba, and Mexico (Bar-Joseph *et al.*, 1989; Batista *et al.*, 1996; Garnsey *et al.*, 1998; Roistacher and Moreno, 1991). Thus, when outbreaks of CTV occurred in Israel, eradication programs were the measure of choice to prevent an epidemic (Bar-Joseph *et al.*, 1974, 1983, 1989; Raccah *et al.*, 1976). However, although eradication was initially sufficient to contain the disease, spread of CTV was observed in the 1970s (Bar-Joseph *et al.*, 1989). Improved detection methods led to further removal of infected trees that were previously missed due to asymptomatic infection (Bar-Joseph *et al.*, 1983). It was concluded that eradication and containment are not by themselves sufficient to control the disease (Bar-Joseph *et al.*, 1989; Ben-Ze'ev *et al.*, 1989).

In larger areas with a high CTV incidence, eradication is not a viable option as the rate of tree removal is not sufficient to offset new infections (Garnsey *et al.*, 1998). Therefore, cross-protection appeared to be a viable option to protect citrus crops from CTV epidemics. The earliest large-scale experiments with cross-protection were carried out in Brazil (reviewed by Costa and Müller, 1980). Grant and Costa (1951) were the first authors to describe a mild CTV strain that would protect sour orange rootstocks against challenge inoculation with a severe CTV strain by either aphids or bud-transmission. At that time CTV was already widespread in Brazil and the introduction of new rootstock material had not prevented infection of lime, grapefruit, and sweet orange plants. This made exploitation of cross-protection on a large-scale acceptable (Costa and Müller, 1980). Many different combinations of rootstock, scion, and mild CTV isolates were tested under controlled conditions and the best combinations supplied to growers for field tests. By the late 1970s, 5 million cross-protected orange and lime trees had been planted in Brazil, this number was exceeded at the beginning of the 1980s when 8 million sweet oranges alone were cross-protected, and by the end of the century there were 80 million cross-protected sweet orange trees (Costa and Müller, 1980; Müller, 1980; Müller and Costa, 1972, 1977; Müller *et al.*, 2000). The search for new potential cross-protecting isolates has continued in order to provide cross-protection against novel CTV strains (Müller *et al.*, 1988, 2000; Roistacher and Moreno, 1991; Salibe *et al.*, 2002; Sambade *et al.*, 2007; Souza *et al.*, 2002). Even though used on large scales, hardly any breakdown of cross-protection has been observed. Those few instances of severely infected trees were explained either by infection with severe CTV too close to the time of mild strain inoculation for cross-protection to be possible, or to insufficient spread of the protective strain to all tissues (Costa and Müller, 1980).

Cross-protection against CTV has also been used outside of Brazil. In Florida, CTV had been widespread but had not caused epidemics in this area as citrus plants had been grown using tristeza-tolerant rootstocks and because the endemic CTV strains had been mild (Brlansky *et al.*, 1986; Rocha-Peña *et al.*, 1991; Yokomi *et al.*, 1991). However, the increasing use of sour orange rootstocks in orchards led to the occurrence of areas with CTV-induced quick decline and severe dwarfing of trees (Lee *et al.*, 1987; Rocha-Peña *et al.*, 1991). Floridian CTV strains could be grouped into two classes: asymptomatic and those causing stunting and/or decline of trees possessing sour orange or *C. macrophylla* Wester rootstocks (Powell *et al.*, 1999).

In Florida, some mild CTV strains were found to cross-protect against challenge CTV strains though their effectiveness varied (Cohen, 1976; Rocha-Peña *et al.*, 1991; Yokomi *et al.*, 1987, 1991). In a long-term evaluation taking 16 years it was found that certain mild isolates would protect grapefruits, but not sweet orange, grafted onto sour orange rootstocks (Pelosi *et al.*, 2000; Powell *et al.*, 1992, 1999). It is interesting to note that the success of cross-protection is host-dependent. However, cross-protection in Florida, even where it had been successful, was compromised by the arrival of a new vector, the brown citrus aphid (*Toxoptera citricida* Kirkaldy). Within 5 years of the appearance of the brown citrus aphid, cross-protected trees showed a high incidence of infection with severe CTV isolates (Powell *et al.*, 2003). Although aphid control measures were not effective in preventing natural CTV spread by aphids, the evaluation of new rootstock/scion combinations and mild CTV strains for cross-protection did meet with considerable success (Dekkers and Lee, 2002;

Powell *et al.*, 1997). Meanwhile, in other investigations, naturally occurring mild strains were screened for their protective potential but even in cross-protected grapefruit trees a high proportion contained mixed infections of mild and severe strains (Lin *et al.*, 2002; Ochoa *et al.*, 2000).

Mixed results were reported from Australia. Here, the success of cross-protection was influenced by climate. Warmer temperatures appeared to ameliorate symptom severity of severe strains and enhanced cross-protection (Broadbent *et al.*, 1991; Cox *et al.*, 1976). Cross-protection appeared not to be successful as long-term control strategy for grapefruits despite good protection during the first 12 years or so of the reported trials (Broadbent *et al.*, 1991; Cox *et al.*, 1976; Thornton and Stubbs, 1976; Thornton *et al.*, 1980) and so the search for effective mild protective strains continued (Zhou *et al.*, 2002b). Interestingly, the breakdown of cross-protection under certain conditions was linked to an uneven virus distribution of the protective strain in the infected plants (Zhou *et al.*, 2002a). The potential of cross-protection as a means to prevent economic losses to the citrus industry was also investigated in South Africa. A search for mild, protective strains was conducted (van Vuuren and Moll, 1987; van Vuuren *et al.*, 1993) and, interestingly, cross-protection was afforded by inocula comprising mixtures of mild strains (van Vuuren *et al.*, 2000). In grapefruit production cross-protection was sometimes less effective, possibly due to segregation of the mild strains and some protective CTV strains from Florida and Israel appeared to perform better (van der Vyver *et al.*, 2002; van Vuuren and van der Vyver, 2000).

Overall positive results with cross-protection were reported from California (Roistacher *et al.*, 1987, 1988; Wallace and Drake, 1976), Japan (Ieki and Yamaguchi, 1988; Koizumi *et al.*, 1991; Miyakawa, 1987; Sasaki, 1979), China (Cui *et al.*, 2005), Peru (Bederski *et al.*, 2005), Venezuela (Ochoa *et al.*, 1993), and India (Balaraman and Ramakrishnan, 1980). However, in California and in contrast to the success of cross-protection seen in Brazil and elsewhere, it proved very difficult to find local mild strains that would cross-protect against severe strains; a screen of 100 local isolates failed to yield any protective strains (Roistacher and Dodds, 1993; Roistacher *et al.*, 1987). For the last decade citrus decline has been affecting Pakistan. Cross-protection has been recommended as a control method for the citrus industry. However, no research on its efficacy had been reported by the time of a recent review on CTV in that country (Abbas *et al.*, 2005).

Cross-protection cannot provide total resistance against severe CTV strains and many factors can influence or limit its success such as climate, tree physiology, rootstock/scion combination, infection by aphid vectors, or from infected budding material, etc. Nevertheless, cross-protection against CTV has been, largely, a success and has been seen for some time as the only viable option to keep the citrus industry alive

(Lee *et al.*, 1987; Müller and Costa, 1987). It would appear that as long as genetic resistance continues to be elusive or until engineered resistance becomes available, cross-protection will remain the most effective means of controlling CTV.

V. MECHANISM(S) OF CROSS-PROTECTION

A. Early explanations

The cross-protection phenomenon has been known for almost a century. Yet the mechanism(s) behind this phenomenon remain mysterious, despite the many different theories proposed over the years. Several previous reviews discuss those theories in detail and we refer the readers to these articles (among others, Bozarth and Ford, 1989; Fraser, 1998; Gal-On and Shiboleth, 2006; Lecoq and Raccah, 2001; Pennazio *et al.*, 2001; Ponz and Bruening, 1986; Valle *et al.*, 1988).

Early "naïve" hypotheses suggested the development of immunizing substances neutralizing the virus in the plant or the formation of antibodies (Chester, 1933; Wallace, 1940, 1944). One must bear in mind that in those days the nature of the animal immune system was not yet understood and in the context of contemporary knowledge, these ideas were not unreasonable. Several authors suggested that viruses needed certain materials (such as amino acids) for replication and that closely related viruses would have a similar requirement for those materials. Therefore, a successful challenge infection would only occur with differently related viruses as they had different requirements for replication materials (Bawden, 1934; Köhler, 1934; Kunkel, 1934; Price, 1940). Alternatively, it was suggested that the protective strain might occupy virus-specific replication sites within a cell leaving no room for the challenging virus (Bawden, 1964; Caldwell, 1935; Hull and Plaskitt, 1970; Kassanis *et al.*, 1974; Kunkel, 1934; Price, 1940).

In a "Letter to the Editor" of *Phytopathology*, de Zoeten and Fulton (1975) proposed a model for cross-protection between TMV strains in which the coat protein (CP) of the protecting strain, already present in the cells of the protected plant, would somehow capture the genomic RNA of the challenging strain, before it was able to begin the process of replication. Zaitlin (1976) refuted these ideas, pointing out that he was able to induce cross-protection with a CP-defective mutant against wild-type TMV but not against unrelated PVX, and arguing that de Zoeten and Fulton (1975) had misinterpreted Jockusch's (1968) work with assembly-defective TMV mutants. It should be remembered that the proposal of de Zoeten and Fulton (1975) came before the discovery that the uncoating of TMV occurs concomitantly with the first rounds of replicase protein synthesis and negative-strand synthesis (Shaw *et al.*, 1986; Wu and Shaw, 1997), and in the light of these later experiments their initially suggested mechanism was in any case untenable.

Further investigations delivered confusing results. Sherwood and Fulton (1982) found that a challenge with a necrosis-inducing TMV strain would only produce lesions in dark-green leaf tissue on previously TMV-inoculated *Nicotiana sylvestris* (which possesses the *N′* gene: Loebenstein, 2009) plants but a challenge with naked RNA from an HR-inducing TMV successfully induced necrotic lesions in both dark-green and light-green tissues. When RNA from an HR-inducing TMV strain was coated in BMV CP, no cross-protection was observed and lesions could be found in both dark- and light-green tissues. They therefore suggested that the CP of the protecting strain would prevent uncoating of the challenge strain (Sherwood and Fulton, 1982). A further study using a CP-deficient mutant seemed to support the requirement for the CP for cross-protection (Sherwood, 1987a, reviewed by Sherwood, 1987b). Similarly, a challenge with CMV RNA overcame cross-protection provided by a protective CMV strain whereas CMV challenge with purified virions could not (Dodds *et al.*, 1985).

Contrasting results were obtained by Sarkar and Smitamana (1981). They used TMV mutants that were either CP-free or not producing functional CPs in cross-protection experiments. They found that when either mutant was coinoculated with a chlorotic patches-inducing TMV mutant, the extent of those patches was limited in Samsun tobacco. This mutant also protected against a variety of other TMV strains (Sarkar and Smitamana, 1981). Also, preinoculation of Samsun-EN tobacco with a CP-free TMV mutant reduced the number of local lesions induced by challenge inoculation with a lesion-inducing TMV strain (Sarkar and Smitamana, 1981). Further experiments confirmed these findings suggesting the cross-protection could be achieved without the involvement of CPs (Gerber and Sarkar, 1989).

The cross-protection interactions of two different tobamoviruses were investigated (Zinnen and Fulton, 1986). Preinoculation with wild-type SHMV protected against infection with a SHMV mutant as well as a challenge with RNA from the SHMV mutant. Wild-type SHMV infection provided a weak protection against challenge with TMV. However, when TMV RNA was encapsidated in SHMV CP, the chimera was less infectious than wild-type TMV virions. Mutant SHMV RNA encapsidated in TMV CP did not superinfect plants preinoculated with wild-type SHMV. The authors concluded that in some cases protection was CP-mediated whereas in other situation protection was dependent on other factors (Zinnen and Fulton, 1986). In some cases, CP may interfere with virus replication. BMV CP was able to inhibit viral RNA synthesis *in vitro*, most likely because the CP had partly coated the template RNA and obscured the replicase recognition site (Horikoshi *et al.*, 1987).

An important development came in 1984 when it was first suggested that RNA homology might be a driving factor behind cross-protection. Palukaitis and Zaitlin (1984) proposed that during replication of the protecting strain, a large amount of positive-sense RNA strands would be formed. Upon challenge with a closely related strain, excessive positive-sense RNA strands of the protective strain would base-pair to the complementary or near-complementary negative-sense strands of the challenge virus, thus inhibiting transcription and replication. At the time this model was compelling, since it provided a convenient explanation of why cross-protection worked only between closely related strains, and possibly why it worked between satellite RNAs, viroids, and CP-defective virus strains. In certain ways, particularly with regard to the central role of nucleic acid sequence homology, this thinking prefigured the lessons that would soon emerge from experiments with virus-resistant transgenic plants and the discovery of RNA silencing.

B. Lessons from pathogen-derived resistance

Transformation of plants became possible in the mid-1980s and enabled the pioneering work of Beachy and his colleagues who carried out the first successful transformation experiments leading to the generation of virus-resistant plants. In their initial work, they engineered tobacco plants to constitutively express the TMV *CP* gene and the resulting plants showed significant levels of resistance to challenge with TMV (Powell-Abel *et al.*, 1986). The transgenic plants possessed resistance to inoculation with the parental TMV strain as well as a different TMV strain but this resistance could be overcome by inoculation with purified TMV RNA or higher concentration of challenge inocula (Nelson *et al.*, 1987; Powell-Abel *et al.*, 1986; Register III and Beachy, 1988). This work demonstrated the feasibility of pathogen-derived resistance: engendering resistance to viruses by expression of a virus-derived gene sequence in the host (see review by Gottula and Fuchs, 2009). However, the inspiration behind their work was cross-protection and the aim of these experiments was to "simulate" the proposed inhibitory effect of preexisting CP on the uncoating of a challenging virus.

An important question that arose from the creation of the first TMV-resistant *CP*-transgenic plants was whether resistance was engendered by CP protein or by the *CP* transcript. Work with plants expressing untranslatable CP constructs, protoplast work, and studies in which resistance was assessed at various temperatures indicated that this example of pathogen-derived resistance was dependent on the presence and level of accumulation of the CP polypeptide accumulation and that the primary, though not the only, mechanism of protection was that the transgene-derived CP interfered the uncoating of the challenge virus RNA (Nejidat and Beachy, 1989; Nelson *et al.*, 1987; Powell-Abel *et al.*, 1986, 1990; Register III and Beachy, 1988; Wu *et al.*, 1990). Additional recent studies confirmed that TMV CP interferes with early stages of infection. Using an inducible promoter system, Koo *et al.* (2004) found that plants expressing transgenic TMV CP prior to infection were resistant to infection and development of disease symptoms whereas plants expressing the CP postinfection were susceptible. To a significant extent, these studies have vindicated the major hypothesis for the mode of action proposed at the inception of studies of TMV CP-mediated protection.

Does the inhibition of uncoating of TMV in *CP*-transgenic plants parallel one of the potential mechanisms of cross-protection between strains of TMV in nontransgenic plants? Culver (1996) used a PVX-derived gene expression vector to express the TMV CP in systemically infected *Nicotiana benthamiana* plants. Normally, TMV can kill this host but when plants infected with the TMV CP-expressing PVX vector were challenged with TMV there was a delay in virus-induced necrosis and TMV accumulation was markedly decreased compared to unprotected plants (Culver, 1996). Interestingly, a PVX vector expressing a nontranslatable version of the TMV *CP* gene also provided some degree of "cross-protection" against TMV challenge but this was not as pronounced as the protection provided by the PVX vector expressing the TMV CP.

The "cross-protection" provided by either the translatable or untranslatable *CP* constructs was overcome at higher challenge inocula concentrations of or challenge with TMV RNA indicating a role for both RNA and protein-mediated protection although CP-mediation appeared to be the main contributing factor. However, it was concluded that the strongest protection was provided by a protein-mediated mechanism involving the inhibition of uncoating or by recoating of challenge virus RNA by TMV CP since expression of assembly- or RNA binding-incompetent CP mutants did not provide "cross-protection" (Lu *et al.*, 1998). Experiments like these suggest that in certain cases, at least, the inhibition of virion uncoating by CP produced by a protective strain, or in the case of the experiments of Culver and colleagues by a viral vector, may underlie cross-protection and further suggest that the effects of expression of CP genes in transgenic plants may mimic what is going on in a cross-protected nontransgenic plant. However, prevention of disassembly cannot explain other instances of cross-protection, where, for example, satellite RNAs or viroids were used as protective agents (Gallitelli *et al.*, 1991; Montasser *et al.*, 1991; Niblett *et al.*, 1978; Sayama *et al.*, 1993, 2001; Tien and Wu, 1991). This model also does not explain why strains of TMV that are CP-deficient can protect plants against challenge infection with closely related TMV strains (Gerber and Sarkar, 1989; Sarkar and

Smitamana, 1981; Zaitlin, 1976). Furthermore, Valkonen and colleagues (2002) found that cross-protection between PVA strains could not be overcome by a challenge with PVA RNA. This contrasts with the properties of cross-protection observed between TMV strains and with the protection against TMV seen in TMV *CP*-transgenic plants. Taken together, these studies suggest that inhibition of disassembly can explain only some instances of cross-protection, and in those cases in which this mechanism does play a role it cannot completely explain all aspects of this resistance phenomenon.

Meanwhile, in the field of transgenic resistance, it had become apparent that not in every case did the degree of virus resistance correlate with the level to which transgene-encoded CP accumulated. These findings initiated one of the lines of study leading to the eventual discovery of RNA silencing (Lindbo and Dougherty, 1992a,b; Lindbo *et al.*, 1993). Resistance to viruses in transgenic plants expressing other, nonstructural viral genes also could not always be attributed firmly to the action of a protein (de Haan *et al.*, 1992; Longstaff *et al.*, 1993), while other experiments also showed that nontranslated sequences or antisense sequences as well as sequences derived from satellite RNAs or defective interfering RNAs were able to provide virus resistance in transgenic plants (e.g., Cillo *et al.*, 2004, 2007; Gerlach *et al.*, 1987; Goregaoker *et al.*, 2000; Harrison *et al.*, 1987; also see Section VII). Subsequently, it was found that many instances of pathogen-derived transgenic resistance are based, at least to some extent, on RNA silencing (Prins *et al.*, 2008).

C. RNA silencing and cross-protection

The discovery that many instances of pathogen-derived transgenic resistance are based on RNA silencing led to new proposals regarding the mechanism of cross-protection and the potentially related phenomenon of recovery (Moore *et al.*, 2001; Ratcliff *et al.*, 1997, 1999; Voinnet, 2001) (Section III). RNA silencing can function as part of plant defense: primarily against viruses, but it can also have a regulatory role in resistance to nonviral pathogens (Goldbach *et al.*, 2003; Navarro *et al.*, 2006; Ratcliff *et al.*, 1997, 1999; Vaucheret *et al.*, 2001).

For a detailed review of RNA silencing, viral silencing suppressors and their roles in plant–virus interactions, the reader is referred to the article by Csorba *et al.* (2009): only a brief overview of the phenomenon will be provided here. In antiviral RNA silencing, double-stranded RNA possibly derived from viral replicative intermediates or single-stranded RNA molecules forming stem loop or hairpin secondary structures, is recognized and cleaved by Dicer-like (DCL) enzymes into small (21–26 nt) RNA molecules usually referred to as small-interfering RNAs (siRNAs) (Hamilton and Baulcombe, 1999; Molnar *et al.*, 2005; Voinnet, 2005).

Double-stranded siRNAs species are unwound and the single-stranded forms can incorporate, together with argonaute (AGO) proteins, into a RNA-induced silencing complex (RISC). The incorporated siRNAs confer sequence-specificity to the RISC, which has RNAse III activity. Single-stranded siRNAs can also be used as primers for synthesis of virus-derived double-stranded RNA by host RNA-directed RNA polymerases. This double-stranded RNA can act as a template for additional DCL activity and this process can amplify RNA silencing against the viral RNA target. A mobile signal can spread throughout the plant inducing systemic silencing and protecting all plant parts against RNA targets which share sequence homology to the RNA which first elicited the silencing (Himber *et al.*, 2003; Mallory *et al.*, 2003; Palauqui *et al.*, 1997; Voinnet and Baulcombe, 1997).

Gal-On and Shiboleth (2006) proposed a revised model of cross-protection that takes RNA silencing into account. They suggested that on entry of the protective virus strain into the initially infected plant cell, the virus uncoats and its RNA is translated into viral proteins (replicases, CP(s), movement protein(s), suppressors of gene silencing, etc.). Viral RNA-dependent RNA polymerase activity replicates the viral genome via the double-stranded replicative RNAs that they suggest may be recognized by plant DCLs (although there are other double-stranded structures in virus-derived RNAs capable of triggering antiviral RNA silencing: Csorba *et al.*, 2009). Meanwhile, since most viruses encode one or more suppressors of gene silencing the early production and utilization of virus-specific siRNAs in antiviral silencing may be limited. Thus, Gal-On and Shiboleth (2006) suggested that a "race" occurs between the infectious viral RNA, and siRNAs derived from it, to enter neighboring cells and tissues. If the siRNAs enter the neighboring cells before the viral RNA, they may "prime" RNA silencing in these cells and defeat the attempt of the virus to spread.

Gal-On and Shiboleth (2006) envisaged three scenarios in which cross-protection might occur. In the first, they considered three ways in which the challenge strain could be hindered from successful infection. Firstly, upon entering an infected cell with a protective strain, the protective strain CP might hinder the incoming strain from uncoating. This is a "traditional" mechanism, which does not involve any role for RNA silencing (see Section V.A). Secondly, however, the uncoated RNA from the challenge strain might be degraded after recognition by activated RISC and, thirdly, this uncoated RNA might hybridize to minus-strand RNA from the protective virus strain, as proposed in the model of Palukaitis and Zaitlin (1984). This double-stranded RNA hybrid might then be subject to DICER-mediated cleavage, resulting in generation of virus-specific siRNAs.

Their second and third scenarios invoked the idea of a race between infectious viral RNA and virus-specific siRNAs (see above). In the second scenario, a challenge virus strain might enter a cell that has been primed by siRNAs but does not contain protective virus strain RNA. An activated RISC will then degrade incoming viral RNA before it can initiate an infection. Their third scenario involves systemic silencing and Gal-On and Shiboleth (2006) argue that it might also explain aspects of the recovery and green island phenomena seen in certain natural infections or in certain lines of transgenic plants expressing virus-derived transgenes (Iyer and Hall, 2000; Ratcliff *et al.*, 1999; Tanzer *et al.*, 1997). In systemic silencing small RNAs amplified by cellular RNA-directed RNA polymerases may travel to locations distant from the inoculated cells and activate RISCs in these uninfected tissues. The RNA of any challenge virus will subject to degradation in those locations.

This model would neatly explain why cross-protection only worked between closely related virus strains, required an interval between inoculations in order to be successful and the breakdown of protection in some cases (Gal-On and Shiboleth, 2006). If the levels of challenge virus RNA infecting a primed cell exceeded the available activated RISC, then a challenge virus might be able to infect those plants after all. Similarly, the occurrence of divergent virus strains might help the challenge virus to evade recognition by RISC.

Several studies support the involvement of RNA silencing in cross-protection (Kubota *et al.*, 2003; Lin *et al.*, 2007; Tsuda *et al.*, 2007; Valkonen *et al.*, 2002). An amino acid change in the HC-Pro of a PVA mutant enhanced its ability to overcome cross-protection (Valkonen *et al.*, 2002). Interestingly, HC-Pro is a silencing suppressor; thus the mutation might have enhanced the silencing suppression activity. However, recent results from our laboratory suggest that cross-protection scenarios other than those proposed by Gal-On and Shiboleth (2006) can occur and may not necessarily involve either RNA silencing or inhibition of uncoating.

We investigated cross-protection between CMV strains using a CMV deletion mutant that is unable to express the 2b silencing suppressor protein (CMVΔ2b). Unlike its parent strain (the Fny strain of CMV) this mutant virus is symptomless in tobacco, *N. benthamiana*, and *Arabidopsis thaliana* (ecotype Col-0) (Lewsey *et al.*, 2009; Soards *et al.*, 2002; Ziebell and Carr, 2009; Ziebell *et al.*, 2007). We reasoned, in line with a suggestion of Gal-On and Shiboleth (2006), that an engineered virus that lacks an RNA silencing suppressor, or one that encodes a mutant version unable to effectively suppress RNA silencing, would be a particularly potent cross-protecting agent. This is because upon infection this mutant would trigger RNA silencing against itself and homologous viral RNA sequences but, lacking the means to inhibit silencing, systemic RNA silencing would protect plants from challenge infection with closely related strains. There was ample evidence from previous work with viral vectors (Ratcliff *et al.*, 1999) and with a mutant of CymRSV unable to express the P19 silencing suppressor protein (Silhavy *et al.*, 2002) to indicate that it would be a viable approach. Indeed, we found that CMVΔ2b did provide limited protection against infection with its parental strain in tobacco and *N. benthamiana*. However, we found no indication that CMVΔ2b induced a strong systemic silencing signal (Ziebell *et al.*, 2007).

Another line of evidence indicating that RNA silencing may not be able to entirely explain cross-protection between CMV strains came from a study by Ziebell and colleagues (2007). Surprisingly, it was found that the CMVΔ2b mutant was able not only to protect plants against its parental (Fny) strain of CMV but also against another CMV strain (TC). Fny-CMV is a subgroup IA CMV strain (Palukaitis and García-Arenal, 2003), whereas TC-CMV belongs to subgroup II. The two strains share only 70% overall RNA sequence similarity and have few regions of homology exceeding 20 nucleotides in extent; making RNA silencing a less convincing explanation for CMVΔ2b-mediated cross-protection (Ziebell *et al.*, 2007). Interestingly, CMVΔ2b-mediated cross-protection proved to be broader in scope than the protection against the virus exhibited by CMV-resistant transgenic plants expressing truncated versions of the 2a replicase gene of Fny strain CMV. These plants showed resistance only to subgroup I strains and were unprotected from infection with CMV strains from subgroup II (Anderson *et al.*, 1992; Carr *et al.*, 1994; Hellwald and Palukaitis, 1995; Zaitlin *et al.*, 1994).

To investigate whether the protection afforded by CMVΔ2b involved localized, rather than systemic, RNA silencing, Ziebell and Carr (2009) used *A. thaliana* mutants compromised in the RNA silencing pathway. *A. thaliana* possesses four DCL enzymes that use small RNAs to direct homology-dependent cleavage of host or viral RNA targets and each DCL is required for the triggering of specific aspects of RNA silencing (Deleris *et al.*, 2006). Primary antiviral defense is provided by DCL4 but DCL2 can to some extent substitute for it in *dcl4* mutants (Deleris *et al.*, 2006). Additionally, DCL3 may also play a minor role in antiviral silencing with DCL1 having an indirect effect through its negative regulation of DCL4 and DCL3 gene expression (Qu *et al.*, 2008). As expected, CMVΔ2b protected against symptoms induced by challenge with the parental CMV strain in wild-type *A. thaliana* (Ziebell and Carr, 2009). However, protection was not abolished in *dcl2, dcl3*, or *dcl4* mutant plants, in *dcl2/4* double-mutant or *dcl2/3/4* triple-mutant plants (Ziebell and Carr, 2009) (Fig. 1). Interestingly, protection against challenge infection with wild-type CMV was strongest in *dcl2/4* where the CMVΔ2b replicated to its highest levels and also induced strong symptoms (Fig. 1). Additionally, it appeared that in wild-type plants DCL4 and, in a subsidiary role, DCL2 were inhibiting accumulation and spread of CMV, which the 2b silencing suppressor

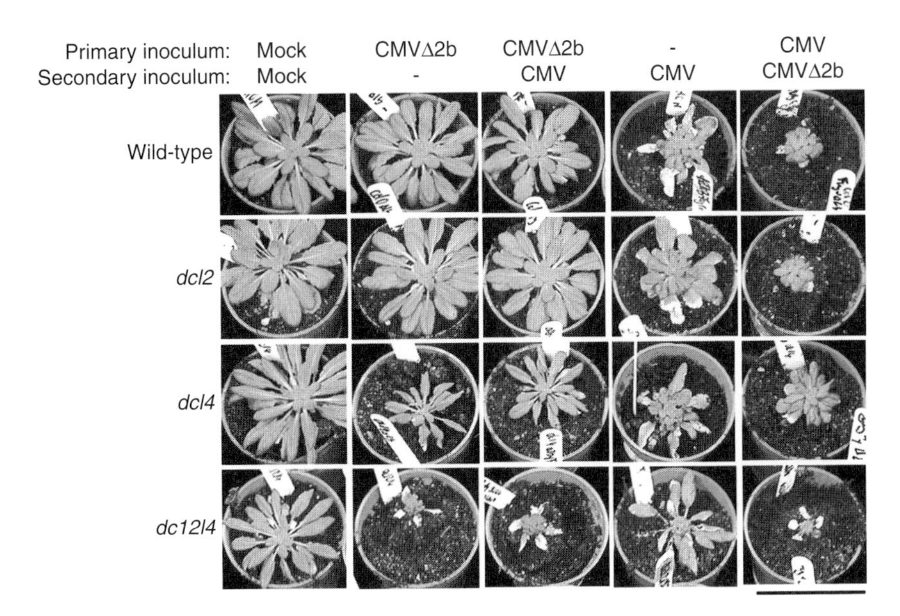

FIGURE 1 Symptoms in cross-protection experiments in wild-type and RNA silencing-deficient mutant *Arabidopsis thaliana* plants. Wild-type plants (Col-0 ecotype) or plants of mutant *dcl2*, *dcl4* or double-mutant *dcl2/4* lines were inoculated (primary inoculum) with Fny-CMVΔ2b (a CMV mutant that does not possess the 2b silencing suppressor protein gene), Fny-CMV (the wild-type CMV strain from which Fny-CMVΔ2b was derived) or were mock-inoculated, or left untreated (–). Ten days later, plants were challenge-inoculated (secondary inoculum) as indicated. On wild-type *A. thaliana* and *dcl2* mutant plants Fny-CMVTΔ2b does not induce symptoms but does cross-protect against Fny-CMV-induced symptoms. On *dcl4* mutants and *dcl2/4* double-mutant plants Fny-CMVΔ2b induces mild and severe symptoms, respectively. Those symptoms could not be distinguished from symptoms induced by Fny-CMV. However, analysis by reverse-transcriptase polymerase chain reaction revealed that Fny-CMVΔ2b protected those mutant plants to a higher proportion from challenge inoculation with Fny-CMV indicating that DCL is not required for sufficient cross-protection. Figure taken from Ziebell and Carr (2009) with kind permission of the copyright holder The Society for General Microbiology.

protein could overcome to some extent: in line with observations by other researchers (Diaz-Pendon and Ding, 2008; Diaz-Pendon *et al.*, 2007). However, in *dcl-2/dcl-4* mutant plants, in which antiviral silencing cannot occur, CMVΔ2b accumulates to higher levels and spreads faster presumably because the lack of the silencing suppressor was no handicap to the mutant virus in this plant background (Ziebell and Carr, 2009). Thus, it appears that DCL-2, DCL-3, and DCL-4 are not required for cross-protection between CMVΔ2b and wild-type CMV, suggesting that

CMVΔ2b provides cross-protection by a mechanism that does not involve any of the currently known pathways of RNA silencing.

D. Exclusion/spatial separation

In CMVΔ2b-infected tobacco plants challenged with wild-type CMV, analysis of inoculated and upper, noninoculated leaves using virus-specific RT-PCR and *in situ* hybridization revealed that the protective strain needs to be present in challenged tissue to provide successful protection (Ziebell *et al.*, 2007). Nakazono-Nagaoka *et al.* (2005) noted similar findings with cross-protection between CMV strains. In tissues inoculated with CMVΔ2b and CMV, the *in situ* hybridization experiments indicated that the CMV and the CMVΔ2b were spatially separated; the two viruses occurred in the same tissues but not within the same cells (Ziebell *et al.*, 2007).

Spatial separation between strains of the same virus appears to be common in doubly infected plant tissues since it has also been described for a number of other viruses, including those belonging to the following genera: *Tobamovirus* (Kunkel, 1934); *Alfamovirus* (Hull and Plaskitt, 1970); *Tritimovirus* (Hall *et al.*, 2001); *Potyvirus* and *Potexvirus* (Dietrich and Maiss, 2003: see Fig. 2); *Cheravirus* (Takahashi *et al.*, 2007), and *Cucumovirus* (Takeshita *et al.*, 2004). In the study of Takeshita *et al.* (2004), it was shown that CMV strains from different subgroups could mutually exclude each other from tissues in both the inoculated and the noninoculated, systemically infected leaves of doubly infected cowpea plants.

Dietrich and Maiss (2003) investigated spatial exclusion between virus strains using confocal scanning laser microscopy to distinguish between cells infected with genetically engineered PPV and PVX variants expressing either GFP or the red fluorescent protein (RFP). They observed mutual exclusion only between closely related viruses (e.g., PPV-GFP and PPV-RFP) but not between different viruses (e.g., PPV-GFP and PVX-RFP) (Dietrich and Maiss, 2003: see Fig. 2). Similar exclusion was observed between strains of ALSV and BYMV (Takahashi and Yoshikawa, 2008; Takahashi *et al.*, 2007) between TNV and TCV (Xi *et al.*, in press) and in a classic study using EM to distinguish between inclusion bodies produced by different AlMV strains (Hull and Plaskitt, 1970).

More recently, González-Jara *et al.* (2009) investigated the multiplicity of infection (i.e., the number of virus particles or genomes that may infect one cell) for TMV using different fluorescent reporter genes. Interestingly, they also found that only a fraction of infected cells were infected simultaneously by two TMV genotypes. They found that infection of one cell by one TMV genotype prevents superinfection with another. This kind of competition may give the initially infecting virus a selective advantage (González-Jara *et al.*, 2009). This observation is line with previous

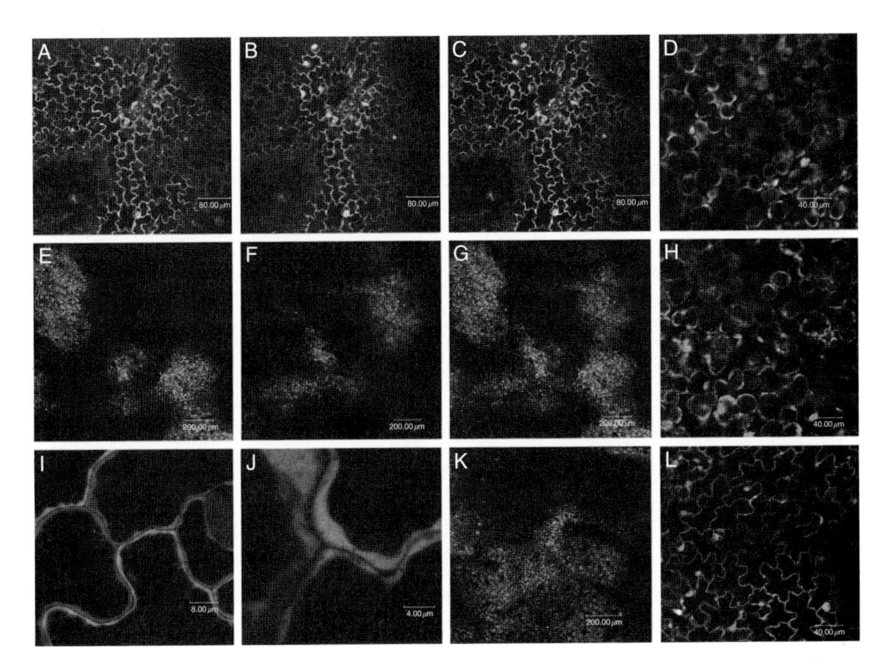

FIGURE 2 Mutual exclusion of closely related viruses. Virus distribution in mixed infections in systemically infected *N. benthamiana* tissues. Confocal imaging of coinfecting PPV-NAT-*AgfpS* (A) and PVX201-*optRed* (B) reveals extensive double infection of epidermal cells. (C) Merged image of (A) and (B). (D) Double infected mesophyll cells. Double infections of PPV-NAT-*AgfpS* (E) and PPV-NAT-*red* (F) result in spatial separation of the two virus populations in epidermal cells as indicated by the differently colored cell clusters in the merged image (G) of (E) and (F). (H) Spatial separation in mesophyll cells. (I) Close-up of (G) at the border region of different fluorescent cell clusters. The coinfected cell appears yellow, whereas the neighboring cells harbor an excess of PPV-NAT-*AgfpS* (green) or PPV-NAT-*red* (red). (J) Close-up of (C) cells coinfected with PPV-NAT-*AgfpS* and PVX201-*optRed* showing an uniform pale yellow color. (K) Coinfecting but spatially separate populations of PPV-NAT-*red* and TVMV-*gfp*. (L) Close-up of (K) reveals that contact of PPV-NAT-*red* and TVMV-*gfp* populations is restricted to a few cells at the border of two different fluorescent cell clusters. Figure and text taken from Dietrich and Maiss (2003) with kind permission of the copy right holder The Society for General Microbiology. (See Page 2 in Color Section at the back of the book.)

observations of highly structured RNA virus populations within plants though this may not be a general feature for all plant viruses, for example, pararetroviruses such as CaMV (Dietrich *et al.*, 2007; García-Arenal *et al.*, 2001; Hall *et al.*, 2001; Jridi *et al.*, 2006; Li and Roossinck, 2004; Monsion *et al.*, 2008). We have theorized that exclusion between strains may be a general mechanism in which virus variants prevent the internecine struggle over host resources, while preserving RNA sequence diversity within a viral population. Among other things, this may provide an additional way of evading RNA silencing (Ziebell and Carr, 2009).

VI. PROTECTION PHENOMENA INVOLVING SUBVIRAL AGENTS

Satellite viruses and satellite RNAs are subviral agents that are hyperparasites of viruses (Hull, 2002). Satellite viruses are dependent upon another virus (called the helper) for replication. Satellite viruses should not be confused with satellite RNAs (or DNAs), which are subviral, hyperparasitic nucleic acid molecules that are dependent on a helper virus for replication and also for encapsidation. The majority of known satellite viruses and RNAs do not share extensive nucleic acid sequence homology with the helper virus genome. The presence of satellite viruses and satellite RNAs can modify the symptoms induced by a helper virus and affect the titer of helper virus in the infected host tissue. The fundamental biology of satellites and their effects on their helper viruses and hosts has been reviewed extensively elsewhere and will not be dealt with in detail here (Collmer and Howell, 1992; Roossinck *et al.*, 1992; Simon *et al.*, 2004).

The effects of satellite viruses and RNAs on helper virus symptoms and pathogenicity can be dramatic and can include ameliorative effects on virus-induced disease (Hull, 2002). Therefore, efforts have been made to use satellites as protective agents or to express them from transgenes in transformed plants to achieve genetically engineered virus resistance (Cillo *et al.*, 2004; Gallitelli *et al.*, 1991; Gerlach *et al.*, 1987; Harrison *et al.*, 1987; Montasser *et al.*, 1991; Sayama *et al.*, 1993, 2001; Tien and Wu, 1991; Yie and Tien, 1993; Yie *et al.*, 1992).

Viroids constitute a third class of subviral infectious agents. They are small, single-stranded, circular RNA molecules that replicate in host plant cells without the aid of a helper virus (Flores *et al.*, 1997; Hull, 2002; Tabler and Tsagris, 2004). Viroids do not encode proteins yet they can cause devastating plant diseases (Ding, 2009). Cross-protection among viroids has been described for PSTVd (Branch *et al.*, 1988; Fernow, 1967; Khoury *et al.*, 1988; Kryczynski and Paduch-Cichal, 1987; Singh *et al.*, 1989, 1990), CEVd (Duran-Vila and Semancik, 1990), for CChMVd (de la Peña and Flores, 2002; de la Peña *et al.*, 1999; Horst, 1975), between PSTVd and HSVd (Branch *et al.*, 1988), between PSTVd and CLVd-N (Singh *et al.*, 1992), and between PSTVd and CEVd (Pallás and Flores, 1989). An extensive study investigated the possibility of cross-protection between different viroids (Niblett *et al.*, 1978). However, as far as we can ascertain cross-protection between viroids has not been utilized commercially on a large scale and will not be further discussed here. For more information, we refer the reader to the reviews by Ding (2009) and Flores *et al.* (2005).

A. Satellite viruses

Few examples of satellite viruses ameliorating the effects of their helper viruses have been described but in some cases interference can be strong enough to obscure the symptoms of the helper virus infection (Scholthof *et al.*, 1999). STNV, the first satellite virus to have been discovered, reduces the accumulation of its helper virus TNV, an effect that has been attributed to the accumulation or action of the STNV CP (Jones and Reichman, 1973). Extensive studies with a different satellite virus, STMV, showed that this satellite virus did not usually change the symptoms induced by the helper virus, TMV (normally strain U5, also known as TMGMV) (Dodds, 1998, 1999; Rodríguez-Alvarado *et al.*, 1994). The presence of STMV in certain pepper varieties led to increased chlorosis relative to mosaic symptoms induced by various TMV helper strains alone (Rodríguez-Alvarado *et al.*, 1994). However, symptoms on Jalapeño peppers induced by one TMV strain were ameliorated after an initial period of severe symptoms (Rodríguez-Alvarado *et al.*, 1994). Different strains of STMV were also able to cross-protect against the other (Mathews and Dodds, 1998). Interestingly, Kurath and Dodds (1994) found that STMV variants with sequence differences of only five nucleotides could interfere with each other in a cross-protection-like manner.

The symptoms induced by PMV on St. Augustinegrass were more dependent on environmental conditions than on the presence or absence of its satellite virus SPMV (Cabrera and Scholthof, 1999). However, on millet plants, the presence of SPMV enhanced the rate of systemic infection and exacerbated the symptoms induced by PMV from mild mottling to severe chlorosis and stunting (Masuta *et al.*, 1987; Scholthof, 1999). Further studies linked the induction of more severe symptoms to domains on the CP of SPMV (Qiu and Scholthof, 2001). When the CP of SPMV was expressed from a PVX vector in a nonhost plant (*N. benthamiana*), SPMV CP did not show any silencing suppression activity itself but was able to interfere with the silencing suppressor protein from PVX (Qiu and Scholthof, 2004). Furthermore, the expression of CP in this way in *N. benthamiana* induced symptoms in those plants (Qiu and Scholthof, 2004). Though the direct mechanisms of the interaction between PMV and SPMV remain to be revealed, it is clear that the properties of the SPMV CP play an important part (Omarov *et al.*, 2005; Qi and Scholthof, 2008).

B. Satellite RNAs

Satellite RNAs are replicated by the helper virus replicase complex and encapsidated by helper virus CP (Hull, 2002; Simon *et al.*, 2004). Typically, they share no nucleotide sequence homology with their helper virus. An exception to this is the satellite RNA of TCV, which contains a sequence domain with homology to TCV RNA. This satellite RNA therefore has some characteristics in common with defective RNAs (Simon and Howell, 1986; Simon *et al.*, 2004). The effects of satellite RNAs on their helper viruses and helper virus-induced symptoms have been well characterized in a number of experimental systems. While a satellite RNA may have no effect at all on symptoms induced by the helper virus (a neutral effect), in many cases they may either ameliorate or exacerbate the symptoms (Collmer and Howell, 1992; Palukaitis and García-Arenal, 2003; Palukaitis *et al.*, 1992; Roossinck *et al.*, 1992).

The effect of satellite RNAs on helper virus-induced symptoms is complicated by nature of the strains of both helper virus and satellite RNA and by the host plant background (species- and cultivar-specific effects) (Cillo *et al.*, 2007; Collmer and Howell, 1992; Collmer *et al.*, 1992; Demler *et al.*, 1996; Devic *et al.*, 1989, 1990; Gonsalves *et al.*, 1982; Grieco *et al.*, 1992, 1997; Jaegle *et al.*, 1990; Jordá *et al.*, 1992; Kaper and Waterworth, 1977; Kosaka *et al.*, 1989; Kuwata *et al.*, 1991; Li and Simon, 1990; Masuta *et al.*, 1988; Mossop *et al.*, 1976; Murant *et al.*, 1988; Oh *et al.*, 1995; Palukaitis, 1988; Raj *et al.*, 2000; Rodríguez-Alvarado and Roossinck, 1997; Roossinck *et al.*, 1992; Sleat, 1990; Sleat and Palukaitis, 1990a,b, 1992; Sleat *et al.*, 1994; Takanami, 1981; Taliansky and Robinson, 1997; Waterworth *et al.*, 1979; Xu and Roossinck, 2000; Xu *et al.*, 2003; Zhang *et al.*, 1994). For authoritative, in-depth reviews on the interactions between satellite RNAs, their helper viruses, hosts, and the various mechanisms underlying these interactions reader is referred to the articles by Collmer and Howell (1992), Palukaitis *et al.* (1992), Palukaitis and García-Arenal (2003), Roossinck *et al.* (1992), and Simon *et al.* (2004).

One of the best characterized satellite RNA and helper virus relationships is that between CMV and its various satellite RNAs. Because of their ability to ameliorate symptoms and/or prevent infection with CMV, certain satellite RNA strains have been used to protect greenhouse and field crops (Gallitelli, 1998; Gallitelli *et al.*, 1991; Montasser *et al.*, 1991; Sayama *et al.*, 1993, 2001; Tien and Wu, 1991; Tien *et al.*, 1987; Wu *et al.*, 1989; Yoshida *et al.*, 1985). In genetically engineered plants, transgenes derived from satellite RNA sequences have been used to engender resistance to CMV (Cillo *et al.*, 2004, 2007; Harrison *et al.*, 1987; Yie and Tien, 1993; Yie *et al.*, 1992).

Several mechanisms have been put forward to explain the effects of satellite RNAs on their helper viruses and on symptom induction. Competition between viral and satellite RNA templates for the viral replicase was an early suggestion to explain how in some cases coinfection with a satellite RNA decreased the accumulation of the helper virus RNA (Habili and Kaper, 1981; Kaper, 1982; Mossop and Francki, 1979). However, the relationships between cucumoviruses and satellite RNA are often more complex and the effects of the satellite RNA on helper virus titer do not necessarily correlate simply with their effects on pathogenesis.

For example, CMV satellite RNAs interfere not only with symptom induction by CMV, but also with symptom induction by another cucumovirus, TAV. But they do not inhibit TAV accumulation (Harrison *et al.*, 1987; Moriones *et al.*, 1994). Additionally, a satellite RNA that induced necrosis during coinfection with CMV did not enhance, but decreased, accumulation of the helper virus (Escriu *et al.*, 2000). Similarly, a satellite RNA of the cucumovirus PSV did not affect accumulation of its helper virus but it did aggravate symptoms engendered by certain PSV isolates (Ferreiro *et al.*, 1996; Militao *et al.*, 1998). Similarly complex helper–satellite relationships were also discovered in a study of the effects of various CMV and CMV satellite RNA combinations in tomato and wild tomato relatives (Cillo *et al.*, 2007). Both benign and necrosis-inducing satellite RNAs were able to downregulate CMV RNA accumulation but this symptom alteration was host-dependent (Cillo *et al.*, 2007). The extent of symptom attenuation in a given host with a given combination of satellite RNA strain and CMV strain correlated to a strong degree with specific downregulated accumulation of the helper virus' RNAs 3 and 4 (Cillo *et al.*, 2007). This work and evidence from a number of earlier studies indicated that host factors must play an important role in determining the nature of helper–satellite relationships and in conditioning symptom expression in hosts coinfected with satellite RNAs and helper viruses (Masuta *et al.*, 1988; Mossop and Francki, 1979; Palukaitis, 1988; Waterworth *et al.*, 1979).

A detailed study on the involvement of host factors was carried out in the Roossinck laboratory using tomato plants coinfected with CMV and the D4 satellite RNA, which induces a form of programmed host cell death (Xu and Roossinck, 2000; Xu *et al.*, 2003). Prior to visible cell death, expression of defense-related host genes increased dramatically. This was not seen in plants infected solely with CMV, which in the absence of D4 does not induce programmed cell death in tomatoes (Xu and Roossinck, 2000). Although plus- and minus-sense D4 satellite RNA strands were found in infected stems, satellite RNA was only found in vascular bundles undergoing programmed cell death. Steady-state levels of minus-strand satellite RNA rose dramatically before the onset of programmed cell death indicating a role for this molecule in pathogenicity (Xu and Roossinck, 2000). However, the exact mechanisms of programmed cell death activation are still unclear (Simon *et al.*, 2004).

Simon and coworkers have characterized the mechanisms of pathogenicity in the TCV–satellite RNA system. Symptoms can be more severe (including necrosis) on host plants infected with both TCV and satellite C RNA of TCV. The helper virus titer is decreased on TCV plus satellite RNA infected plants compared to those in plants infected only with TCV (Li and Simon, 1990). The satellite RNA can bind to viral CP (P38) and this inhibits virion formation. This leads to an increase in the amount of free

CP and, because the TCV CP is also a suppressor of RNA silencing and a pathogenicity determinant, leads to intensified disease symptoms (Kong *et al.*, 1997; Manfre and Simon, 2008; Simon *et al.*, 2004; Wang and Simon, 1999; Zhang and Simon, 2003).

As mentioned earlier, sequences derived from ameliorative satellite RNAs have been expressed from transgenes with the aim of engendering resistance or tolerance in the transformed plants (Cillo *et al.*, 2004, 2007; Gerlach *et al.*, 1987; Harrison *et al.*, 1987; Yie and Tien, 1993; Yie *et al.*, 1992). Transgenic tomatoes expressing a transgene encoding a benign CMV satellite RNA sequence showed increased tolerance to various CMV strains and supported less viral replication (Cillo *et al.*, 2004). These plants were protected from necrosis induction if challenged with inocula comprising CMV and a necrogenic satellite RNA. Interestingly, this was not because of a competition of transgenic and necrogenic satellite RNA for viral replicase but due to sequence-specific degradation of the necrogenic satellite RNA, indicating a role for RNA silencing (Cillo *et al.*, 2004). Thus, various mechanisms condition the relationship between satellite RNAs and their helper viruses: some of these may resemble those involved in cross-protection, while others differ considerably.

Regardless of whether a satellite RNA is generated from a transgene or applied as an inoculum, the use of satellite RNAs to protect crops must be treated with caution. Unfortunately, when satellite RNAs undergo replication by the replicase of a helper virus, mutant satellite RNA sequences are likely to be generated because RNA-dependent RNA polymerases are intrinsically error-prone. Mutant RNAs that retain viability may not have the same effects on the helper virus as the wild-type satellite RNA. For example, serial passage of a benign D-satellite RNA of CMV produced a heterogeneous population of satellite RNAs (Kurath and Palukaitis, 1989). In addition, spontaneous mutation of a benign satellite RNA into a pathogenic variant can occur over time, raising concerns that the use of satellite RNAs as means of protection could seriously backfire (Palukaitis and Roossinck, 1996; Tepfer and Jacquemond, 1996).

VII. CONCLUDING THOUGHTS

The mechanism or mechanisms of cross-protection remain mysterious and somewhat confusing and perhaps there is no simple, single explanation for the phenomenon. Nevertheless, research into cross-protection and its causes has been and continues to be fruitful in terms of the theoretical insights and practical benefits it has yielded. Against the background of so many emerging viral threats to agriculture (see Section I), cross-protection still has a potentially important role to play in plant protection, even if it is used as a stopgap method against a novel viral disease pending the development and testing of resistant crop varieties developed through conventional breeding or transgenesis.

Work in recent years from a number of groups has shown that viral suppressors of RNA silencing are major pathogenicity determinants (Csorba *et al.*, 2009; Ding *et al.*, 1995; Gal-On, 2000; Kubota *et al.*, 2003; Lewsey *et al.*, 2007; Lin *et al.*, 2007; Tsuda *et al.*, 2007). Several studies have shown that viral mutants expressing altered versions of these factors, or unable to produce them at all, can induce protection against the parental viral strain and/or homologous viral sequences expressed from an unrelated viral vector (Lin *et al.*, 2007; Ratcliff *et al.*, 1999; Silhavy *et al.*, 2002; Szittya *et al.*, 2002; Ziebell *et al.*, 2007). Engineering and testing of virus variants containing specific mutations or deletions in the genes encoding silencing suppressors, perhaps coupled with insights from developing methods to mathematically model the behavior of viruses in mixed populations, at the population and individual plant levels (Martín and Elena, 2009; Zhang and Holt, 2001), offers a rapid way of screening for potential cross-protection agents.

It is clear, however, that while some engineered constructs may induce protection by triggering RNA silencing (Ratcliff *et al.*, 1999; Silhavy *et al.*, 2002), others may not (Lin *et al.*, 2007; Ziebell and Carr, 2009; Ziebell *et al.*, 2007) (see Sections V.C and V.D). Before agents of this sort could be used in crop protection, the mode of action of each new example would need to be determined before deployment in order to assist with risk assessment. But from a practical point of view the exact nature of the mechanism (e.g., RNA silencing induction vs. exclusion or some other mode of action) is probably not important. If a mutant can no longer induce symptoms, but can inhibit infection by a severe strain, it will be potentially useful.

While work with precisely engineered mutant viruses will aid in developing new cross-protection strategies, it may not be acceptable or allowable within certain jurisdictions to deploy these agents directly in the field. This is, perhaps ironic, given the success of the pioneering work of Rast (1972), who isolated protective TMV strains by random chemical mutagenesis by treating crude preparations of TMV with nitrous acid that were used safely for many years in commercial tomato cultivation in the United Kingdom, Netherlands, and Japan (see Yeh and Gonsalves, 1984 and references therein). However, given the advances in high throughput sequencing it will be possible to exploit the vast natural variation occurring in natural virus populations and the small RNA populations they engender in their host plants (Donaire *et al.*, 2009). With methods such as pyrosequencing that yield thousands of reads it will be possible to detect naturally occurring mutants or appropriate small RNA target sites, with the desired sequence variations within the silencing suppressor genes, that can be isolated and utilized in the development of new cross-protection agents.

ACKNOWLEDGMENTS

We are grateful to colleagues for useful discussions over the years on cross-protection and allied topics including John Walsh, Nicola Spence, Roger Hull, Peter Palukaitis, and Alex Murphy, as well as past and present members of the Cambridge University Plant Virology Group. We are very grateful to our Librarian, Christine Alexander, for excellent assistance in obtaining the more rare bibliographical materials. The Walter Grant Scott Fellowship of Trinity Hall, Cambridge supported H.Z. Research in the J.P.C. lab is funded by grants from the Biotechnological and Biological Sciences (BB/D008204/1, BB/F014376/1), The Leverhulme Trust, the European Union, and the Isaac Newton Trust.

REFERENCES

Aapola, A. I. E., and Rochow, W. F. (1971). Relationships among three isolates of barley yellow dwarf virus. *Virology* **46**:127–141.

Abbas, M., Khan, M. M., Mughal, S. M., and Khan, I. A. (2005). Prospects of classical cross protection technique against *Citrus tristeza closterovirus* in Pakistan. *Hort. Sci. (Prague)* **32**:74–83.

Aguilar, I., Sánchez, F., and Ponz, F. (2000). Different forms of interference between two tobamoviruses in two different hosts. *Plant Pathol.* **49**:659–665.

Ahoonmanesh, A., and Shalla, T. A. (1981). Feasibility of cross-protection for control of tomato mosaic virus in fresh market field-grown tomatoes. *Plant Dis.* **65**:56–58.

Anderson, J. M., Palukaitis, P., and Zaitlin, M. (1992). A defective replicase gene induces resistance to cucumber mosaic virus in transgenic tobacco plants. *Proc. Natl. Acad. Sci. USA* **89**:8759–8763.

Anderson, P. K., Cunningham, A. A., Patel, N. G., Morales, F. J., Epstein, P. R., and Daszak, P. (2004). Emerging infectious diseases of plants: Pathogen pollution, climate change and agrotechnology drivers. *Trends Ecol. Evol.* **19**:535–544.

Anonymous, (2004). FAO report 2004. http://www.fao.org/fileadmin/user_upload/foodclimate/HLCdocs/declaration-E.pdf, accessed 23.07.2009.

Balaraman, K., and Ramakrishnan, K. (1980). Strain variation and cross-protection in citrus tristeza virus on acid lime. *In* (E. C. Calavan, S. M. Garnsey, and L. W. Timmer, eds.), *Proc. 8th Conf. IOCV*, pp. 60–68.

Bar-Joseph, M. (1978). Cross protection incompleteness: A possible cause for natural spread of citrus tristeza virus after a prolonged period in Israel. *Phytopathology* **68**:1110–1111.

Bar-Joseph, M., Loebenstein, G., and Oren, Y. (1974). Use of electron microscopy in eradication of tristeza sources recently found in Israel. *In* (L. G. Weathers and M. Cohen, eds.), *Proc. 6th Conf. IOCV*, pp. 83–85.

Bar-Joseph, M., Roistacher, C. N., Garnsey, S. M., and Gumpf, D. J. (1981). A review on tristeza, an ongoing threat to citriculture. *Proc. Int. Soc. Citriculture* **1**:419–423.

Bar-Joseph, M., Roistacher, C. N., and Garnsey, S. M. (1983). The epidemiology and control of citrus tristeza disease. *In* "Plant Virus Epidemiology" (R. T. Plumb and J. M. Thresh, eds.), Blackwell Scientific Publications, Oxford.

Bar-Joseph, M., Marcus, R., and Lee, R. F. (1989). The continuous challenge of citrus tristeza virus control. *Ann. Rev. Phytopathol.* **27**:291–316.

Bateson, M., Lines, R., Revill, P., Chaleeprom, W., Ha, C., Gibbs, A., and Dale, J. (2002). On the evolution and molecular epidemiology of the potyvirus *Papaya ringspot virus*. *J. Gen. Virol.* **83**:2575–2585.

Batista, L., Porras, D. N., Gutiérrez, A., Peña, I., Rodriguez, J., Fernandez del Amo, O., Pérez, R., Morera, J. L., Lee, R. F., and Niblett, C. L. (1996). Tristeza and *Toxoptera citricida* in Cuba: Incidence and control strategies. *In* (J. V. da Graça, P. Moreno, and R. K. Yokomi, eds.), *Proc. 13th Conf. IOCV*, pp. 104–111.

Baulcombe, D. C. (2004). RNA silencing in plants. *Nature* **431:**356–363.

Bawden, F. C. (1934). Studies on a virus causing foliar necrosis of the potato. *Proc. R. Soc. Lond. B* **116:**375–395.

Bawden, F. C. (1964). *Plant Viruses and Virus Diseases*, 4th Edn. Ronald Press, New York.

Bederski, K., Roistacher, C. N., and Müller, G. W. (2005). Cross protection against the severe citrus tristeza virus stem pitting in Peru. *In* (M. E. Hilf, N. Duran-Vila, and M. A. Rocha-Peña, eds.), *Proc. 16th Conf. IOCV*, pp. 117–126.

Bennet, C. W. (1951). Interference phenomena between plant viruses. *Ann. Rev. Microbiol.* **5:**295–308.

Ben-Ze'ev, I. S., Bar-Joseph, M., Nitzan, Y., and Marcus, R. (1989). A severe citrus tristeza virus isolate causing the collapse of trees of sour orange before the virus is detectable throughout the canopy. *Ann. Appl. Biol.* **114:**293–300.

Bitterlin, M. W., and Gonsalves, D. (1988). Serological grouping of tomato ringspot virus isolates: Implications for diagnosis and cross-protection. *Phytopathology* **78:**278–285.

Bodaghi, S., Mathews, D. M., and Dodds, J. A. (2004). Natural incidence of mixed infections and experimental cross protection between two genotypes of tobacco mild green mosaic virus. *Phytopathology* **94:**1337–1341.

Bodine, E. W. (1942). Antagonism between strains of the peach-mosaic virus in Western Colorado. *Phytopathology* **32:**1.

Boyle, J. S., and Bergman, E. L. (1969). The prevention of blotchy ripening in tomato. *Phytopathology* **49:**397.

Bozarth, R. F., and Ford, R. E. (1989). Viral interactions: Induced resistance (cross-protection) and viral interference among plant viruses. *In* "Experimental and Conceptual Plant Pathology, Vol. 3" (W. M. Hess, R. S. Sing, U. S. Singh, and D. J. Weber, eds.), pp. 551–567. Gordon & Breach Scientific Publications, New York.

Branch, A. D., Benenfeld, B. J., Franck, E. R., Shaw, J. F., Lee Varban, M., Willis, K. K., Rosen, D. L., and Robertson, H. D. (1988). Interference between coinoculated viroids. *Virology* **163:**538–546.

Brlansky, R. H., Pelosi, R. R., Garnsey, S. M., Youtsey, C. O., Lee, R. F., Yokomi, R. K., and Sonoda, R. M. (1986). Tristeza quick decline epidemic in South Florida. *Proc. Fla. State Hort. Soc.* **99:**66–69.

Broadbent, L. (1975). Epidemiology and control of tomato mosaic virus. *Ann. Rev. Phytopathol.* **14:**75–96.

Broadbent, P., Bevington, K. B., and Coote, B. G. (1991). Control of stem pitting of grapefruit in Australia by mild strain protection. *In* (R. H. Brlansky, R. F. Lee, and L. W. Timmer, eds.), *Proc. 11th Conf. IOCV*, pp. 64–70.

Burgyán, J., and Gáborjányi, R. (1984). Cross-protection and multiplication of mild and severe strains of TMV in tomato plants. *Phytopathol. Z.* **110:**156–167.

Cabrera, O., and Scholthof, K.-B. G. (1999). The complex viral etiology of St. Augustine decline. *Plant Dis.* **83:**902–904.

Caldwell, J. (1935). On the interactions of two strains of a plant virus; Experiments on induced immunity in plants. *Proc. R. Soc. Lond. B* **117:**120–139.

Canto, T., Aranda, M. A., and Fereres, A. (2009). Climate change effects on physiology and population processes of hosts and vectors that influence the spread of hemipteran-borne plant viruses. *Global Change Biol.* **15:**1884–1894.

Carr, J. P., Gal-On, A., Palukaitis, P., and Zaitlin, M. (1994). Replicase-mediated resistance to cucumber mosaic virus in transgenic plants involves suppression of both virus replication in the inoculated leaves and long-distance movement. *Virology* **199:**439–447.

Cassells, A. C., and Herrick, C. C. (1977). Cross-protection between mild and severe strains of tobacco mosaic virus in doubly inoculated tomato plants. *Virology* **78:**253–260.

Castle, S., Palumbo, J., and Prabhaker, N. (2009). Newer insecticides for plant virus disease management. *Virus Res.* **141:**131–139.

Chamberlain, E. E., Atkinson, J. D., and Hunter, J. A. (1969). Cross-protection between strains of apple mosaic virus. *N. Z. J. Agric. Res.* **7:**480–490.

Chatchawankanphanich, O., Jamboonsri, W., Kositrana, W., and Attathom, S. (2000). Screening for mild strains of papaya ringspot virus for cross protection. *Thai J. Agric. Sci.* **33:**147–152.

Chester, K. S. (1933). The problem of acquired physiological immunity in plants. *Q. Rev. Biol.* **8:**275–324.

Chiang, C.-H., Lee, C.-Y., Wang, C.-H., Jan, F.-J., Lin, S.-S., Chen, T.-C., Raja, J. A. J., and Ye, S.-D. (2007). Genetic analysis of an attenuated *Papaya ringspot virus* strain applied for cross-protection. *Eur. J. Plant Pathol.* **118:**333–348.

Cho, J. J., Ullman, D. E., Wheatley, E., Holly, J., and Gonsalves, D. (1992). Commercialization of ZYMV cross protection for zucchini production in Hawaii. *Phytopathology* **82:**1073.

Cillo, F., Finetti-Sialer, M. M., Papanice, M. A., and Gallitelli, D. (2004). Analysis of mechanisms involved in the *Cucumber mosaic virus* satellite RNA-mediated transgenic resistance in tomato plants. *Mol. Plant–Microbe Interact.* **17:**98–108.

Cillo, F., Pasciuto, M. M., De Giovanni, C., Finetti-Sialer, M. M., Ricciardi, L., and Gallitelli, D. (2007). Response of tomato and its wild relatives in the genus *Solanum* to cucumber mosaic virus and satellite RNA combinations. *J. Gen. Virol.* **88:**3166–3176.

Cochran, L. C. (1954). The origin and interaction of forms of the peach mosaic virus. *Congr. Int. Bot.* 202–204.

Cohen, M. (1976). A comparison of some tristeza isolates and a cross-protection trial in Florida. *In* (E. C. Calavan, ed.), *Proc. 7th Conf. IOCV*, pp. 50–54.

Collmer, C. W., and Howell, S. H. (1992). Role of satellite RNA in the expression of symptoms caused by plant viruses. *Ann. Rev. Phytopathol.* **30:**419–442.

Collmer, C. W., Stenzler, L., Chen, X., Fay, N., Hacker, D., and Howell, S. H. (1992). Single amino acid change in the helicase domain of the putative RNA replicase of turnip crinkle virus alters symptom intensification by virulent satellites. *Proc. Natl. Acad. Sci. USA* **89:**309–313.

Costa, A. S., and Müller, G. W. (1980). Tristeza control by cross protection: A U.S.–Brazil cooperative success. *Plant Dis.* **64:**538–541.

Covey, S. N., Al-Kaff, N. S., Lángara, A., and Turner, D. S. (1997). Plants combat infection by gene-silencing. *Nature* **385:**781–782.

Cox, J. E., Fraser, L. R., and Broadbent, P. (1976). Stem pitting of grapefruit: Field protection by the use of mild strains, an evaluation of trials in two climatic districts. *In* (E. C. Calavan, ed.), *Proc. 7th Conf. IOCV*, pp. 68–70.

Crowdy, S. H., and Posnette, A. F. (1947). Virus diseases of cacao in West Africa. II. Cross-immunity experiments with viruses 1A, 1B and 1C. *Ann. Appl. Biol.* **34:**403–411.

Csorba, T., Pantaleo, V., and Burgyán, J. (2009). RNA silencing, an antiviral mechanism. *Adv. Virus Res.* **75:**35–71.

Cui, B.-F., Cui, S.-W., Wang, H.-X., Weng, F.-L., and Gong, J.-Q. (2005). Study of cross protection of citrus tristeza virus disease on Bendizao mandarin (*Citrus succosa*) [in Chinese]. *J. Zhejiang Univ. (Agric.Life Sci.)* **31:**433–438.

Culver, J. N. (1996). Tobamovirus cross protection using a potexvirus vector. *Virology* **226:**228–235.

de Haan, P., Gielen, J. J., Prins, M., Wijkamp, I. G., van Schepen, A., Peters, D., van Grinsven, M. Q., and Goldbach, R. (1992). Characterization of RNA-mediated resistance to tomato spotted wilt virus in transgenic tobacco plants. *BioTechnology* **10:**1133–1137.

Dekkers, M. G. H., and Lee, R. F. (2002). Evaluation of recently selected mild isolates of citrus tristeza virus for cross-protection of Hamlin sweet orange on smooth flat Seville rootstock. *In* (N. Duran-Vila, R. G. Milne, and J. V. da Graça, eds.), *Proc. 15th Conf. IOCV*, pp. 136–150.

de la Peña, M., and Flores, R. (2002). Chrysanthemum chlorotic mottle viroid RNA: Dissection of the pathogenicity determinant and comparative fitness of symptomatic and non-symptomatic variants. *J. Mol. Biol.* **321:**411–421.

de la Peña, M., Navarro, B., and Flores, R. (1999). Mapping the molecular determinant of pathogenicity in a hammerhead viroid: A tetraloop within the *in vivo* branched RNA conformation. *Proc. Natl. Acad. Sci. USA* **96:**9960–9965.

Deleris, A., Gallego-Bartolome, J., Bao, J., Kasschau, K. D., Carrington, J. C., and Voinnet, O. (2006). Hierarchical action and inhibition of plant Dicer-like proteins in antiviral defence. *Science* **313:**68–71.

Demler, S. A., Rucker, D. G., de Zoeten, G. A., Ziegler, A., Robinson, D. J., and Murant, A. F. (1996). The satellite RNAs associated with the groundnut rosette disease complex and pea enation mosaic virus: Sequence similarities and ability of each other's helper virus to support their replication. *J. Gen. Virol.* **77:**2847–2855.

Desbiez, C., and Lecoq, H. (1997). Zucchini yellow mosaic virus. *Plant Pathol.* **46:**809–829.

Devic, M., Jaegle, M., and Baulcombe, D. (1989). Symptom production on tobacco and tomato is determined by two distinct domains of the satellite RNA of cucumber mosaic virus (strain Y). *J. Gen. Virol.* **70:**2765–2774.

Devic, M., Jaegle, M., and Baulcombe, D. C. (1990). Cucumber mosaic virus satellite RNA (strain Y): Analysis of sequences which affect systemic necrosis on tomato. *J. Gen. Virol.* **71:**1443–1449.

de Zoeten, G. A., and Fulton, R. W. (1975). Understanding generates possibilities. *Phytopathology* **65:**221–222.

Dias, P. R. P., and Rezende, J. A. M. (2001). Problemas na premunização de melancia para o controle do mosaico causado pelo *Papaya ringspot virus*. *Fitopathol. Bras.* **26:**651–654.

Diaz-Pendon, J. A., and Ding, S.-W. (2008). Direct and indirect roles of viral suppressors of RNA silencing in pathogenesis. *Ann. Rev. Phytopathol.* **46:**303–326.

Diaz-Pendon, J. A., Li, F., Li, W.-X., and Ding, S.-W. (2007). Suppression of antiviral silencing by cucumber mosaic virus 2b protein in *Arabidopsis* is associated with drastically reduced accumulation of three classes of viral small interfering RNAs. *Plant Cell* **19:**2053–2063.

Dietrich, C., and Maiss, E. (2003). Fluorescent labelling reveals spatial separation of potyvirus populations in mixed infected *Nicotiana benthamiana* plants. *J. Gen. Virol.* **84:**2871–2876.

Dietrich, C., Al Abdallah, Q., Lintl, L., Pietruszka, A., and Maiss, E. (2007). A chimeric plum pox virus shows reduced spread and cannot compete with its parental wild-type viruses in a mixed infection. *J. Gen. Virol.* **88:**2846–2851.

Ding, B. (2009). The biology of viroid–host interactions. *Ann. Rev. Phytopathol.* **47:**105–131.

Ding, S. W., Li, W. X., and Symons, R. H. (1995). A novel naturally occurring hybrid gene encoded by a plant RNA virus facilitates long distance virus movement. *EMBO J.* **14:**5762–5772.

Dodds, J. A. (1982). Cross-protection and interference between electrophoretically distinct strains of cucumber mosaic virus in tomato. *Virology* **118:**235–240.

Dodds, J. A. (1998). Satellite tobacco mosaic virus. *Ann. Rev. Phytopathol.* **36:**295–310.

Dodds, J. A. (1999). Satellite tobacco mosaic virus. *In* "Satellites and Defective Viral RNAs, Vol. 239" (P. K. Vogt and A. O. Jackson, eds.), pp. 145–157. Springer, Berlin.

Dodds, J. A., Lee, S. Q., and Tiffany, M. (1985). Cross protection between strains of cucumber mosaic virus: Effect of host and type of inoculum on accumulation of virions and double-stranded RNA of the challenge strain. *Virology* **144:**301–309.

Donaire, L., Wang, Y., Gonzalez-Ibeas, D., Mayer, K. F., Aranda, M. A., and Llave, C. (2009). Deep-sequencing of plant viral small RNAs reveals effective and widespread targeting of viral genomes. *Virology* **392:**203–214.

Duran-Vila, N., and Semancik, J. S. (1990). Variations in the "cross protection" effect between two strains of citrus exocortis viroid *Ann. Appl. Biol.* **117:**367–377.

Escriu, F., Fraile, A., and García-Arenal, F. (2000). Evolution of virulence in natural populations of the satellite RNA of cucumber mosaic virus. *Phytopathology* **90:**480–485.

Fernández-Rodríguez, T., Rubio, L., Carballo, O., and Marys, E. (2008). Genetic variation of papaya ringspot virus in Venzuela. *Arch. Virol.* **153:**343–349.

Fernow, K. H. (1967). Tomato as a test plant for detecting mild strains of potato spindle tuber virus. *Phytopathology* **57:**1347–1352.

Ferreira, S. A., Pitz, K. Y., Manshardt, R., Zee, F., Fitch, M., and Gonsalves, D. (2002). Virus coat protein transgenic papaya provides practical control of *Papaya ringspot virus* in Hawaii. *Plant Dis.* **86:**101–105.

Ferreiro, C., Ostrowaka, K., Lopéz-Moya, J. J., and Diaz-Ruiz, J. R. (1996). Nucleotide sequence and symptom modulating analysis of a peanut stunt virus-associated satellite RNA from Poland: High level of sequence identity with the American PSV satellites. *Eur. J. Plant Pathol.* **102:**779–786.

Fletcher, J. T. (1978). The use of avirulent virus strains to protect plants against the effects of virulent strains. *Ann. Appl. Biol.* **89:**110–114.

Fletcher, J. T., and Rowe, J. M. (1975). Observations and experiments on the use of an avirulent mutant strain of tobacco mosaic virus as a means of controlling tomato disease. *Ann. Appl. Biol.* **81:**171–179.

Flores, R., Di Serio, F., and Hernádez, C. (1997). Viroids: The noncoding genomes. *Sem. Virol.* **8:**65–73.

Flores, R., Hernández, C., Martínez de Alba, A. E., Daròs, J.-A., and Di Serio, F. (2005). Viroids and viroid–host interactions. *Ann. Rev. Phytopathol.* **43:**117–139.

Fraile, A., and García-Arenal, F. (2010). The coevolution of plants and viruses: Resistance and pathogenicity. *Adv. Virus Res.* **76:**1–32.

Fraser, R. S. S. (1998). Introduction to classical crossprotection. *In* "Plant Virology Protocols" (G. D. Foster and S. C. Taylor, eds.), pp. 13–24. Humana Press, Totowa.

Fuchs, M., Ferreira, S., and Gonsalves, D. (1997). Management of virus diseases by classical and engineered protection. *Mol. Plant Pathol.* http://www.bspp.org.uk/mppol/1997/0116fuchs.

Fuji, S., Iida, T., and Nakamae, H. (2000). Selection of an attenuated strain of Japanese yam mosaic virus and its use for protecting yam plants against severe strains [in Japanese]. *Jpn. J. Phytopathol.* **66:**35–39.

Fulton, R. W. (1986). Practices and precautions in the use of cross protection for plant virus disease control. *Ann. Rev. Phytopathol.* **24:**67–81.

Gaba, V., Zelcer, A., and Gal-On, A. (2004). Invited review: Cucurbit biotechnology—The importance of virus resistance. *In Vitro Cell. Dev. Biol. Plant* **40:**346–358.

Gallitelli, D. (1998). Present status of controlling cucumber mosaic virus. *In* "Plant Virus Disease Control" (A. Hadidi, R. K. Khetarpal, and H. Koganezawa, eds.), pp. 507–523. APS Press, St. Paul, MN.

Gallitelli, D. (2000). The ecology of *Cucumber mosaic virus* and sustainable agriculture. *Virus Res.* **71:**9–21.

Gallitelli, D., Vovlas, C., Martelli, G., Montasser, M. S., Tousignant, M. E., and Kaper, J. M. (1991). Satellite-mediated protection of tomato against cucumber mosaic virus: II Field test under natural epidemic conditions in southern Italy. *Plant Dis.* **75:**93–95.

Gal-On, A. (2000). A point mutation in the FRNK motif of the *Potyvirus* helper component-protease gene alters symptom expression in cucurbits and elicits protection against the severe homologous virus. *Phytopathology* **90:**467–473.

Gal-On, A. (2007). *Zucchini yellow mosaic virus*: insect transmission and pathogenicity—The tails of two proteins. *Mol. Plant Pathol.* **8:**139–150.

Gal-On, A., and Shiboleth, Y. M. (2006). Cross-protection. *In* "Natural Resistance Mechanisms of Plants to Viruses" (G. Loebenstein and J. P. Carr, eds.), pp. 261–288. Springer, Dordrecht.

Gal-On, A., Katsir, P., and Yongzang, W. (2000). Genetic engineering of attenuated viral cDNA of zucchini yellow mosaic virus for protection of cucurbits. *Acta Hort.* **510:**343–347.
García-Arenal, F., Fraile, A., and Malpica, J. M. (2001). Variability and genetic structure of plant virus populations. *Ann. Rev. Phytopathol.* **39:**157–186.
Garnsey, S. M., Gottwald, T. R., and Yokomi, R. K. (1998). Control strategies for citrus tristeza virus. *In* "Plant Virus Disease Control" (A. Hadidi, R. K. Khetarpal, and H. Koganezawa, eds.), pp. 639–658. APS Press, St. Paul, MN.
Gerber, M., and Sarkar, S. (1989). The coat protein of tobacco mosaic virus does not play a significant role for cross-protection. *J. Phytopathol.* **124:**323–331.
Gerlach, W. L., Llewellyn, D., and Haseloff, J. (1987). Construction of a plant disease resistance gene from the satellite RNA of tobacco ringspot virus. *Nature* **328:**802–805.
Goldbach, R., Bucher, E., and Prins, M. (2003). Resistance mechanisms to plant viruses: An overview. *Virus Res.* **92:**207–212.
Gonsalves, D. (1998). Control of papaya ringspot virus in papaya: A case study. *Ann. Rev. Phytopathol.* **36:**415–437.
Gonsalves, D. (2006). Transgenic papaya development, release, impact and challenges. *Adv. Virus Res.* **67:**317–354.
Gonsalves, D., and Garnsey, S. M. (1989). Cross protection techniques for control of plant virus diseases in the tropics. *Plant Dis.* **73:**592–597.
Gonsalves, D., and Ishii, M. (1980). Purification and serology of papaya ringspot virus. *Phytopathology* **70:**1028–1032.
Gonsalves, D., Provvidenti, R., and Edwards, M. C. (1982). Tomato white leaf: The relation of an apparent satellite RNA and cucumber mosaic virus. *Phytopathology* **72:**1533–1538.
González-Jara, P., Fraile, A., Canto, T., and García-Arenal, F. (2009). The multiplicity of infection of a plant virus varies during colonization of its eukaryotic host. *J. Virol.* **83:**7487–7494.
Goregaoker, S. P., Eckhardt, L. G., and Culver, J. N. (2000). Tobacco mosaic virus replicase-mediated cross-protection: Contributions of RNA and protein-derived mechanisms. *Virology* **273:**267–275.
Goto, T., Komochi, S., and Oshima, N. (1966). Study on control of plant virus diseases by vaccination with attenuated virus. (2). Effects of concentration and time elapsed after inoculation of tomato with attenuated TMV against infection with virulent parent strain. *Ann. Phytopathol. Soc. Jpn.* **32:**221–226.
Gottula, J., and Fuchs, M. (2009). Towards a quarter century of pathogen-derived resistance and practical approaches to plant virus disease control. *Adv. Virus Res.* **75:**161–183.
Grant, T. J., and Costa, A. S. (1951). A mild strain of the tristeza virus of citrus. *Phytopathology* **41:**114–122.
Grieco, F., Cillo, F., Barbarossa, L., and Gallitelli, D. (1992). Nucleotide sequence of a cucumber mosaic virus satellite RNA associated with a tomato top stunting. *Nucleic Acids Res.* **20:**6733.
Grieco, F., Lanave, C., and Gallitelli, D. (1997). Evolutionary dynamics of cucumber mosaic virus satellite RNA during natural epidemics in Italy. *Virology* **229:**166–174.
Habili, N., and Kaper, J. M. (1981). Cucumber mosaic virus-associated RNA 5 VII. Double-stranded form accumulation and disease attenuation in tobacco. *Virology* **112:**250–261.
Hall, J. S., French, R., Hein, G. L., Morris, T. J., and Stenger, D. C. (2001). Three distinct mechanisms facilitate genetic isolation of sympatric wheat streak mosaic virus lineages. *Virology* **282:**230–236.
Hamilton, A. J., and Baulcombe, D. C. (1999). A species of small antisense RNA in posttranscriptional gene silencing in plants. *Science* **286:**950–952.
Hammond, J., Lecoq, H., and Raccah, B. (1999). Epidemiological risks from mixed virus infections and transgenic plants expressing viral genes. *Adv. Vir. Res.* **54:**189–314.

Harrison, B. D., Mayo, M. A., and Baulcombe, C. (1987). Virus resistance in transgenic plants that express cucumber mosaic virus satellite RNA. *Nature* **328:**799–802.
Hellwald, K. H., and Palukaitis, P. (1995). Viral RNA as a potential target for two independent mechanisms of replicase-mediated resistance against cucumber mosaic virus. *Cell* **83:**937–946.
Herold, F., and Weibel, J. (1962). Electron microscopic demonstration of papaya ringspot virus. *Virology* **18:**302–311.
Himber, C., Dunoyer, P., Moissiard, G., Ritzenthaler, C., and Voinnet, O. (2003). Transitivity-dependent and -independent cell-to-cell movement of RNA silencing. *EMBO J.* **22:**4523–4533.
Horikoshi, M., Nakayama, M., Yamaoka, N., Furusawa, I., and Shishiyama, J. (1987). Brome mosaic virus coat protein inhibits viral RNA synthesis *in vitro*. *Virology* **158:**15–19.
Horst, R. K. (1975). Detection of a latent infectious agent that protects against infection by chrysanthemum chlorotic mottle viroid. *Phytopathology* **65:**1000–1003.
Hughes, J. d'A., and Ollennu, L. A. A. (1994). Mild strain protection of cocoa in Ghana against cocoa swollen shoot virus—A review. *Plant Pathol.* **43:**442–457.
Hull, R. (2002). Matthews' Plant Virology. 4th Edn. Academic Press, London and San Diego.
Hull, R., and Plaskitt, A. (1970). Electron microscopy on the behavior of two strains of alfalfa mosaic virus in mixed infections. *Virology* **42:**773–776.
Huss, B., Walter, B., and Fuchs, M. (1989). Cross-protection between arabis mosaic virus and grapevine fanleaf virus isolates in *Chenopodium quinoa*. *Ann. Appl. Biol.* **114:**45–60.
Ichiki, T. U., Nagaoka, E. N., Hagiwara, K., Uchikawa, K., Tsuda, S., and Omura, T. (2005). Integration of mutations responsible for the attenuated phenotype of *Pepper mild mottle virus* strains results in a symptomless cross-protecting strain. *Arch. Virol.* **150:**2009–2020.
Ieki, H., and Yamaguchi, A. (1988). Protective interference of mild strains of citrus tristeza virus against a severe strain in Morita navel orange. *In* (L. W. Timmer, S. M. Garnsey, and L. Navarro, eds.), *Proc. 10th Conf. IOCV*, pp. 86–90.
Inoue-Nagata, A. K., de Mello Franco, C., Martin, D. P., Marques Rezende, J. A., Ferreira, G. B., Dutra, L. S., and Nagata, T. (2007). Genome analysis of a severe and a mild isolate of *Papaya ringspot virus*-type W found in Brazil. *Virol. Gen.* **35:**119–127.
Iyer, L. M., and Hall, T. C. (2000). Virus recovery is induced in brome mosaic virus p2 transgenic plants showing synchronous complementation and RNA-2-specific silencing. *Mol. Plant–Microbe Interact.* **13:**247–258.
Jaegle, M., Devic, M., Longstaff, M., and Baulcombe, D. C. (1990). Cucumber mosaic virus satellite RNA (Y strain): Analysis of sequences which affect yellow mosaic symptoms on tobacco. *J. Gen. Virol.* **71:**1905–1912.
Jedlinski, H., and Brown, C. M. (1965). Cross-protection and mutual exclusion by three strains of barley yellow dwarf virus in *Avena sativa* L. *Virology* **26:**613–621.
Jockusch, H. (1968). Two mutants of tobacco mosaic virus temperature-sensitive in two different functions. *Virology* **35:**94–101.
Jones, I. M., and Reichman, M. E. (1973). The protein synthesized in tobacco leaves infected with tobacco necrosis virus and satellite tobacco necrosis virus. *Virology* **52:**49–56.
Jordá, C., Alfaro, A., Aranda, M. A., Moriones, E., and García-Arenal, F. (1992). Epidemic of cucumber mosaic virus plus satellite RNA in tomatoes in eastern Spain. *Plant Dis.* **76:**363–366.
Jridi, C., Martin, J.-F., Marie-Jeanne, V., Labonne, G., and Blanc, S. (2006). Distinct viral populations differentiate and evolve independently in a single perennial host plant. *J. Virol.* **80:**2349–2357.
Kajihara, H., Muramoto, K., Inoue, T., Kameya-Iwaki, M., Suyama, N., Sumida, Y., and Matsumoto, O. (2000). Production of attenuated strain of Japanese yam mosaic virus of *Dioscorea opposita* and its effectivenes of cross protection [in Japanese]. *Bull. Yamaguchi Agric. Exp. Stat.* **51:**33–38.

Kameya-Iwaki, M., Tochihara, H., Hanada, K., and Torigoe, H. (1992). Attenuated isolate of watermelon mosaic virus (WMV-2) and its cross protection against virulent isolate [in Japanese]. *Ann. Phytopathol. Soc. Jpn.* **58:**491–494.

Kaper, J. M. (1982). Rapid synthesis of double-stranded cucumber mosaic virus-associated RNA 5: Mechanism controlling viral pathogenesis? *Biochem. Biophys. Res. Commun.* **105:**1014–1022.

Kaper, J. M., and Waterworth, H. E. (1977). Cucumber mosaic virus associated RNA 5: Causal agent for tomato necrosis. *Science* **196:**429–431.

Kassanis, B., Gianinazzi, S., and White, R. F. (1974). A possible explanation of the resistance of virus-infected tobacco plants to second infection. *J. Gen. Virol.* **23:**11–16.

Kasschau, K. D., and Carrington, J. C. (1998). A counterdefensive strategy of plant viruses: Suppression of posttranscriptional gene silencing. *Cell* **95:**461–470.

Khoury, J., Singh, R. P., Boucher, A., and Coombs, D. H. (1988). Concentration and distribution of mild and severe strains of potato spindle tuber viroid in cross-protected tomato plants. *Phytopathology* **78:**1331–1336.

Köhler, E. (1934). Untersuchungen über die Viruskrankheiten der Kartoffel III. Weitere Versuche mit Viren aus der Mosaikgruppe. *Phytopathol. Z.* **7:**1–30.

Koizumi, M., Kuhara, S., Ieki, H., Kano, T., Tanaka, A., and Iwanami, T. (1991). A report of preinoculation to control stem pitting disease of Naval orange in fields up to 1989. *In* (R. H. Brlansky, R. F. Lee, and L. W. Timmer, eds.), *Proc. 11th Conf. IOCV*, pp. 125–127.

Komar, V., Vigne, E., Demangeat, G., Lemaire, J. M., and Fuchs, M. (2008). Cross-protection as control strategy against grapevine fanleaf virus. *Plant Dis.* **92:**1689–1694.

Kondo, T., Kasai, K., Yamashita, K., and Ishitani, M. (2007). Selection and discrimination of an attenuated strain of Chinese yam necrotic mosaic virus for cross-protection. *J. Gen. Plant Pathol.* **73:**152–155.

Kong, Q., Oh, J. W., Carpenter, C. D., and Simon, A. E. (1997). The coat protein of turnip crinkle virus is involved in subviral RNA-mediated symptom modulation and accumulation. *Virology* **238:**478–485.

Koo, J. C., Asurmendi, S., Bick, J., Woodford-Thomas, T., and Beachy, R. N. (2004). Ecdysone agonist-inducible expression of a coat protein gene from tobacco mosaic virus confers viral resistance in transgenic *Arabidopsis*. *Plant J.* **37:**439–448.

Kosaka, Y., and Fukunishi, T. (1993). Attenuated isolates of soybean mosaic virus derived at a low temperature. *Plant Dis.* **77:**882–886.

Kosaka, Y., and Fukunishi, T. (1997). Multiple inoculation with three attenuated viruses for the control of cucumber virus disease. *Plant Dis.* **81:**733–738.

Kosaka, Y., Hanada, K., Fukunishi, T., and Tochihara, H. (1989). Cucumber mosaic virus isolate causing tomato necrotic disease in Kyoto prefecture. *Ann. Phytopathol. Soc. Jpn.* **55:**229–232.

Kosaka, Y., Ryang, B.-S., Kobori, T., Shiomi, H., Yasuhara, H., and Kataoka, M. (2006). Effectiveness of an attenuated zucchini yellow mosaic virus isolate for cross-protecting cucumber. *Plant Dis.* **90:**67–72.

Kryczynski, S., and Paduch-Cichal, E. (1987). A comparative study of four viroids of plants. *J. Phytopathol.* **120:**121–129.

Kubota, K., Tsuda, S., Tamai, A., and Meshi, T. (2003). *Tomato mosaic virus* replication potein suppresses virus-targeted posttranscriptional gene silencing. *J. Virol.* **77:**11016–11026.

Kunkel, L. O. (1934). Studies on acquired immunity with tobacco and aucuba mosaics. *Phytopathology* **24:**437–466.

Kurath, G., and Dodds, J. A. (1994). Satellite tobacco mosaic virus sequence variants with only five nucleotide differences can interfere with each other in a cross protection-like phenomenon in plants. *Virology* **202:**1065–1069.

Kurath, G., and Palukaitis, P. (1989). RNA sequence heterogeneity in natural populations of three satellite RNAs of cucumber mosaic virus. *Virology* **173:**231–240.

Kurihara, Y., and Watanabe, Y. (2003). Cross-protection in *Arabidopsis* against crucifer *Tobamovirus* Cg by an attenuated strain of the virus. *Mol. Plant Pathol.* **4:**259–269.

Kuti, J. O., and Moline, H. E. (1986). Effects of inoculation with a mild strain of tomato aspermy virus on the growth and yield of tomatoes and the potential for cross protection. *J. Phytopathol.* **115:**56–60.

Kuwata, S., Masuta, C., and Takanami, Y. (1991). Reciprocal phenotype alterations between two satellite RNAs of cucumber mosaic virus. *J. Gen. Virol.* **72:**2385–2389.

Latorre, B. A., and Flores, V. (1985). Strain identification and cross-protection of potato virus Y affecting tobacco in Chile. *Plant Dis.* **69:**930–932.

Lecoq, H. (1998). Control of plant virus diseases by cross-protection. *In* "Plant Virus Disease Control" (A. Hadidi, R. K. Khetarpal, and H. Konganezawa, eds.), pp. 33–40. APS Press, St. Paul, MN.

Lecoq, H., and Raccah, B. (2001). Cross-protection: Interactions between strains exploited to control plant virus diseases. *In* "Biotic Interactions in Plant–Pathogen Associations" (M. J. Jeger and N. J. Spence, eds.), pp. 177–192. CABI Publishing, Wallingford, Oxon.

Lecoq, H., Lemaire, J. M., and Wipf-Scheibel, C. (1991). Control of zucchini yellow mosaic virus in squash by cross-protection. *Plant Dis.* **75:**208–211.

Lee, R. F., Brlansky, R. H., Garnsey, S. M., and Yokomi, R. K. (1987). Traits of citrus tristeza virus important for mild strain cross protection of citrus: The Florida approach. *Phytophylactica* **19:**215–218.

Legg, J. P., Owor, B., Sseruwagi, P., and Ndunguru, J. (2006). Cassava mosaic virus disease in East and Central Africa: Epidemiology and management of a regional pandemic. *Adv. Virus Res.* **67:**355–418.

Lesley, J. W., and Wallace, J. M. (1938). Acquired tolerance to curly top in the tomato. *Phytopathology* **28:**548–553.

Lewsey, M., Robertson, F. C., Canto, T., Palukaitis, P., and Carr, J. P. (2007). Selective targeting of miRNA-regulated plant development by a viral counter-silencing protein. *Plant J.* **50:**240–252.

Lewsey, M., Surette, M., Robertson, F. C., Ziebell, H., Choi, S. H., Ryu, K. H., Canto, T., Palukaitis, P., Payne, T., Walsh, J. A., and Carr, J. P. (2009). The role of the *Cucumber mosaic virus* 2b protein in viral movement and symptom induction. *Mol. Plant–Microbe Interact.* **22:**642–654.

Li, H., and Roossinck, M. J. (2004). Genetic bottlenecks reduce population variation in an experimental RNA virus population. *J. Virol.* **78:**10582–10587.

Li, X. H., and Simon, A. E. (1990). Symptom intensification on cruciferous hosts by the virulent sat-RNA of turnip crinkle virus. *Phytopathology* **80:**238–242.

Liefting, L., Pearson, M., and Pone, S. (1992). The isolation and evaluation of two naturally occuring mild strains of vanilla necrosis potyvirus for control by cross-protection. *J. Phytopathol.* **136:**9–15.

Lin, Y., Rundell, P. A., and Powell, C. A. (2002). *In situ* immunoassay (ISIA) of field grapefruit trees inoculated with mild isolates of *Citrus tristeza virus* indicates mixed infections with severe isolates. *Plant Dis.* **86:**458–461.

Lin, S.-S., Wu, H.-W., Jan, F.-J., Hou, R. F., and Ye, S.-D. (2007). Modifications in the helper component-protease of zucchini yellow mosaic virus for generation of attenuated mutants for cross protection against severe infection. *Phytopathology* **97:**287–296.

Lindbo, J. A., and Dougherty, W. G. (1992a). Pathogen-derived resistance to a potyvirus: Immune and resistant phenotypes in transgenic tobacco expressing altered forms of a potyvirus coat protein nucleotide sequence. *Mol. Plant–Microbe Interact.* **5:**144–153.

Lindbo, J. A., and Dougherty, W. G. (1992b). Untranslatable transcripts of the tobacco etch virus coat protein gene sequence can interfere with tobacco etch virus replication in transgenic plants and protoplasts. *Virology* **189:**725–733.

Lindbo, J. A., Silva-Rosales, L., Proebsting, W. M., and Dougherty, W. G. (1993). Induction of a highly specific antiviral state in transgenic plants: Implications for regulation of gene expression and virus resistance. *Plant Cell* **5:**1749–1759.

Loebenstein, G. (2009). Local lesions and induced resistance. *Adv. Virus Res.* **75:**73–117.

Loebenstein, G., Cohen, J., Shabtai, S., Coutts, R. H., and Wood, K. R. (1977). Distribution of cucumber mosaic virus in systemically infected tobacco leaves. *Virology* **81:**117–125.

Longstaff, M., Brigneti, G., Boccard, F., Chapman, S., and Baulcombe, D. (1993). Extreme resistance to potato virus X infection in plants expressing a modified component of the putative viral replicase. *EMBO J.* **12:**379–386.

Lu, B., Stubbs, G., and Culver, J. N. (1998). Coat protein interactions involved in tobacco mosaic tobamovirus cross-protection. *Virology* **248:**188–198.

Madden, L. V., and Wheelis, M. (2003). The threat of plant pathogens as weapons against U.S. crops. *Ann. Rev. Phytopathol.* **41:**155–176.

Mahmood, T., and Rush, C. M. (1999). Evidence of cross-protection between beet soilborne mosaic virus and beet necrotic yellow vein virus in sugar beet. *Plant Dis.* **83:**521–526.

Maia, I. G., Haenni, A.-L., and Bernardi, F. (1996). Potyviral HC-Pro: A multifunctional protein. *J. Gen. Virol.* **77:**1335–1341.

Mallory, A. C., Mlotshwa, S., Bowman, L. H., and Vance, V. B. (2003). The capacity of transgenic tobacco to send a systemic RNA silencing signal depends on the nature of the inducing transgene locus. *Plant J.* **35:**82–92.

Manfre, A. J., and Simon, A. E. (2008). Importance of coat protein and RNA silencing in satellite RNA/virus interactions. *Virology* **379:**161–167.

Marenaud, C., Dunez, J., and Bernhard, R. (1976). Identification and comparison of different strains of apple chlorotic leaf spot virus and possibilities of cross protection. *Acta Hort.* **67:**219–225.

Martín, S., and Elena, S. F. (2009). Application of game theory to the interaction between plant viruses during mixed infections. *J. Gen. Virol.* **90:**2815–2820.

Masuta, C., Zuidema, D., Hunter, B. G., Heaton, L. A., Sopher, D. S., and Jackson, A. O. (1987). Analysis of the genome of satellite panicum mosaic virus. *Virology* **159:**329–338.

Masuta, C., Kuwata, S., and Takanami, Y. (1988). Disease modulation on several plants by cucumber mosaic virus satellite RNA (Y strain). *Ann. Phytopathol. Soc. Jpn.* **54:**332–336.

Mathews, D. M., and Dodds, J. A. (1998). Naturally occurring variants of satellite tobacco mosaic virus. *Phytopathology* **88:**514–519.

Matthews, R. E. F. (1949). Studies on potato virus X II. Criteria of relationships between strains. *Ann. Appl. Biol.* **36:**460–474.

McKinney, H. H. (1929). Mosaic diseases in the Canary Islands, West Africa and Gibraltar. *J. Agric. Res.* **39:**557–578.

McKinney, H. H. (1941). Virus-antagonism tests and their limitations for establishing relationship between mutants, and non-relationship between distinct viruses. *Am. J. Bot.* **28:**770–778.

McMillan, R. T., and Gonsalves, D. (1987). Effectiveness of cross-protection by a mild mutant of papaya ringspot virus for control of ringspot disease of papaya in Florida. *Proc. Fla. State Hort. Soc.* **100:**294–296.

Militao, V., Moreno, I., Rodríguez-Cerezo, E., and García-Arenal, F. (1998). Differential interactions among isolates of peanut stunt cucumovirus and its satellite RNA. *J. Gen. Virol.* **79:**177–184.

Miyakawa, T. (1987). Protection against citrus tristeza seedling yellows infection in citrus by pre-inoculation with stem pitting isolates. *Phytophylactica* **19:**193–195.

Molnar, A., Csorba, T., Lakatos, L., Varallyay, E., Lacomme, C., and Burgyan, J. (2005). Plant virus-derived small interfering RNAs originate predominantly from highly structured single-stranded viral RNAs. *J. Virol.* **79:**7812–7818.

Monsion, B., Froissart, R., Michalakis, Y., and Blanc, S. (2008). Large bottleneck size in cauliflower mosaic virus populations during host plant colonization. *PLoS Pathog.* **4:** e1000174.

Montasser, M. S., Tousignant, M. E., and Kaper, J. M. (1991). Satellite-mediated protection of tomato against cucumber mosaic virus: I Greenhouse experiments and simulated epidemic conditions in the field. *Plant Dis.* **75:**86–92.

Moore, C. J., Sutherland, P. W., Forster, R. L. S., Gardner, R. C., and MacDiarmid, R. M. (2001). Dark green islands in plant virus infection are the result of posttranscriptional gene silencing. *Mol. Plant–Microbe Interact.* **14:**939–946.

Moreno, P., Ambrós, S., Albiach-Martí, M. R., Guerri, J., and Peña, L. (2008). *Citrus tristeza virus*: A pathogen that changed the course of the citrus industry. *Mol. Plant Pathol.* **9:**251–268.

Moriones, E., Diaz, I., Fernandez-Cuartero, B., Fraile, A., Burgyan, J., and García-Arenal, F. (1994). Mapping helper virus functions for cucumber mosaic virus satellite RNA with pseudorecombinants derived from cucumber mosaic and tomato aspermy viruses. *Virology* **205:**574–577.

Mossop, D. W., and Francki, R. I. B. (1979). Comparative studies on two satellite RNAs of cucumber mosaic virus. *Virology* **95:**395–404.

Mossop, D. W., Francki, R. I., and Grivell, C. J. (1976). Comparative studies on tomato aspermy and cucumber mosaic viruses V. Purification and properties of a cucumber mosaic virus inducing severe chlorosis. *Virology* **74:**544–546.

Müller, G. W. (1980). Use of mild strains of citrus tristeza virus (CTV) to reestablish commercial production of 'Pera' sweet orange in Sao Paulo, Brazil. *Proc. Fla. State Hort. Soc.* **93:**62–64.

Müller, G. W., and Costa, A. S. (1972). Reduction in yield of Galego lime avoided by preimmunization with mild strains of tristeza virus. *In* (W. C. Price, ed.), *Proc. 5th Conf. IOCV*, pp. 171–175.

Müller, G. W., and Costa, A. S. (1977). Tristea control in Brazil by preimmunization with mild strains. *Proc Int. Soc. Citriculture* **3:**868–872.

Müller, G. W., and Costa, A. S. (1987). Search for outstanding plants in tristeza infected citrus orchards: The best approach to control the disease by preimmunization. *Phytophylactica* **19:**197–198.

Müller, G. W., Costa, A. S., Castro, J. L., and Guirado, N. (1988). Results from preimmunization tests to control the Capão Bonito strain of tristeza. *In* (L. W. Timmer, S. M. Garnsey, and L. Navarro, eds.), *Proc. 10th Conf. IOCV*, pp. 82–85.

Müller, G. W., Targon, M. L. P. N., and Machado, M. A. (2000). Thirty years of preimmunized sweet orange in the citriculture in São Paulo State, Brazil. *In* (J. V. da Graça, R. F. Lee, and R. K. Yokomi, eds.), *Proc. 14th Conf. IOCV*, pp. 400–402.

Murant, A. F., Rajeshwari, R., Robinson, D. J., and Raschke, J. H. (1988). A satellite RNA of groundnut rosette virus is largely responsible for symptoms of roundnut rosette disease. *J. Gen. Virol.* **69:**1479–1486.

Nakazono-Nagaoka, E., Sato, C., Kosaka, Y., and Natsuaki, T. (2004). Evaluation of cross-protection with an attenuated isolate of *Bean yellow mosaic virus* by differential detection of virus isolates using RT-PCR. *J. Gen. Plant Pathol.* **70:**359–362.

Nakazono-Nagaoka, E., Suzuki, M., Kosaka, Y., and Natsuaki, T. (2005). RT-PCR-RFLP analysis for evaluating cross protection by an attenuated isolate of *Cucumber mosaic virus*. *J. Gen. Plant Pathol.* **71:**243–246.

Nakazono-Nagaoka, E., Takahashi, T., Shimizu, T., Kosaka, Y., Natsuaki, T., Omura, T., and Sasaya, T. (2009). Cross-protection against bean yellow mosaic virus (BYMV) and clover yellow vein virus by attenuated BYMV isolate M11. *Phytopathology* **99:**251–257.

Navarro, L., Dunoyer, P., Jay, F., Arnold, B., Dharmasiri, N., Estelle, M., Voinnet, O., and Jones, J. D. G. (2006). A plant miRNA contributes to antibacterial resistance by repressing auxin signaling. *Science* **312:**436–439.

Nejidat, A., and Beachy, R. N. (1989). Decreased levels of TMV coat protein in transgenic tobacco plants at elevated temperatures reduce resistance to TMV infection. *Virology* **173**:531–538.
Nelson, M. R., and Wheeler, R. E. (1978). Biological and serological characterization and separation of potyviruses that infect peppers. *Phytopathology* **68**:979–984.
Nelson, R. S., Powell-Abel, P., and Beachy, R. N. (1987). Lesions and virus accumulation in inoculated transgenic tobacco plants expressing the coat protein gene of tobacco mosaic virus. *Virology* **158**:126–132.
Niblett, C. L., Dickson, E., Fernow, K. H., Horst, R. K., and Zaitlin, M. (1978). Cross protection among four viroids. *Virology* **91**:198–203.
Nishiguchi, M. (2007). Basic studies on attenuated viruses. *J. Gen. Plant Pathol.* **73**:418–420.
Novaes, Q. S., and Rezende, J. A. M. (2005). Protection between strains of *Passion fruit woodiness virus* in sunnhemp. *Fitopathol. Bras.* **30**:307–311.
Ochoa, F., Carballo, O., Trujillo, G., Mayoral de Izaquirre, M. L., and Lee, R. F. (1993). Biological characterization and evaluation of cross protection potential of citrus tristeza virus isolates in Venzuela. *In* (P. Moreno, J. V. da Graça, and L. W. Timmer, eds.), *Proc. 12th Conf. IOCV*, pp. 1–7.
Ochoa, F. M., Cevik, B., Febres, V. J., Niblett, C. L., and Lee, R. F. (2000). Molecular characterization of Florida citrus tristeza virus isolates with potential use in mild strain cross protection. *In* (J. V. da Graça, R. F. Lee, and R. K. Yokomi, eds.), *Proc. 14th Conf. IOCV*, pp. 94–102.
Oh, J. W., Kong, Q., Song, C., Carpenter, C. D., and Simon, A. E. (1995). Open reading frames of turnip crinkles virus involved in satellite symptom expression and incompatibility with *Arabidopsis thaliana* ecotype Dijon. *Mol. Plant–Microbe Interact.* **8**:979–987.
Olson, E. O. (1956). Mild and severe strains of tristeza virus in Texas citrus. *Phytopathology* **46**:336–341.
Olson, E. O. (1958). Responses of lime and sour orange seedlings and four scion-rootstock combinations to infection by strains of the tristeza virus. *Phytopathology* **48**:454–459.
Omarov, R. T., Qi, D., and Scholthof, K.-B. G. (2005). The capsid protein of satellite panicum mosaic virus contributes to systemic invasion and interacts with its helper virus. *J. Virol.* **79**:9756–9764.
Oshima, N. (1981). Control of tomato mosaic virus by attenuated virus. *Jpn. Agric. Res. Q.* **14**:222–228.
Owor, B., Legg, J. P., Okao-Okuja, G., Obonyo, R., Kyamanywa, S., and Ogenga-Latigo, M. W. (2004). Field studies of cross protection with *Cassava mosaic geminiviruses* in Uganda. *J. Phytopathol.* **152**:243–249.
Palauqui, J. C., Elmayan, T., Pollien, J.-M., and Vaucheret, H. (1997). Systemic acquired silencing: transgene specific post-transcriptional silencing is transmitted by grafting from silenced stocks to non-silenced scions. *EMBO J.* **16**:4738–4745.
Pallás, V., and Flores, R. (1989). Interactions between citrus exocortis and potato spindle tuber viroids in plants of *Gynura aurantiaca* and *Lycopersicon esculentum*. *Intervirology* **30**:10–17.
Palukaitis, P. (1988). Pathogenicity regulation by satellite RNAs of cucumber mosaic virus: Minor nucleotide sequence changes alter host responses. *Mol. Plant–Microbe Interact.* **1**:175–181.
Palukaitis, P., and García-Arenal, F. (2003). Cucumoviruses. *Adv. Virus Res.* **62**:241–323.
Palukaitis, P., and Roossinck, M. J. (1996). Spontaneous change of a benign satellite RNA of cucumber mosaic virus to a pathogenic variant. *Nat. Biotech.* **14**:1264–1268.
Palukaitis, P., and Zaitlin, M. (1984). A model to explain the 'cross-protection' phenomenon shown by plant viruses and viroids. *In* ''Plant–Microbe Interactions, Vol. 1'' (T. Kosuge and E. W. Nester, eds.), pp. 420–429. Macmillan, New York.
Palukaitis, P., Roossinck, M. J., Dietzgen, R. G., and Francki, R. I. (1992). Cucumber mosaic virus. *Adv. Virus Res.* **41**:281–348.
Pelosi, R. R., Rundell, P. A., Cohen, M., and Powell, C. A. (2000). Evaluation of a sixteen-year citrus tristeza virus cross-protection trial in Florida. *In* (J. V. da Graça, R. F. Lee, and R. K. Yokomi, eds.), *Proc. 14th Conf. IOCV*, pp. 111–114.

Pennazio, S., Roggero, P., and Conti, M. (2001). A history of plant virology. Cross-protection. *Microbiologica* **24:**99–114.

Perring, T. M., Farrar, C. A., Blua, M. J., Wang, H. L., and Gonsalves, D. (1995). Cross protection of cantaloupe with a mild strain of zucchini yellow mosaic virus: Effectiveness and application. *Crop Prot.* **14:**601–606.

Pimentel, D., Lach, L., Zuniga, R., and Morrison, D. (2000). Environmental and economic costs of non-indigenous species in the United States. *BioScience* **50:**53–65.

Ponz, F., and Bruening, G. (1986). Mechanisms of resistance to plant viruses. *Ann. Rev. Phytopathol.* **24:**355–381.

Posnette, A. F., and Cropley, R. (1956). Apple mosaic viruses host reactions and strain interference. *J. Hort. Sci.* **31:**119–133.

Posnette, A. F., and Todd, J. M. A. (1951). Virus disease of cacao in West Africa. VIII. The search for virus-resistant cacao. *Ann. Appl. Biol.* **38:**785.

Posnette, A. F., and Todd, J. M. A. (1955). Virus disease of cacao in west Africa. IX. Strain variation and interference in virus 1A. *Ann. Appl. Biol.* **43:**433–453.

Powell, C. A., Pelosi, R. R., and Cohen, M. (1992). Superinfection of orange trees containing mild isolates of citrus tristeza virus with severe Florida isolates of citrus tristeza virus. *Plant Dis.* **76:**141–144.

Powell, C. A., Pelosi, R. R., and Bullock, R. C. (1997). Natural field spread of mild and severe isolates of citrus tristeza virus in Florida. *Plant Dis.* **81:**18–20.

Powell, C. A., Pelosi, R. R., Rundell, P. A., Stover, E., and Cohen, M. (1999). Cross-protection of grapefruit from decline-inducing isolates of citrus tristeza virus. *Plant Dis.* **83:**989–991.

Powell, C. A., Pelosi, R. R., Rundell, P. A., and Cohen, M. (2003). Breakdown of cross-protection of grapefruit from decline-inducing isolates of citrus tristeza virus following introduction of the brown citrus aphid. *Plant Dis.* **87:**1116–1118.

Powell-Abel, P., Nelson, R. S., De, B., Hoffmann, N., Rogers, S. G., Fraley, R. T., and Beachy, R. N. (1986). Delay of disease development in transgenic plants that express the tobacco mosaic virus coat protein gene. *Science* **232:**738–743.

Powell-Abel, P., Sanders, P. R., Tumer, N., Fraley, R. T., and Beachy, R. N. (1990). Protection against tobacco mosaic virus infection in transgenic plants requires accumulation of coat protein rather than coat protein RNA sequences. *Virology* **175:**124–130.

Price, W. C. (1932). Acquired immunity to ring-spot in *Nicotiana. Contrib. Boyce Thompson Inst.* **4:**359–403.

Price, W. C. (1935). Acquired immunity from cucumber mosaic in *Zinnia. Phytopathology* **25:**776–789.

Price, W. C. (1936). Specificity of acquired immunity from tobacco-ring-spot diseases. *Phytopathology* **26:**665–675.

Price, W. C. (1940). Acquired immunity from plant virus diseases. *Q. Rev. Biol.* **15:**338–361.

Prins, M., Laimer, M., Noris, E., Schubert, J., Wassenegger, M., and Tepfer, M. (2008). Strategies for antiviral resistance in transgenic plants. *Mol. Plant Pathol.* **9:**73–83.

Purcifull, D., Edwardson, J., Hiebert, E., and Gonsalves, D. (1984). Papaya ringspot virus. *In* "CMI/AAB Description of Plant Viruses, DPV 292" (D. Robinson, R. Mumford, M. Stevens, and M. Adams, eds.), http://www.dpvweb.net/index.php.

Qi, D., and Scholthof, K.-B. G. (2008). Multiple activities associated with the capsid protein of satellite panicum mosaic virus are controlled separately by the N- and C-terminal regions. *Mol. Plant–Microbe Interact.* **21:**613–621.

Qiu, W., and Scholthof, K.-B. G. (2001). Genetic identification of multiple biological roles associated with the capsid protein of satellite panicum mosaic virus. *Mol. Plant–Microbe Interact.* **14:**21–30.

Qiu, W., and Scholthof, K.-B. G. (2004). Satellite panicum mosaic virus capsid protein elicits symptoms on a nonhost plant and interferes with a suppressor of virus-induced gene silencing. *Mol. Plant–Microbe Interact.* **17:**263–271.

Qu, F., Ye, X., and Morris, T. J. (2008). *Arabidopsis* DRB4, AGO1, AGO7 and RDR6 participate in a DCL4-initiated antiviral RNA silencing pathway negatively regulated by DCL1. *Proc. Natl. Acad. Sci. USA* **105:**14732–14737.

Raccah, B., Loebenstein, G., Bar-Joseph, M., and Oren, Y. (1976). Transmission of tristeza by aphids prevelant on citrus, and operation of the tristeza suppression programe in Israel. *In* (E. C. Calavan, ed.), *Proc. 7th Conf. IOCV*, pp. 47–49.

Raj, S. K., Srivastava, A., Chandra, G., and Singh, B. P. (2000). Role of satellite RNA of an Indian isolate of cucumber mosaic virus in inducing lethal necrosis of tobacco plants. *Indian J. Exp. Biol.* **38:**613–616.

Ram, R. D., Verma, R., Tomer, S. P. S., and Prakash, S. (2006). Management of papaya ring spot virus through cross-protection strategy. *J. Maharashtra Agric. Univ.* **31:**92–95.

Rancovic, M., and Paunovic, S. (1988). Further studies on the resistance of plums to sharka (plum pox) virus. *Acta Hort.* **235:**283–290.

Rast, A. T. B. (1972). M II-16, an artificial symptomless mutant of tobacco mosaic virus for seedling inoculation of tomato crops. *Netherlands J. Plant Pathol.* **78:**110–112.

Rast, A. T. B. (1975). Variability of tobacco mosaic virus in relation to control of tomato mosaic in glasshouse tomato crops by resistance breeding and cross protection. *Agric. Res. Rep.* **834:**1–76.

Ratcliff, F., Harrison, B. D., and Baulcombe, D. C. (1997). A similarity between viral defense and gene silencing in plants. *Science* **276:**1558–1560.

Ratcliff, F. G., MacFarlane, S. A., and Baulcombe, D. C. (1999). Gene silencing without DNA. RNA-mediated cross-protection between viruses. *Plant Cell* **11:**1207–1216.

Ravelonandro, M., Briard, P., Glasa, M., and Adam, S. (2008). The ability of a mild isolate of plum pox virus to cross-protect against sharka virus. *Acta Hort.* **781:**281–286.

Reddy, D. V. R., Sudarshana, M. R., Fuchs, M., Rao, N. C., and Tottappilli, G. (2009). Genetically engineered virus-resistant plants in developing countries: Current status and future prospects. *Adv. Virus Res.* **75:**185–220.

Register III, J. C., and Beachy, R. N. (1988). Resistance to TMV in transgenic plants results from interference with an early event in infection. *Virology* **166:**524–532.

Rezende, J. A. M., and Pacheco, D. A. (1998). Control of papaya ringspot virus-type W in zucchini squash by cross-protetion in Brazil. *Plant Dis.* **82:**171–175.

Rezende, J. A. M., and Sherwood, J. L. (1991). Breakdown of cross protection between strains of tobacco mosaic virus due to susceptibility of dark green areas to superinfection. *Phytopathology* **81:**1490–1496.

Rezende, J. A. M., Urban, L., Sherwood, J. L., and Melcher, U. (1992). Host effect on cross protection between two strains of tobacco mosaic virus. *J. Phytopathol.* **136:**147–153.

Rocha-Peña, M. A., Lee, R. F., Permar, T. A., Yokomi, R. K., and Garnsey, S. M. (1991). Use of enzyme-linked immunosorbent and dot-immunobinding assays to evaluate two mild-strain cross protection experiments after challenge with a severe citrus tristeza virus isolate. *In* (R. H. Brlansky, R. F. Lee, and L. W. Timmer, eds.), *Proc. 11th Conf. IOCV*, pp. 93–102.

Rodini, B. (2009). The role of plant biosecurity in preventing and controlling emerging plant virus disease epidemics. *Virus Res.* **141:**150–157.

Rodríguez-Alvarado, G., and Roossinck, M. J. (1997). Structural analysis of a necrogenic strain of cucumber mosaic cucumovirus satellite RNA in planta. *Virology* **236:**155–166.

Rodríguez-Alvarado, G., Kurath, G., and Dodds, J. A. (1994). Symptom modification by satellite tobacco mosaic virus in pepper types and cultivars with helper tobamoviruses. *Phytopathology* **84:**617–621.

Rodríguez-Alvarado, G., Kurath, G., and Dodds, J. A. (2001). Cross-protection between and within subgroup I and II of cucumber mosaic virus isolates from pepper. *Agrociencia* **35:**563–573.

Roistacher, C. N., and Dodds, J. A. (1993). Failure of 100 mild citrus tristeza viurs isolates from California to cross protect against a challenge by severe sweet orange stem pitting isolates. *In* (P. Moreno, J. V. daGraça, and L. W. Timmer, eds.), *Proc. 12th Conf. IOCV*, pp. 100–107.

Roistacher, C. N., and Moreno, P. (1991). The worldwide threat from destructive isolates of citrus tristeza virus—A review. *In* (R. H. Brlansky, R. F. Lee, and L. W. Timmer, eds.), *Proc. 11th Conf. IOCV*, pp. 7–19.

Roistacher, C. N., Dodds, J. A., and Bash, J. A. (1987). Means of obtaining and testing protective strains of seedling yellows and stem pitting tristeza virus: A preliminary report. *Phytophylactica* **19:**199–203.

Roistacher, C. N., Dodds, J. A., and Bash, J. A. (1988). Cross protection against citrus tristeza seedling yellows and stem pitting viruses by protective isolates developed in greenhouse plants. *In* (L. W. Timmer, S. M. Garnsey, and L. Navarro, eds.), *Proc. 10th Conf. IOCV*, pp. 91–100.

Román, M. P., Cambra, M., Juárez, J., Moreno, P., Duran-Vila, N., Tanaka, F. A. O., Alves, E., Kitajima, E. W., Yamamoto, P. T., Bassanezi, R. B., Teixeira, D. C., Junior, W. C. J., *et al.* (2004). Sudden death of citrus in Brazil: A graft-transmissible bud union disease. *Plant Dis.* **88:**453–467.

Roossinck, M. J., Sleat, D., and Palukaitis, P. (1992). Satellite RNAs of plant viruses: Structures and biological effects. *Microbiol. Rev.* **56:**265–279.

Salaman, R. N. (1933). Protective inoculation against a plant virus. *Nature* **131:**468.

Salaman, R. N., Smith, K. M., MacClement, W. D., Bawden, F. C., Bernal, J. D., McFarlane, A. S., Findlay, G. M., Watson, M. A., Murphy, P. A., and Elford, W. J. (1938). A discussion on new aspects of virus disease. *Proc. R. Soc. Lond. B* **125:**291–310.

Salibe, A. A., Souza, A. A., Targon, M. L. P. N., Müller, G. W., Coletta Filho, H. D., and Machado, M. A. (2002). Selection of a mild sub-isolate of citrus tristeza virus for preimmunization of Pera sweet orange. *In* (N. Duran-Vila, R. G. Milne, and J. V. da Graça, eds.), *Proc. 15th Conf. IOCV*, pp. 348–351.

Sambade, A., Ambrós, S., López, C., Ruiz-Ruiz, S., Hermoso de Mendoza, A., Flores, R., Guerri, J., and Moreno, P. (2007). Preferential accumulation of severe variants of *Citrus tristeza virus* in plants co-inoculated with mild and severe variants. *Arch. Virus* **152:**1115–1126.

Sarkar, S., and Smitamana, P. (1981). A proteinless mutant of tobacco mosaic virus: Evidence against the role of a viral coat protein for interference. *Mol. Genet. Genomics* **184:**158–159.

Sasaki, A. (1979). Control of Hassaku dwarf by preimmunization with mild strain. *Rev. Plant Protec. Res.* **12:**80–87.

Sayama, H., Sato, T., Kominato, M., Natsuaki, T., and Kaper, J. M. (1993). Field testing of a satellite-containing attenuated strain of cucumber mosaic virus for tomato protection in Japan. *Phytopathology* **83:**405–410.

Sayama, H., Kominato, M., Ubukata, M., and Sato, T. (2001). Three-year risk assessment of the practical application of cross-protection in processing tomato fields using an attenuated cucumber mosaic virus (CMV) strain containing an ameliorative satellite RNA. *Acta Hort.* **542:**47–53.

Scholthof, K.-B. G. (1999). A synergism induced by satellite panicum mosaic virus. *Mol. Plant–Microbe Interact.* **12:**163–166.

Scholthof, K.-B. G., Jones, R. W., and Jackson, A. O. (1999). Biology and structure of plant satellite viruses activated by icosahedral helper viruses. *In* "Satellites and Defective Viral RNAs, Vol. 239" (P. K. Vogt and A. O. Jackson, eds.), pp. 124–143. Springer, Berlin.

Sclavounos, A. P., Voloudakis, A. E., Arabatzis, C., and Kyriakopoulou, P. E. (2006). A severe Hellenic CMV tomato isolate: Symptom variability in tobacco, characterization and discrimination of variants. *Eur. J. Plant Pathol.* **115:**163–172.

Shaw, J. G., Plaskitt, K. A., and Wilson, T. M. A. (1986). Evidence that tobacco mosaic virus particles disassemble cotranslationally *in vivo*. *Virology* **148:**326–336.

Sheen, T.-F., Wang, H.-L., and Wang, D.-N. (1998). Control of papaya ringspot virus by cross protection and cultivation techniques. *J. Jpn. Soc. Hort. Sci.* **67:**1232–1235.
Sherwood, J. L. (1987a). Demonstration of the specific involvement of coat protein in tobacco mosaic virus (TMV) cross protection using a TMV coat protein mutant. *J. Phytopathol.* **118:**358–362.
Sherwood, J. L. (1987b). Mechanisms of cross-protection between plant virus strains. *In* "Plant Resistance to Viruses" (D. Evered and S. Harnett, eds.), pp. 136–150. Wiley, Chichester.
Sherwood, J. L., and Fulton, R. W. (1982). The specific involvement of coat protein in *Tobacco mosaic virus* cross protection. *Virology* **119:**150–158.
Silhavy, D., Molnár, A., Lucioli, A., Szittya, G., Hornyik, C., Tavazza, M., and Burgyán, J. (2002). A viral protein suppresses RNA silencing and binds silencing-generated, 21- to 25-nucleotide double-stranded RNAs. *EMBO J.* **21:**3070–3080.
Simmonds, J. H. (1959). Mild strain protection as a means of reducing losses from the Queensland woodiness virus in the passion vine. *Qld. J. Agric. Sci.* **16:**371–380.
Simon, A. E., and Howell, S. H. (1986). The virulent satellite RNA of turnip crinkle virus has a major domain homologous to the 3′ end of the helper virus genome. *EMBO J.* **5:**3423–3428.
Simon, A. E., Roossinck, M. J., and Havelda, Z. (2004). Plant virus satellite and defective interfering RNAs: New paradigms for a new century. *Ann. Rev. Phytopathol.* **42:**415–437.
Singh, M., and Singh, R. P. (1995). Host dependent cross-protection between PVYN, PVYO and PVA in potato cultivars and *Solanum brachycarpum*. *Can. J. Plant Pathol.* **17:**82–86.
Singh, R. P., Khoury, J., Boucher, A., and Somerville, T. H. (1989). Characteristics of cross-protection with potato spindle tuber viroid strains in tomato plants. *Can. J. Plant Pathol.* **11:**263–267.
Singh, R. P., Boucher, A., and Somerville, T. H. (1990). Cross-protection with strains of potato spindle tuber viroid in the potato plant and other solanaceous hosts. *Phytopathology* **80:**246–250.
Singh, R. P., Lakshman, D. K., Boucher, A., and Tavantzis, S. M. (1992). A viroid from *Nematanthus wettsteinii* plants closely related to the *Columnea* latent viroid. *J. Gen. Virol.* **73:**2769–2774.
Sleat, D. E. (1990). Nucleotide sequence of a new satellite RNA of cucumber mosaic virus. *Nucleic Acids Res.* **18:**3416.
Sleat, D. E., and Palukaitis, P. (1990a). Induction of tobacco chlorosis by certain cucumber mosaic virus satellite RNAs is specific to subgroup II helper strains. *Virology* **176:**292–295.
Sleat, D. E., and Palukaitis, P. (1990b). Site-directed mutagenesis of a plant viral satellite RNA changes its phenotype from ameliorative to necrogenic. *Proc. Natl. Acad. Sci. USA* **87:**2946–2950.
Sleat, D. E., and Palukaitis, P. (1992). A single nucleotide change within a plant virus satellite RNA alters the host specificity of disease induction. *Plant J.* **2:**43–49.
Sleat, D. E., Zhang, L., and Palukaitis, P. (1994). Mapping determinants within cucumber mosaic virus and its satellite RNA for the induction of necrosis in tomato plants. *Mol. Plant–Microbe Interact.* **7:**189–195.
Smith, H. C. (1963). Interaction between isolates of barley yellow dwarf virus. *N. Z. J. Agric. Res.* **6:**343–353.
Soards, A. J., Murphy, A. M., Palukaitis, P., and Carr, J. P. (2002). Virulence and differential local and systemic spread of *Cucumber mosaic virus* in tobacco are affected by the CMV 2b protein. *Mol. Plant–Microbe Interact.* **15:**647–653.
Souza, A. A., Müller, G. W., Targon, M. L. P. N., Takita, M. A., and Machado, M. A. (2002). Stability of the mild protective 'PIAC' isolate of citrus tristeza virus. *In* (N. Duran-Vila, R. G. Milne, and J. V. da Graça, eds.), *Proc. 15th Conf. IOCV*, pp. 131–135.
Spence, N. J., Mead, A., Miller, A., Shaw, E. D., and Walkey, D. G. A. (1996). The effect on yield in courgette and marrow of the mild strain of zucchini yellow mosaic virus used for cross-protection. *Ann. Appl. Biol.* **129:**247–259.

Storey, H. H., and Nichols, R. F. W. (1938). Mosaic diseases of cassava. *Ann. Appl. Biol.* **25:**790–806.

Strange, R. N., and Scott, P. R. (2005). Plant disease: A threat to global food security. *Ann. Rev. Phytopathol.* **43:**83–116.

Stubbs, L. L. (1964). Transmission and protective inoculation studies with viruses of the citrus tristeza complex. *Aust. J. Agric. Res.* **15:**752–770.

Szittya, G., Molnár, A., Silhavy, D., Hornyik, C., and Burgyán, J. (2002). Short defective interfering RNAs of tombusviruses are not targeted but trigger post-transcriptional gene silencing against their helper virus. *Plant Cell* **14:**359–372.

Tabler, M., and Tsagris, M. (2004). Viroids: Petite RNA pathogens with distinguished talents. *Trends Plant Sci.* **9:**339–348.

Takahashi, T., and Yoshikawa, N. (2008). Analysis of cell-to-cell and long-distance movement of apple latent spherical virus in infected plants using green, cyan and yellow fluorescent proteins. *In* "Plant Virology Protocols, Vol. 451" (G. D. Foster, I. E. Johansen, Y. Hong, and P. D. Nagy, eds.), pp. 545–554. Humana Press, Totowa.

Takahashi, T., Sugawara, T., Yamatsuta, T., Isogai, M., Natsuaki, T., and Yoshikawa, N. (2007). Analysis of the spatial distribution of identical and two distinct virus populations differently labeled with cyan and yellow fluorescent proteins in coinfected plants. *Phytopathology* **97:**1200–1206.

Takanami, Y. (1981). A striking change in symptoms on cucumber mosaic virus-infected tobacco plants induced by a satellite RNA. *Virology* **109:**120–126.

Takeshita, M., Shigemune, N., Kikuhara, K., Furuya, N., and Takanami, Y. (2004). Spatial analysis for exclusive interactions between subgroups I and II of *Cucumber mosaic virus* in cowpea. *Virology* **328:**45–51.

Taliansky, M. E., and Robinson, D. J. (1997). Trans-acting untranslated elements of groundnut rosette virus satellite RNA are involved in symptom production. *J. Gen. Virol.* **78:**1277–1285.

Tan, S.-H., Nishiguchi, M., Sakamoto, W., Ogura, Y., Murata, M., Ugaki, M., Tomiyama, M., and Motoyoshi, F. (1997). Molecular analysis of the genome of an attenuated strain of cucumber green mottle mosaic virus. *Ann. Phytopathol. Soc. Jpn.* **63:**470–474.

Tanzer, M. M., Thompson, W. F., Law, M. D., Wernsman, E. A., and Uknes, S. (1997). Characterization of post-transcriptionally suppressed transgene expression that confers resistance to tobacco etch virus infection in tobacco. *Plant Cell* **9:**1411–1423.

Tennant, P. F., Gonsalves, C., Ling, K.-S., Fitch, M., Manshardt, R., Slightom, J. L., and Gonsalves, D. (1994). Differential protection against papaya ringspot virus isolates in coat protein gene transgenic papaya and classically cross-protected papaya. *Phytopathology* **84:**1359–1366.

Tepfer, M., and Jacquemond, M. (1996). Sleeping satellites: A risky prospect. *Nat. Biotech.* **14:**1226.

Thompson, J. R., and Tepfer, M. (2010). Assessment of the benefits and risks for engineered virus resistance. *Adv. Virus Res.* **76:**33–56.

Thornton, I. R., and Stubbs, L. L. (1976). Control of tristeza decline of grapefruit on sour orange rootstock by preinduced immunity. *In* (E. C. Calavan, ed.), *Proc. 7th Conf. IOCV*, pp. 55–57.

Thornton, I. R., Emmett, R. W., and Stubbs, L. L. (1980). A further report on the grapefruit tristeza preimmunization trial at Mildura, Victoria. *In* (E. C. Calavan, S. M. Garnsey, and L. W. Timmer, eds.), *Proc. 8th Conf. IOCV*, pp. 51–53.

Thung, T. H. (1931). Smetstof en plantecel bij enkele virusziekten van de Tabaksplant. *Hand. 6de Ned.-Ind. Natuurwet. Congress* 450–463. Abstract (1932) in *Rev. Appl. Mycol.* **11:**750–751.

Thung, T. H. (1935). Infective principle and plant cell in some virus diseases of the tobacco plant II. *Hand. 7de Ned.-Ind. Natuurwet. Congress* 496–507.

Tian, Z., Qiu, J., Yu, J., Han, C., and Liu, W. (2009). Competition between cucumber mosaic virus subgroup I and II isolates in tobacco. *J. Phytopathol.* **157:**457–464.

Tien, P., and Wu, G. S. (1991). Satellite RNA for the biocontrol of plant disease. *Adv. Virus Res.* **39:**321–339.

Tien, P., Zhang, X., Qiu, B., Qin, B., and Wu, G. (1987). Satellite RNA for the control of plant diseases caused by cucumber mosaic virus. *Ann. Appl. Biol.* **111:**143–152.

Tomaru, K., Hidaka, Z., and Udagawa, A. (1967). Strains of cucumber mosaic virus isolated from tobacco plants. V. Studies on interference among cucumber mosaic virus strains [in Japanese]. *Bull. Hatano Tobacco Exp. Stat.* **58:**79–88.

Tomlinson, J. A., and Shepherd, R. J. (1978). Studies on mutagenesis and cross-protection of cauliflower mosaic virus. *Ann. Appl. Biol.* **90:**223–231.

Tripathi, S., Suzuki, J. Y., Ferreira, S. A., and Gonsalves, D. (2008). *Papaya ringspot virus*-P: Characteristics, pathogenicity, sequence variability and control. *Mol. Plant Pathol.* **9:**269–280.

Tsuda, S., Kubota, K., Kanda, A., Ohki, T., and Meshi, T. (2007). Pathogenicity of pepper mild mottle virus is controlled by the RNA silencing suppression activity of its replication protein but not the viral accumulation. *Phytopathology* **97:**412–420.

Tumer, N. E., O'Connell, K. M., Nelson, R. S., Sanders, P. R., Beachy, R. N., Fraley, R. T., and Shah, D. M. (1987). Expression of alfalfa mosaic virus coat protein gene confers cross-protection in transgenic tobacco and tomato plants. *EMBO J.* **6:**1181–1188.

Uga, H., Kobayshi, Y. O., Hagiwara, K., Honda, Y., and Omura, T. (2004). Selection of an attenuated isolate of *Bean yellow mosaic virus* for protection of dwarf gentian plants from viral infection in the field. *J. Gen. Plant Pathol.* **70:**54–60.

Urban, L. A., Sherwood, J. L., Rezende, J. A. M., and Melcher, U. (1990). Examination of mechanisms of cross protection with non-transgenic plants. *In* "Recognition and Response in Plant–Virus Interactions" (R. S. S. Fraser, ed.), pp. 415–426. Springer-Verlag, Berlin.

Urcuqui-Inchima, S., Haenni, A.-L., and Bernardi, F. (2001). Potyvirus proteins: A wealth of functions. *Virus Res.* **74:**157–175.

Valenta, V. (1959a). Interference studies with yellows-type plant viruses. I. Cross protection tests with European viruses. *Acta Virol.* **3:**65–72.

Valenta, V. (1959b). Interference studies with yellows-type plant viruses. II. Cross protection tests with European and American viruses. *Acta Virol.* **3:**145–152.

Valkonen, J. P. T., Rajamäki, M.-L., and Kekarainen, T. (2002). Mapping of viral genomic regions important in cross-protection between strains of a potyvirus. *Mol. Plant–Microbe Interact.* **15:**683–692.

Valle, R. P., Skrzeczkowski, J., Morch, M. D., Joshi, R. L., Gargouri, R., Drugeon, G., Boyer, J. C., Chapeville, F., and Haenni, A. L. (1988). Plant viruses and new perspectives in cross-protection. *Biochimie* **70:**695–703.

van der Vyver, J. B., van Vuuren, S. P., Luttig, M., and daGraça, J. V. (2002). Changes in the *Citrus tristeza virus* status of pre-immunized grapefruit field trees. *In* (N. Duran-Vila, R. G. Milne, and J. V. daGraça, eds.), *Proc. 15th Conf. IOCV*, pp. 175–185.

van Vuuren, S. P., and Moll, J. N. (1987). Glasshouse evaluation of citrus tristeza virus isolates. *Phytophylactica* **19:**219–221.

van Vuuren, S. P., and van der Vyver, J. B. (2000). Comparison of South African pre-immunizing citrus tristeza virus isolates with foreign isolates in three grapefruit selections. *In* (J. V. da Graça, R. F. Lee, and R. K. Yokomi, eds.), *Proc. 14th Conf. IOCV*, pp. 50–56.

van Vuuren, S. P., Collins, R. P., and da Graça, J. V. (1993). Evaluation of citrus tristeza virus isolates for cross protection of grapefruit in South Africa. *Plant Dis.* **77:**24–28.

van Vuuren, S. P., van der Vyver, J. B., and Luttig, M. (2000). Diversity among sub-isolates of cross-protecting citrus tristeza virus isolates in South Africa. *In* (J. V. da Graça, R. F. Lee, and R. K. Yokomi, eds.), *Proc. 14th Conf. IOCV*, pp. 103–110.

Vaucheret, H., Beclin, C., and Fagard, M. (2001). Post-transcriptional gene silencing in plants. *J. Cell Sci.* **114:**3083–3091.

Vigne, E., Marmonier, A., Komar, V., Lemaire, O., and Fuchs, M. (2009). Genetic structure and variability of virus populations in cross-protected grapevines superinfected by *Grapevine fanleaf virus*. *Virus Res.* **144:**154–162.

Voinnet, O. (2001). RNA silencing as a plant immune system against viruses. *Trends Genet.* **17:**449–459.

Voinnet, O. (2005). Induction and suppression of RNA silencing: Insights from viral infections. *Nat. Rev. Genet.* **6:**206–220.

Voinnet, O., and Baulcombe, D. C. (1997). Systemic signalling in gene silencing. *Nature* **389:**553.

Walkey, D. G. A. (1992). Zucchini yellow mosaic virus: Control by mild strain protection. *Phytoparasitica* **20:**99S–103S.

Walkey, D. G. A., Lecoq, H., Collier, R., and Dobson, S. (1992). Studies on the control of zucchini yellow mosaic virus in courgettes by mild strain protection. *Plant Pathol.* **41:**762–771.

Wallace, J. M. (1939). Recovery from and acquired tolerance of curly top in *Nicotiana tabacum*. *Phytopathology* **29:**743–749.

Wallace, J. M. (1940). Evidence of passive immunization of tobacco, *Nicotiana tabacum*, from the virus of curly top. *Phytopathology* **30:**673–679.

Wallace, J. M. (1944). Acquired immunity from curly top in tobacco and tomato. *J. Agric. Res.* **69:**187–214.

Wallace, J. M., and Drake, R. J. (1976). Studies on the recovery of citrus plants from seedling yellows and the resulting protection against reinfection. *In* (W. C. Price, ed.), *Proc. 5th Conf. IOCV*, pp. 127–136.

Wang, M., and Gonsalves, D. (1992). Artificial induction and evaluation of a mild isolate of tomato spotted wilt virus. *J. Phytopathol.* **135:**233–244.

Wang, J., and Simon, A. E. (1999). Symptom attenuation by a satellite RNA *in vivo* is dependent on reduced levels of virus coat protein. *Virology* **259:**234–245.

Wang, H. L., Yeh, S.-D., Chiu, R. J., and Gonsalves, D. (1987). Effectiveness of cross-protection by mild mutants of papaya ringspot virus for control of ringspot disease of papaya in Taiwan. *Plant Dis.* **71:**491–497.

Wang, H. L., Gonsalves, D., and Provvidenti, R. (1991). Effectiveness of cross protection by a mild strain of zucchini yellow mosaic virus in cucumber, melon, and squash. *Plant Dis.* **75:**203–207.

Wang, W. Q., Natsuaki, T., Kosaka, Y., and Okuda, S. (2006). Comparison of the nucleotide and amino acid sequences of parental and attenuated isolates of *Zucchini yellow mosaic virus*. *J. Gen. Plant Pathol.* **72:**52–56.

Waterworth, H. E., Kaper, J. M., and Tousignant, M. E. (1979). CARNA 5, the small cucumber mosaic virus-dependent replicating RNA, regulates disease expression. *Science* **204:**845–847.

Webb, R. E., Larson, R. H., and Walker, J. C. (1952). Relationships of potato leaf roll virus strains. *Res. Bul. Agric. Exp. Stat., Coll. Agric. Univ. Wisconsin* **178:**1–38.

Wen, H., Lister, R. M., and Fattouh, F. A. (1991). Cross-protection among strains of barley yellow dwarf virus. *J. Gen. Virol.* **72:**791–799.

Wingard, S. A. (1928). Hosts and symptoms of ring spot, a virus disease of plants. *J. Agric. Res.* **37:**127–153.

Wu, X., and Shaw, J. G. (1997). Evidence that a viral replicase protein is involved in the disassembly of tobacco mosaic virus particles *in vivo*. *Virology* **239:**426–434.

Wu, G., Kang, L., and Tien, P. (1989). The effect of satellite RNA on cross-protection among cucumber mosaic virus strains. *Ann. Appl. Biol.* **114:**489–496.

Wu, X. J., Beachy, R. N., Wilson, T. M. A., and Shaw, J. G. (1990). Inhibition of uncoating of tobacco mosaic virus particles in protoplasts from transgenic tobacco plants that express the viral coat protein gene. *Virology* **179:**893–895.

Xi, D., Yang, H., Jiang, Y., Xu, M., Shang, J., Zhang, Z., Cheng, S., Sang, L., and Lin, H. (in press). Interference between *Tobacco necrosis virus* and *Turnip crinkle virus* in *Nicotiana benthamiana. J. Phytopathol.*

Xu, P., and Roossinck, M. J. (2000). Cucumber mosaic virus D satellite RNA-induced programmed cell death in tomato. *Plant Cell* **12:**1079–1092.

Xu, P., Blancaflor, E. B., and Roossinck, M. J. (2003). In spite of induced multiple defense responses, tomato plants infected with *Cucumber mosaic virus* and D satellite RNA succumb to systemic necrosis. *Mol. Plant–Microbe Interact.* **16:**467–476.

Yamaya, J., Yoshioka, M., Meshi, T., Okada, Y., and Ohno, T. (1988). Cross protection in transgenic tobacco plants expressing a mild strain of tobacco mosaic virus. *Mol. Gen. Genomics* **215:**173–175.

Yarden, G., Hemo, R., Livne, H., Maoz, E., Lev, E., Lecoq, H., and Raccah, B. (2000). Cross-protection of *Cucurbitaceae* from zucchini yellow mosaic potyvirus. *Acta Hort.* **510:**349–359.

Ye, J., Quing, L., and Zhou, X.-P. (2006). Cross-protection mediated by a ßC1 deletion DNAß associated with *Tomato yellow leaf curl China virus* [in Chinese]. *J. Zhejiang Univ. (Agric. Life Sci.)* **32:**479–482.

Yeh, S.-D., and Gonsalves, D. (1984). Evaluation of induced mutants of papaya ringspot virus for control by cross protection. *Phytopathology* **74:**1086–1091.

Yeh, S.-D., and Gonsalves, D. (1994). Practices and perspective of control of papaya ringspot virus by cross protection. *In* "Advances in Disease Vector Research, vol. 10" (K. F. Harris, ed.), pp. 237–257. Springer, New York.

Yeh, S.-D., and Kung, Y.-J. (2007). The past and current approaches for control of papaya ringspot virus in Taiwan. *Acta Hort.* **740:**235–243.

Yeh, S.-D., Gonsalves, D., Wang, H. L., Namba, R., and Chiu, R. J. (1988). Control of papaya ringspot virus by cross-protection. *Plant Dis.* **72:**375–380.

Yie, Y., and Tien, P. (1993). Plant virus satellite RNAs and their role in engineering resistance to virus diseases. *Sem. Virol.* **4:**363–368.

Yie, Y., Zhao, F., Zhao, S. Z., Liu, Y. Z., Liu, Y. L., and Tien, P. (1992). High resistance to cucumber mosaic virus conferred by satellite RNA and coat protein in transgenic commercial tobacco cultivar G-140. *Mol. Plant–Microbe Interact.* **5:**460–465.

Yokomi, R. K., Garnsey, S. M., Lee, R. F., and Cohen, M. (1987). Use of insect vectors to screen for protecting effects of mild citrus tristeza virus isolates in Florida. *Phytophylactica* **19:**183–185.

Yokomi, R. K., Garnsey, S. M., Permar, T. A., Lee, R. F., and Youtsey, C. O. (1991). Natural spread of severe citrus tristeza virus isolates in citrus preinfected with mild CTV isolates. *In* (R. H. Brlansky, R. F. Lee, and L. W. Timmer, eds.), *Proc. 11th Conf. IOCV*, pp. 86–92.

Yoon, J. Y., Ahn, H. I., Kim, M., Tsuda, S., and Ryu, K. H. (2006). *Pepper mild mottle virus* pathogenicity determinants and cross protection effect of attenuated mutants in pepper. *Virus Res.* **118:**23–30.

Yoshida, K., Goto, T., and Iizuka, N. (1985). Attenuated isolates of cucumber mosaic virus produced by satellite RNA and cross-protection between attenuated isolates and virulent ones. *Ann. Phytopathol. Soc. Jpn.* **51:**238–242.

You, B.-J., Chiang, C.-H., Chen, L.-F., Su, W.-C., and Ye, S.-D. (2005). Engineered mild strains of papaya ringspot virus for broader cross protection in cucurbits. *Phytopathology* **95:**533–540.

Zaitlin, M. (1976). Viral cross-protection: More understanding is needed. *Phytopathology* **66:**382–383.

Zaitlin, M., Anderson, J. M., Perry, K. L., Zhang, L., and Palukaitis, P. (1994). Specificity of replicase-mediated resistance to cucumber mosaic virus. *Virology* **201**:200–205.

Zhang, X.-S., and Holt, J. (2001). Mathematical models of cross protection in the epidemiology of plant-virus diseases. *Phytopathology* **91**:924–934.

Zhang, F., and Simon, A. E. (2003). Enhanced viral pathogenesis associated with a virulent mutant virus or a virulent satellite RNA correlates with reduced virion accumulation and abundance of free coat protein. *Virology* **312**:8–13.

Zhang, L., Kim, C. H., and Palukaitis, P. (1994). The chlorosis-induction domain of the satellite RNA of cucumber mosaic virus: Identifying sequences that affect accumulation and the degree of chlorosis. *Mol. Plant–Microbe Interact.* **7**:208–213.

Zhou, C. Y., Broadbent, P., Hailstones, D. L., Bowyer, J., and Connor, R. (2002a). Movement and titre of *Citrus tristeza virus* (pre-immunizing isolate PB61) within seedlings and field trees. *In* (N. Duran-Vila, R. G. Milne, and J. V. da Graça, eds.), *Proc. 15th Conf. IOCV*, pp. 39–47.

Zhou, C. Y., Hailstones, D. L., Broadbent, P., Connor, R., and Bowyer, J. (2002b). Studies on mild strain cross protection against stem-pitting citrus tristeza virus. *In* (N. Duran-Vila, R. G. Milne, and J. V. da Graça, eds.), *Proc. 15th Conf. IOCV*, pp. 151–157.

Ziebell, H. (2008). Mechanisms of cross-protection. *CAB Rev.: Perspect. Agric. Vet. Sci. Nutr. Nat. Resour.* **3**(049). http://www.cababstractsplus.org/cabreviews/index.asp, 13pp.

Ziebell, H., and Carr, J. P. (2009). Effects of dicer-like endoribonucleases 2 and 4 on infection of *Arabidopsis thaliana* by cucumber mosaic virus and a mutant virus lacking the 2b counter-defence protein gene. *J. Gen. Virol.* **90**:2288–2292.

Ziebell, H., Payne, T., Berry, J. O., Walsh, J. A., and Carr, J. P. (2007). A cucumber mosaic virus mutant lacking the 2b counter-defence protein gene provides protection against wild-type strains. *J. Gen. Virol.* **88**:2862–2871.

Zinnen, T. M., and Fulton, R. W. (1986). Cross-protection between sunn-hemp mosaic and tobacco mosaic viruses. *J. Gen. Virol.* **67**:1679–1687.

INDEX

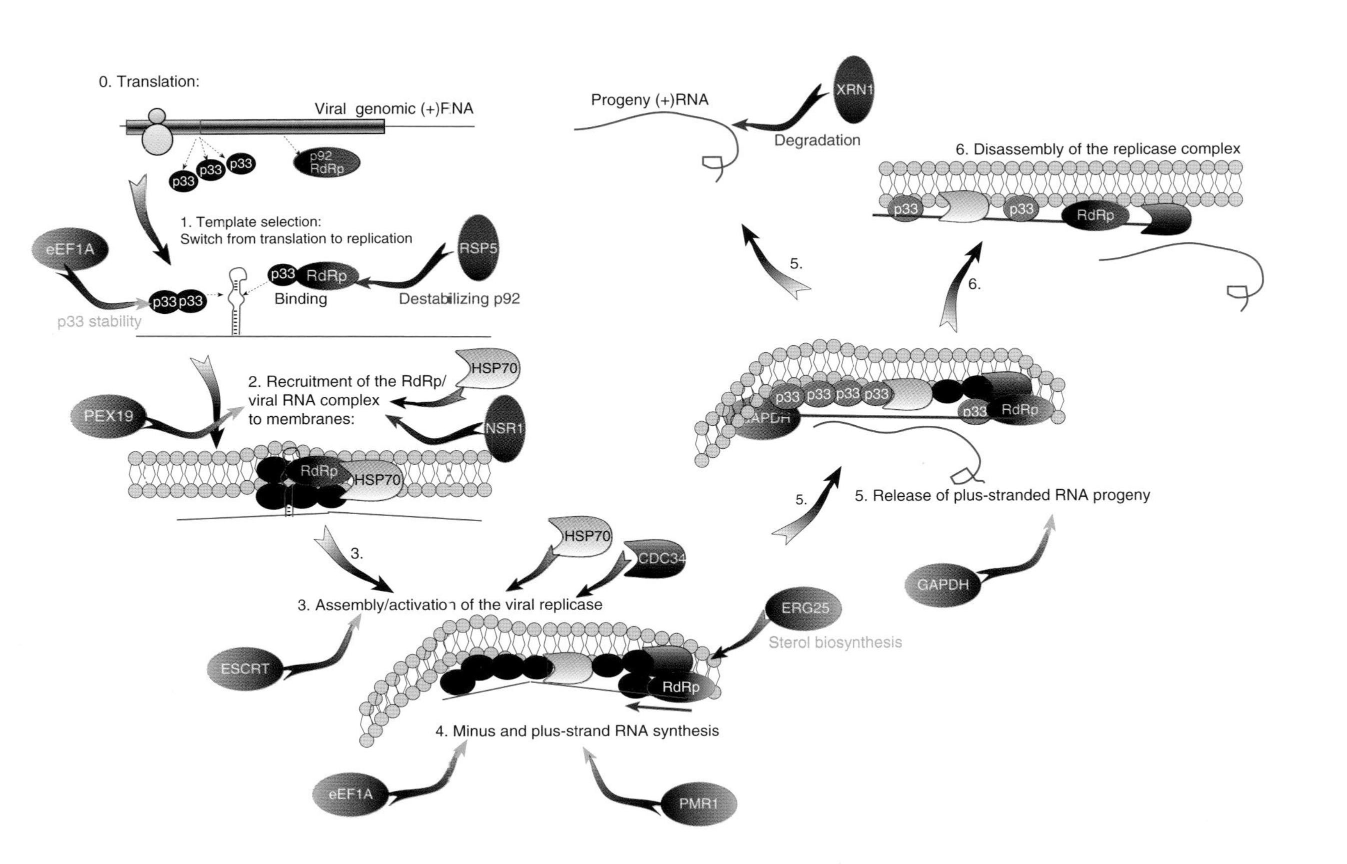

Figure 5, Peter D. Nagy and Judit Pogany (See Page 156 of this Volume)

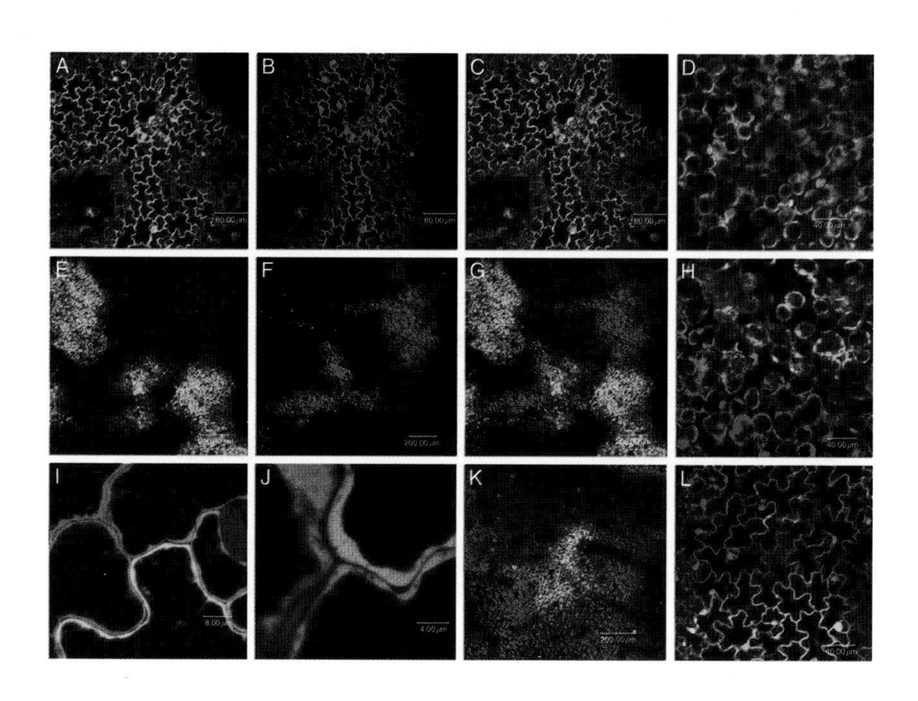

Figure 2, Heiko Ziebell and John Peter Carr (See Page 237 of this Volume)